TOWARD AN ANTHROPOLOGY OF GRAPHING

Toward an Anthropology of Graphing

Semiotic and Activity-Theoretic Perspectives

by

Wolff-Michael Roth
University of Victoria, British Columbia, Canada

KLUWER ACADEMIC PUBLISHERS
DORDRECHT / BOSTON / LONDON

A C.I.P. Catalogue record for this book is available from the Library of Congress.

ISBN 1-4020-1374-4 (HB)
ISBN 1-4020-1376-0 (PB)

Published by Kluwer Academic Publishers,
P.O. Box 17, 3300 AA Dordrecht, The Netherlands.

Sold and distributed in North, Central and South America
by Kluwer Academic Publishers,
101 Philip Drive, Norwell, MA 02061, U.S.A.

In all other countries, sold and distributed
by Kluwer Academic Publishers,
P.O. Box 322, 3300 AH Dordrecht, The Netherlands.

Printed on acid-free paper

Contents

Preface

During the summer of 1990, while taking my holidays to teach a university course of physics for elementary teachers, I also tutored one of the tenth-grade students at my school in physics, chemistry, and mathematics. In return for working with him for free, I had requested permission to audiotape our sessions; I wanted to use the transcripts as data sources for a chapter that I had been invited to write. It so happened that I discovered and read Jean Lave's *Cognition in Practice* that very summer, which inspired me to read other books on mathematics in everyday situations. Two years later, while conducting a study with my teacher colleague G. Michael Bowen on eighth-grade students' learning during an open-inquiry ecology unit, I discovered these students' tremendous data analysis skills that appeared to be a function of the deep familiarity with the objects and events that they had studied and mathematized earlier in the unit. I reported my findings in two articles, 'Mathematization of experience in a grade 8 open-inquiry environment: An introduction to the representational practices of science' and 'Where is the context in contextual word problems?: Mathematical practices and products in Grade 8 students' answers to story problems'.[1] Beginning with that study, I developed a research agenda that focused on mathematical knowing in science and science-related professions.

During the early 1990s, I was also interested in the notion of *authentic practice* as a metaphor for planning school science curriculum. To achieve the goal of a more 'authentic' curriculum, I wanted to find out what scientists actually did at various moments of their work and how their actions and experiences related to their scientific and mathematical competencies. I was particularly interested in the representations scientists used in their journals and that somehow were the outcomes of complex processes in their workplace. Because of the prevalence of graphs in science, I wanted to find out how scientists interpreted them so that I could develop a normative framework for the appropriation of graph-related competencies in my school-related work. This initially led me to a comparative investigation of graphs and other mathematical representations in scientific journals and high school biology textbooks and subsequently to a study of graph interpretations by scientists.[2]

During the initial graph interpretation sessions, I was quite stunned noting that scientists were not at all the experts that the science education and expert-novice literatures portrayed them to be. I found scientists to make errors similar in structure and kind to those children had been reported to make. But, because of the experience and background that these scientists have had, I could not plausibly seek recourse to the same deficit explanations (variously labeled 'misconceptions', 'mental deficits', or 'misunderstandings') that had been used with students. I therefore decided to study the phenomenon more exhaustively by seeking situations where I could follow scientists or technicians in their work to understand their work-related graphing expertise and how it related to the problems that scientists faced when asked to interpret introductory-level graphs in their own discipline. This book is the result of several years of studying the development and use of graphs by scientists and of the interview study in which scientists responded to several graph interpretation tasks.

A book like this does not flow from an author's mind but is the outcome of an agency-structure dialectic. Structure includes the social relations that an author maintains with colleagues, research assistants, research participants, and granting agencies. Without such relations, I could not possibly be the researcher and author that I am. I therefore want to thank all those who, in direct and indirect ways, allowed me to write this book (my agency). First of all, I am grateful to the scientists who took time out of their busy schedules to comply with my interview request and to those who allowed me (or my graduate students in some cases) to spend time in their laboratory or field research sites to observe them in their activities and, sometimes, to participate as an apprentice in their research in ways that fit the need of the ongoing work.

I am also grateful to those individuals who assisted me in collecting the data by interviewing scientists, videotaping the encounters, or transcribing the video- and audiotapes. These individuals include G. Michael Bowen, who completed his doctoral dissertation under my supervision during the time that these studies were conducted, and my wife Sylvie Boutonné, who contributed as a research assistant, videotaping sessions, and transcribing the tapes. In the same breath, I acknowledge the Social Sciences and Humanities Research Council of Canada, which supported the work presented here through two regular grants and one major collaborative research initiative grant. Michelle K. McGinn was an early companion in my research on graphing. In our discussions, which also related to her own MA work on mathematics in the kitchen, I developed my understanding of the situated nature of mathematics in science. Domenico Masciotra and Daniel Lawless, too, were valuable discussion partners while I attempted to think about developmental aspects in the relationship between individuals and graphs. In many discussions with Kenneth Tobin, I have had opportunities for articulating and elaborating issues surrounding activity theory and its application to concrete settings. Tracy Noble, Julian Williams, Jim Kaput, Celia Hoyles, and Richard Noss have been gracious listeners and respondents to various aspects of this work.

My ideas and understanding also developed as I attempted to articulate them in articles submitted to a variety of research journals. I attempted to communicate how scientists used graphs, how they responded to my graphing tasks, and how they arrived at and interpreted graphs that issued from their own engagement in research. But I did not always arrive at making myself clear. The anonymous reviewers of various journals, representing my community of practice, have pushed me to articulate my ideas more clearly and therefore deserve special thanks. The resulting publications in *Journal for Research in Mathematics Education, Cognition and Instruction, Journal of Curriculum Studies, Social Studies of Science* (Chapter 7), *International Journal for Computers in Mathematical Learning, International Journal of Learning Technologies,* and *Science, Technology, and Human Values* were important starting points for the ideas presented here.

ONE

Toward an Anthropology of Graphing

An Introduction

> The consensual nature of mathematics is expressed and described mathematically; that is, it is available in the actions of doing intelligible mathematics. To say this does not imply that mathematicians' practices are given a complete and determinate representation by mathematical formulae but that no such representation can be constructed and none is missing.[1]

This is a book about graphs and graphing both 'in captivity', that is, when scientists do graphing tasks posed by a social scientist, and 'in the wild', that is, when scientists use, construct, collect data for, and interpret graphs in their everyday work. I had come to do this research because early on in my work on scientific representations I wanted to have some samples of expertise to be used as reference for the graphing practices that I observed among high school and university students. It turned out, and I can say this much without anticipating the results reported in the first part of this book, that a number of practicing scientists did not exhibit expertise on my tasks in the way that this is often defined in the literature on expertise. The scientists did not exhibit expertise although the tasks were taken from introductory university textbooks in their own field. On the other hand, when the scientists explained graphs that they had produced themselves, they talked a lot about the contextual details of the object they studied, methodology they used, typical problems they encountered, and so on. I had the hunch that expertise in graph interpretation was a function of the familiarity with the 'system' that the graph referred to and with the particular way of constructing graphs in that context. Following this hunch, I subsequently conducted several ethnographic studies of graphing in the workplace; in this book, I report on graph use and graph interpretation by a water technician working on a farm where she monitored the water levels in the creek, by ecologists concerned with understanding the life history of reptiles, and vision biologists investigating the absorption of light in salmonid retina. My analyses show that scientists engage much less in making inference from graphs as articulating what they already understand and establishing coherence with features that they detect in the graphs. In this chapter, I present a general introduction to the field of scientific representation and outline some details pertaining to the study of graphing 'in captivity'. I want to lay the ground that makes it plausible for taking an anthropological approach to graphs and graphing.

1.1. GRAPHING IS PERVASIVE

Graphical representations, which in the sociology of science and in postmodern discourse have come to be known as *inscriptions*, are central to scientific practice. They are tools used for analyzing and understanding many scientific phenomena and are central to the rhetoric of scientific communication.[2] In fact, scientists and engineers become dependent on graphical representations so that, in their absence, they fail to accomplish tasks, interrupt meetings in order to fetch some representation, or at least use gestures to reproduce transient facsimile in the air.[3] It is virtually impossible to find a science textbook or scientific journal without graphs and diagrams. A recent survey of more than 2,500 pages from ecology research journals showed that there are 14 graphical representations per 10 pages.[4] A similar survey of chapters and journal articles from different disciplines reported about 11 representations.[5] We can say that graphing practices, which includes producing, reading, and interpreting graphs, are central to not only to scientific communication but also to the scientific enterprise more broadly.

Graphing is an interesting practice because people of all ages experience difficulties when asked to produce or interpret graphs.[6] Even those who have graduated with a college or university science degree have difficulties interpreting graphs and data more generally,[7] so that some central question pose themselves: 'What are the graph-related practices of science?', 'Are scientists equally competent reading familiar and unfamiliar graphs?', and 'If scientists are much more competent than recent university graduates, what are the experiences that allow the competencies to develop?' These are some of the questions that led me to the research reported here.

In the opening quote, Michael Lynch makes the point that any understanding of intelligible mathematical practice is expressed and described by observable mathematical activity; and no description of mathematical practice can ever be complete. In contrast, many researchers interested in graphing seek mathematical understandings exclusively in individuals' heads rather than in their public activity of doing math. There is no problem with such an approach if it is considered to be part of a more general research agenda towards understanding graphs and graphing. However, there are problems with such an approach if the research concerns itself *only* with mental processes as inferred from behaviors on narrowly specified tasks while omitting to investigate graphing in the natural settings where it occurs. Much of what we do is situated and we take the resources we need directly from the context and use it in task-relevant work.[8] Thus, to read and interpret a graph, a reader does not have to represent it (for example, by visualizing it) in mind in its entirety together with the caption. Rather, attention and short-term memory can be understood as pointer systems that reference objects in the world to pick those aspects that are then picked and represented for explicit reasoning.

Another aspect that is not captured by research on mental processes pertains to those activities (sometimes referred to as 'routines') that we engage in because this is the way we learned them without questioning or understanding the structure. For example, most people in a given society speak its language, but few know the grammar said to underlie the language in any explicit form. We do not need an explicit representation of grammar (the underlying rule-based systems), and know our way around the world.[9] The research concerned with this domain generally does not speak of skills and explicit knowledge, but about the practices that are characteristic of particular socio-cultural groupings, communities, and so on. Human beings can be competent even without having theoretical that is rule-based understandings of their practices.[10] We do have practical understandings, which, however, normally obtain only local coherence rather than global coherence. Pertaining to graphs and graphing, there too may be 'know how' that is not available in any explicit way. This book is about articulating features of graph-relating expertise not normally covered in the literature.

A final motive for engaging in the research reported here and in writing this book is my dissatisfaction with the deficit thinking that pervade the (especially educational) literature as it pertains scientific and mathematical practices. Because existing models are often based on research with 'experts' and 'novices', they are normative in the sense that novices are said to lack what the experts have. However, I have not seen clearly being untangled personal experience—familiarity with the lived world, its representations, and translations between the two—and structures that are said to characterize 'expert' reasoning. It could be that the changes individuals undergo arise from their familiarity with the topic and an associated change in the relation between the signifiers they use and what they want to do with them. Learning is simply associated with changing ways of doing things in a continuously changing world. Most importantly, it could be that sign (representation, inscription) use is irremediably mediated by

the current salient purposes for the sake of which an individual engages in signing in the first place. It is plausible that 'novices' find it difficult to know what a
graph refers to because of (a) the arbitrary nature of signs and, more importantly, (b) because they do not participate in the same relevance structures as the
normal users of the graph. Even the simplest graph can be perceived in different
ways, depending on its contingent relevance, which then determines how one
elaborates it. Whether it is the height or slope that needs to be salient at a particular point in a line graph depends on the context and determines the correctness of an answer. However, as an individual gets to work in some workplace or
within a community more broadly, which include other graph users, she may
ultimately become so familiar with texts and graphs that she makes no more
distinctions between the graphs as signs and the things that they are said to signify. Such transparency can be observed, for example, when people stop at a red
light or at a stop sign; they do not *interpret* the sign but engage in appropriate
action. In this case, signifier and signified fuse leading many to confuse the two.
At work, scientists do use graphs in such a fused way; to understand the phenomenon and how it comes about, we need to take an anthropological approach.

At this point, then, I have articulated the starting point and endpoint of my
itinerary: scientists' readings and interpretations of selected graphs constitute the
first part of this book, whereas scientists' and technicians' use of graphs and the
building of graphical competence constitutes the second part. I am therefore
moving from a laboratory-based ('in captivity') approach of studying graphing
to studying graphing 'in the wild'.

In the first part of this book, I present the results of my studies on scientists'
responses to graphing tasks. Some graphs were culled from those used in undergraduate training of biologists and ecologists; others were brought by the scientists to the interviews I conducted with them; still other graph use was recorded
at the 'workplace' of scientists and technicians. My studies show that it would
be a mistake to attempt to explain graphing competence as a psychological phenomenon only, for it has been recognized that cognitive ability cannot be read
out directly. Rather, responses are contingent on question format and context.[11]
Previous research has not dealt sufficiently with the problems of getting at the
structures it has presumed to underlie observable graphing actions. A second
problem arises because research has the potential to affect educational policy.
Thus, a focus on cognitive abilities said to reside between the ears and underneath the skull easily leads to the conclusion that some individuals have innate
problems learning graphs.

The studies in this book were not designed to reject or supplant psychological theories of competence in graphing. Rather, my intent is to provide suffi-

ciently detailed descriptions of graph interpretations that can serve as testbeds for my own theory and the theories of others. Thus, unless theories of graphing can deal with the kind of data that are presented here, they are much more limited than is generally assumed. Second, my sense is that we should do with minimal theoretical frameworks and seek recourse to hidden variables only when we have exhausted describing and understanding those aspects of expertise directly available to observation. This is the approach that Bruno Latour has encapsulated in his seventh rule of method for social scientists interested in studying scientists at work. Thus, 'Before attributing any special quality to the mind or to the method of people, let us examine first the many ways through which inscriptions are gathered, combined, tied together, and sent back. Only if there is something unexplained once the networks have been studied shall we start to speak of cognitive factors'.[12] This approach leads us to the practices observable while people go about their business, whether this is related to doing one's best on the tasks in a psychological laboratory or going about the activities at one's workplace.

1.2. NATURE OF PRACTICE

Recent theorizing of everyday activity has focused on the notion of 'practice'. This focus is evident in cognitive anthropology and sociology, but has also had some currency in the more 'hardcore' disciplines of cognitive science and artificial intelligence.[13] Practice pervades everyday activity and precedes subject-object relations. In praxis, we produce both the world of objects and the world of our knowledge about these objects. Practices are patterned actions produced in and through social activity, that is, they are produced in cooperation with others. To share practice is also to share an understanding of the world with others. This 'sharing' arises from the mediating function our bodies play in relating cognition to world: The (social and physical) world structures cognition from early on so that it comes as no surprise that cognition later reflects the structures of the world.[14]

> The agent engaged in practice knows the world but in a way that, as Merleau-Ponty showed, does not establish itself through the exteriority of a knowing consciousness. In some sense, he understands the world too well, without objectivizing distance, as going without saying, precisely because he finds himself there, because he is part of it, because he inhabits it like a habit [clothing] or familiar habitat.[15]

The notion of practice decenters the discussion of knowing and learning by focusing on observable features of members' activities in the process of accomplishing their goals. The difference between cognitive ability and practice perspectives is exemplified by another difference. This other difference is that between the meaning of signs (words, graphs, or formulas) residing in someone's head and the situated use of signs as part of a community's everyday discursive practice and the different role signs (representations) play in activity.[16] Representations are often taken to be general and context independent, whereas they are relational in practice-oriented theories. Traditionally, words (concepts) have been treated as if they had meaning (i.e., referents, sense, or both). However, in the approach I am taking here, meaning is an existential not a property that is attached to signs (graphs, words), lying behind or floating between them.[17] In practice, meaning exists only in so far as these words can take particular places in discourse (or, as Wittgenstein would say, in language games). For example, John Seely Brown and his colleagues discuss the case of children who learned words through dictionary definitions.[18] When these children constructed novel sentences, these sentences were inappropriate. That is, the knowledge of these children was literally use*less*. The children 'knew' these words only in as far as they could recognize them, but placed these words into inappropriate contexts. Words are use*ful* only when we know their places in language games in addition to knowing their particular shape.

There exists a close relationship between practices and the goals and intentions of those who enacted them.[19] This also holds for the use of graphs. Outside schools, people use graphs to achieve certain ends. A newspaper editor may illuminate the message of an article by graphing the relationship between alcohol consumption and cholesterol levels. A physicist may enhance an article with a graph that underscores a claim about the signal to noise ratio of a new detector. Finally, a cognitive scientist may use plots to highlight the different correlations between posttest scores and prior knowledge for different types of learners. In Chapters 6–8, I show how the everyday activity of a person situates and provides context for the knowing and learning of the inscriptions that they are working with.

'In captivity' (school, psychology laboratory), however, people make graphs for the purpose of making graphs. That is, they have few (and in some contexts no) opportunities to populate graphs with their own intentions and therefore experience them in the fundamental mode of 'in-order-to'. Related to this is a restriction in the range of choices available to individuals working with graphs. For example, while it is quite normal for scientists to abandon the interpretation of a graph (see Chapter 8), doing so at school or in a formal study on graphing leads to the attribution of failure (see Chapter 4). Such failure, as attributed to activity by an outside researcher, is therefore an artifact of method. Failure as viewed by a practitioner, in contrast, is much part of ongoing activity as success. Therefore, if we want to understand competent graphing and how people become competent, we really need to observe graphing practices and how they change in everyday situations. This is what I do in Chapters 6–8, which show competent activity that provides a striking contrast to what scientists do on graphing tasks.

One of my ultimate concerns for doing research on graphing among scientist lies with the design of learning environments for students in general and for science students in particular. Taking a developmental perspective on practice has significant implications for the way in which we conceive of school-related activities designed to allow the appropriation of graphing-related competencies. In the public debate about education, one often hears the claim that individuals need to learn knowledge and skills before they can legitimately participate in a practice. However, complex scientific practices such as conducting physics research are appropriated as newcomers with little prior knowledge legitimately but peripherally participate with more experienced members.[20] These trajectories of individuals within communities of scientific practice share a lot in common with those that individuals traverse in traditional practices such as tailoring or midwifery.[21] This is because we bodily participate in the world, we are formed by the structures of this world without being aware of it, being immersed in this practice. Through increasing participation and learning, newcomers turn into old-timers and core participants in the practice as they increasingly reflect the explicit and, more importantly, the tacit relevance structures of the field.

Everyday mathematical practices such as using numbers, graphing, scaling, or timing are inseparably submerged, tied up, and dispersed within a dense network of ordinary conversational activity.[22] Thus, 'the knowing and learning of mathematics is situated in social and intellectual communities of practice, and for their knowing of mathematical knowing to be active and useful, individuals must learn to act and reason mathematically in the settings of their practice'.[23] Consequently, learning mathematics means that a person acquires aspects of an intellectual practice, rather than just acquiring any information and skills. Participating with others affords acquiring a practice without knowing explicit rules for doing what we do. Participation in mathematical practices allows discursive and material practices to be acquired through mutual observation, emulation, and correction in shared situations. In the process, any mathematical expression achieves meaning as part of observable practices and their circumstances because the consensual culture of mathematics is available only in the actions of doing intelligible mathematics. That is, we can understand a lot about mathematical practices such as graphing when we know how and for what reasons these are employed in the particular contexts that we can observe them. When observation is not possible or intelligibility not immediately apparent, such as when proofs are complex—mathematical proofs and solutions to complex theoretical problems in physics involve the computers—the legitimacy of the work done by a mathematician or theoretical physicist can be seriously questioned.[24]

Graphing consists of an array of signing practices that include talking, writing, gesturing, and drawing. These practices are co-deployed so that any single one can only be understood within the network of practices, that is, in its relationships to other practices. To do science, one has to be able to juggle and combine the various practices that are co-deployed to make scientific 'concepts'. In the following section, I describe graph interpretation as a semiotic practice. These practices relate to natural phenomena not because of some logical necessity but because they are associated with the conventions established in each field. In scientific and engineering communities, graphs have three major purposes: graphs are semiotic objects that constitute and represent other aspects of reality, graphs serve a rhetorical function in scientific communication, and graphs act as conscription devices that mediate collective scientific activities (talking, constructing facts). In Sections 1.4–1.6, I focus on each of these aspects by weaving together findings from science studies and my own work.

1.3. READING GRAPHS AS SEMIOTIC PRACTICE

> When Detective Bertrand entered the room, the curtains filled with air by the wind, the stereo was playing Arvo Pärt's *Stabat Mater*, a couple of whiskey glasses on the coffee table, a single cigarette butt in the ash tray. Professor Ashmore was lying in the corner; a trickle of blood had left a trace from his ear down his neck and behind the collar of his Baytown University sweatshirt. Asking the officers who accompanied him not to disturb anything until he tells them otherwise, Bertrand begins to inspect the site…

Many a detective novel begins in this or similar way, or has some such situation as its major component. We can see from the evening news that police officers engage in such work every day, picking up some bloodied glove and, without touching it, placing it in a plastic bag. As we read on in the novel, or follow the news, we learn that the core of the detective and police activity is to secure anything that could be a clue to what had gone on, which could help in reconstructing the events that had led to Professor Ashmore's death. In this work, each object, its placement and state, or its attributes can potentially be viewed as a sign that tells something about the events leading up to the present state of the room. That is, because detectives do not have access to the event as they happen they take the next best things, signs, to tell them about events past. Signs are standing in for, re-present or make present again, worldly objects and events. In his effort, Bertrand cannot simply induce the explanation from the 'facts', for it is not clear at this point what the relevant facts are. Given that the music is playing, it is possible that the time of death predates Bertrand's arrival by less than the 60 minutes of the CD—unless, of course, the CD player was placed on repeated play of the same CD. But this fact itself only comes to life if it is constructed by someone, that is, the CD player set on 'repeat play' has to be perceived before it can be a sign of something else. Is the nature of the music itself—a curious transformation of medieval plainchant and twentieth-century minimalist tendencies—a clue to the visitor? Who smoked the cigarette? Who drank the whiskey? Was there more than one visitor? Is it relevant that the window was open? Are there things that we do not (not yet) perceive that provide clues to the events? Above all, there is the question whether one could look differently at the situation and therefore see something that could serve as a clue.

In his work as a detective, Bertrand has learned to iterate the evidence. He identifies possible clues in the room, and then tries to relate them to the narratives Professor Ashmore's relatives, his colleagues, or his club mates provided him with. By going back and forth between the narratives and the room at the time of his first entering, Bertrand reconstructs a network of relations between the objects and the life of Professor Ashmore (as far as he knows it through the narratives). In this reconstruction, he draws a lot on his experience as a detective and as a real person inhabiting this world, knowing how people behave, and some of the deviant behaviors that are often associated with crimes.[25]

Reading graphs has a lot of similarities with the activities of my fictional detective Bertrand. Readers of unfamiliar graphs are faced with traces and clues that they can use to reconstruct possible scenarios that may have led someone else to construct the graph as a representation. This work, as the present book shows, involves a lot of understanding of local particulars and of the modes of data acquisition. It turns out that even experienced scientists have problems when the graphs and the situations they represent are not familiar, even when the graphs are from the scientists' own domain and part of the staple in undergraduate courses and introductory textbooks (see especially Chapter 4).

In the first part of this book, I follow the routes marked by semioticians and philosophers concerned with understanding and meaning and make the assumption that the lines making graphs and all additional textual information constitute text.[26] These texts are therefore complexes or networks of signs that readers have to perceive in such a way that it allows them to reconstruct a consistent and coherent narrative of a situation that the graph can be said to describe. If the graph is the result of an experiment, we can imagine that the reader considers her work as trying to find out what the original situation was like in which the data were collected. Or, if we are dealing with a graphical model, the reader might consider her work as trying to reconstruct possible scenarios that would lead to the patterns expressed in the graph and accompanying text. Thus, reading and interpreting graphs are concerned with texts (signs) and with the entities that these texts are said to signify (referent). Understanding the cognition relative to this process of reading therefore involves developing an understanding of the processes that are observable while an individual (or group) textually elaborates the graphical representation such as to provide a description of the referents.

The studies in this book are concerned with the performative dimensions (praxis) of sign use, specifically with an understanding of how highly educated individuals construct meaning from graphs. Here, each individual is understood as being engaged in a process of semiosis, identifying signs and establishing sign-referent relations by elaborating interpretants. The starting point for this investigation is a triadic relationship between sign, its referent (content), and all the interpretants that mediate the relationship between sign and referent.[27] For example, when one of the scientists in my database used the word 'caribou' (uttered or written) s/he in fact made use of a sign (expression) /caribou/. The referent (content) of /caribou/ consists in a concept, itself expressed as «caribou».[28] The content segments can correspond to physical entities (caribou, reindeer moss, birch), abstract concepts (habitat, food web, energy), actions (feeding, giving birth), genera and species (animal, plant) as well as directions and relations (above, before, after, below).

The interpretant can take many forms, usually other signs, including outline drawings of caribou, photographs of caribou in the Arctic, translations into a synonym (/reindeer/) or another language (Fr. /renne/, Ger. /Ren/), or any metaphoric translation (in conjunction with c = /Santa/, to signify «Christmas»). Every interpretant (sign, expression, or sequence of expression that translate a previous expression), besides translating the referent (content) of the sign, increasingly articulates our understanding of it. The sign-referent (S-R) relation is such that another relation between sign-interpretant (S-I) can be grafted onto the first.

Sign, referent (content), and interpretant are all to be understood as segmentations of matter. For example, we use mouth and throat to produce the sounds ['kæribu:] (International Phonetic Association) that others hear as /caribou/ or we use a pen or keyboard to produce a trace /caribou/ on paper; in each case, the sign requires matter. Signs are not limited to the spoken and written but can be entirely natural; traces in the sand that a biologist 'reads' as the passing of a lizard are also signs. The content of a sign, that is, its referent also consists of matter segmented in a particular way from the continuum of matter that is the world. This segmentation of matter may be another (written or verbal) expression, a natural object, or any other form of re-presentation that lies between the extremes of nature and word.[29] To interpret, then, means to define one portion of the continuum (sign), which serves as a vehicle for another segmentation of the continuum (referent, content). To define the relationship between the sign and referent portions, we make use of yet further portions (interpretants). In this way, we end up with signification as the result of networks, in which each node represents a segmentation of matter, without however having made a postmodern declaration that 'everything is text'. In other words, underlying all existence and signing processes is a continuum or 'dynamic object'. However, this object is always conveyed through and therefore mediated by some sign system. We therefore perceive and experience the objective world through the contents of signs or 'immediate objects'. But sign systems are culturally specific ways of organizing the immediate world into fields, axes, and subsystems which leads to incoherent ways of cutting the world across academic disciplines, cultures, subcultures, etc.

The content (referent) and sign are not simply related but part of a four-parameter relationship that binds and constrains the values of each parameter. Thus, content R and sign S are related via the relation $R = f_r(S,c)$ where c are the contextual constituents of the sign and r the social conventions that regulate sign use.[30] It is notable that there exist many, perhaps an infinite number of sets $\{S, R, c, r\}$ that fit the relation. It is evident that this relation changes over time and as a function of the disciplinary developments. Thus, both the interpretants of /atom/ have changed significantly over the past 2000 years, and with it the content «atom»—as a look into any physics, chemistry, or physical science textbook will show.

Each sign is modulated by articulated and non-articulated (non-) constituents c. For example, in the context of a debt-income graph (Figure 1.1), the signs /debt/ and /income/ are articulated constituents of the respective lines. Each articulated constituent assists the reader to establish referents external to the graph. Thus /amount/, /income/, and debt all are articulated constituents of the primary signs, the two line graphs.

Unarticulated constituents (also conventions for use of axes) are aspects of the graph that are constitutive of its content, but not made explicit in the representation. For example, the fact that the abscissa values increase from left to

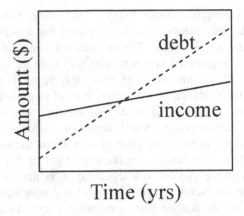

Figure 1.1. Confounding of mathematical and linguistic conventions.

right, the ordinate values from the bottom to top is central to the accepted con-tent of the graph, but is not made available. Also, it is not made explicit that debt really refers to a negative amount in terms of accounting although it has a posi-tive value in the present graph. In banking, for example, and embodied in spreadsheets, losses are 'signed' negative; thus total movement of accounts are evaluated by *adding* losses and gains. There are also articulated non-constituents (e.g., existing grid lines do not constitute the meaning of the graph) and unar-ticulated non-constituents (e.g., orthogonal projections of each curve point onto axes that attribute specific values to the point). Because graphs are used to re-duce rather than provide extraneous information, I am less concerned with the non-constituents of a sign.

1.4. GRAPHS AS SIGN OBJECTS

An important point seldom addressed in the literature on graphing is the rela-tionship between a graph and the reality it constitutes. In a frequently used test item, research participants are required to imagine walking across the room, and then are given a choice between various graphs, or are asked to draw a graph, re-presenting the imagined walk.[31] Here, researchers assume a simple isomorphic relationship that can be expressed {'physical experience of [or imagined] walk' ↔ graph}. In such research it is often assumed that a graphically literate person should be able to read specifics of the walk from the graph or construct a graph after making (or imagining) a walk *because* of an isomorphism between graph and world. However, such an approach overlooks that signs are arbitrary in their relationship to that which they signify. Even iconic relationships such as between the drawing of a cat and an instance of a living cat (furry creature that meows) have to be learned; iconic relationships are strongly influenced by cultural conventions.[32] Furthermore, signs do not form independent conglomerates but interact according to more or less specified grammars. These grammars, too, are conventional rather than derivable from an inherent logic in the representing expressive form.

One can often find graphs that confound linguistic and mathematical conventions such that the sign of a quantity is sometimes embodied in the variable and sometimes in the operation. Figure 1.1 exemplifies graphs of the type often found in the media; here income and debt are plotted into the same direction of the Cartesian axes by using the labels 'debt' and 'income' (or birth rate and death rate) to indicate the direction. However, the conventions followed in such plots are different from those in mathematics observed when plotting the functions $f_1(x) = x$ and $f_2(x) = -x$. Here, the expectation is that one function has a positive slope whereas the other is plotted with a negative slope down from left to right. Furthermore, plotting an increasing debt as a curve that moves right and up is also contrary to the everyday conception of 'going into the hole', 'business is going down', and so on. We can learn two lessons from these considerations. First, such opposing conventions are likely to make it more difficult for newcomers to learn at what point one or the other convention has to be followed. Second, the grammar of the convention cannot be taken from the representation itself, unless it is explicitly included in the graph/caption/main text information.

Recent studies in philosophy, history of science, and ethnomethodology provide descriptions of the considerable work that it takes to relate independent semiotic objects such as graphs to the phenomena they come to represent. These relationships are eventually accepted because the stability of the system of conventions regulating this work, rather than because of a priori ontological connections. As part of this work, semiotic objects, perceived phenomena, and available tools (technical, linguistic) change and are mutually adjusted until they can be regarded as isomorphic.

In a study of physics lectures, Ken Tobin and I have shown the complexity of the process that translated a simple phenomenon (a ball rolling down an inclined plane) into different sets of tables and graphs.[33] Ultimately, the professor arrived at sets of sketches, which to him stood for other more cumbersome verbal and mathematical descriptions of motion. Most importantly, these translations were not evident to the students attending the lecture. When tested, these university students could not reproduce the graphs, the conceptual underpinning

a.

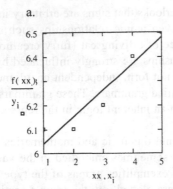

b.

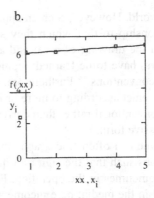

Figure 1.2. Two representations of the same data set. a. If there is theoretical reason for a relationship, the author would emphasize this by bringing out the slope. b. If there are strong theoretical reasons that the relationship is spurious, a re-presentation of the data would be chosen that de-emphasizes the appearance of a relationship.

of the domain the graphs referred to (Newtonian motion), or to the translations between phenomenon and its different re-representations in data tables and graphs. Thus, rather than assuming a simple mapping of a walk across the room onto a graph, it would be more appropriate to study the translations, transformations, and transpositions required to construct the relationship between the event and its re-representation. Even more curious, the professor incorrectly mapped an example, which he intended to help students understand the graphs, onto the graphs that he and his students were looking at.[34] When he used an analogy from work, 'receiving income' and 'receiving raises', he made inappropriate analogies between the distance-time, velocity-time, and acceleration-time graphs and the corresponding notions of salary, hourly wages, increase in wages, etc. In a cognitive paradigm, we would describe this as a failure to 'transfer'. Yet the physicist involved was very successful in publishing and getting grants, and had volunteered to teach the particular class where we conducted the research.

Graphing is not a simple practice. But it is frequently not evident at all (even more so for novices) whether something is a sign (semiotic object) that points to an interesting' henomenon or merely to an artifact. Thus, whether the set of paired numbers $\{(x, y)|(1, 6.1)\ (2, 6.1),\ (3, 6.2),\ (4, 6.3)\}$, which can be represented by an x-y line graph represents a phenomenon of interest or an artifact of measurement is a matter of context. If there were strong evidence (e.g., from theory) that there should be a phenomenon, then a researcher would choose to

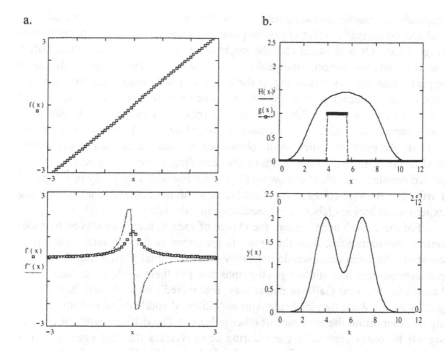

Figure 1.3. Two techniques for coaxing signals from otherwise inconspicuous data. a. The original signal appears to be a straight line [top]. Applying first and second derivative gives 'clear' indications of a signal [bottom]. b. The original signal appears a single hump [top]. Using the technique of 'unfolding', and assuming that the instrument had a square aperture function [top], reveals that there is a double peak [bottom].

scale the ordinate from 6 to 6.5, a steeper curve results (*Fig. 1.2.a*). When plotted with the abscissa scaled from 0 to 7, there is hardly anything to see (*Fig. 1.2.b*).

Physicists frequently use the mathematical practices of integration (or spline functions) to turn sequences of data points into smooth graphs and give them a more familiar look, or use differentiation to obtain a dominant peak where data show a continuous and smooth function. How one can coax a clear 'fact' from a seemingly straight-line signal is simulated in Figures 1.3. For example, a physicist might get a signal from an electronic counter that is almost straight. However, because she assumes that the straight line really hides a phenomenon, she might manipulate the original electronic response or the count in a way that corresponds to a mathematical differentiation. New signals can be constructed (first and second derivative) that appear to point to a clear signal from a phenomenon (*Fig. 1.3.a*). Or a physical chemist might observe a broad peak coming off a spectrometer, but expects two peaks to be 'hidden' underneath it. Such hiding appears, thus goes the reasoning, if the instrument had an aperture function (accepted more radiation than at any single frequency), for example, in the form of a square function (*Fig. 1.3.b.top*). Making a reasonable estimate as to this aperture function and using a mathematical procedure called 'unfolding', he can separate the peak attributed to the phenomenon from the characteristics attributed to the of measurement width of the recording device. As a result, the expected double peaks clearly show up (*Fig. 1.3.b.bottom*). Especially in Chapter 8 I articulate such practices in considerable detail in the context of a cutting-edge experimental biology laboratory specializing in fish vision.

For the experienced person, the choice of axes is therefore driven by theoretical considerations; how the graph is presented is furthermore a rhetorical question. But even scientists do not easily separate signal from noise. For example, astronomers had studied photographs and put them aside as uninteresting. Later, after the first Galilean pulsar was 'discovered', they studied their photographs again. What they had first read as 'blotches' suddenly turned into 'good' signals supporting the existence of other pulsars.[35] The difficult work of reading signals becomes especially clear during controversies that are eventually declared as hoaxes.[36] Thus, serious scientists had supported N-rays and cold fusion until someone else constructed the phenomena as artifacts: they turned 'good' signals' into 'blotches' and 'wiggles'. This issue is not a trivial matter, but part of the daily work of scientists, politicians, and economists. How does one come to make such distinctions?

1.5. GRAPHING AS RHETORICAL PRACTICE

Research in scientific laboratories has shown repeatedly how scientists use graphs as rhetorical devices that, in conjunction with their captions, elaborate the main texts.[37] Graphs are used to highlight certain features of researchers' constructions about nature. All construction of graphs necessitates translations that produce nature-representation relationships, which are then moved through a cascade of representations until the final inscription has been produced. Rarely are the graphs obtained in some direct way. As I already exemplified in the previous section, scientists have developed many techniques to coax those graphs that are constitutive of the phenomena that they are after from the original, noisy data. Scientists make assumptions not only about nature but also about their instruments used to constitute proof of the phenomena.

In addition to using mathematical techniques described above, researchers employ a variety of other processes to make these features stand out, and to eliminate anything that could be distracting. Geneticists cut and paste audioradiographs used in the 'identification' of certain DNA sequences to obtain graphical re-representations on which DNA sequences are distributed along the y-axis, and various samples constitute the categorical x-axis.[38] Change of scale, change relative to some reference value, and the spatial representation of time are yet other techniques employed to construct phenomena through graphical representations. Howard Wainer illustrated an interesting example of how a journalist's claim that was supported by a graph could be inverted.[39] The original claim that student achievement remained constant over a ten-year period despite increasing education expenditures was supported by a double y-axis graph. By manipulating the scales for both y-axes separately, Wainer created a graph suggesting that student achievement gains 'soared' while education expenditures remained stable; a simple change of scale supported opposite interpretations.

Seldom does the literature on graphing clarify that much of scientists' work involves differentiating blotches and wiggles into just that and 'real signals'.[40] Here we can see the close relationship between graphs as a means to constitute a phenomenon (the scientific experimental manipulations that culminate in a graph through which the phenomenon comes to life), and graphs as clear and unmediated evidence of a phenomenon that enters everyday discourse as 'fact'. The purpose of manipulations is to help others with less or no experience see what a researcher has seen. But in these manipulations, phenomena are constructed which otherwise do not exist; these graphs not only bring out phenomena, being constituted in and through scientists' descriptive work but are used as evidence for the phenomena.[41] The precarious relationship between phenomena of the physical world and the representational practices of science has been repeatedly pointed out, and described as a sequence of *translations* and *evidence-fixations*.[42]

Whether an inscription is recognized as a legitimate representation of a natural phenomenon within a community depends on the degree to which the practices are accounted for and measure up to shared, legitimate communal procedures. The degree of an inscription's credibility therefore depends on the extent to which the associated data transformation practices correspond to explicit and implicit communal norms. For example, astronomical data are often acquired by using an optical sensor that converts light intensities into a digital signal.[43] These signals are initially recorded on magnetic tape in a serial fashion.

Each data point is later translated into a number on a 256-step scale. Depending on the ultimate purpose, astronomers and specialists in digital image processing produce different two-dimensional visual displays and thereby different 'facts'. The results of visual image processing differ according to the *purpose* for the sake of which they are produced. Images that are accepted by astronomers for one context (e.g., publication in *Scientific American*) are not accepted in another (e.g., conferences, astronomy journals) because the requisite transformations are associated with different degrees of credibility within the community. In this way, order is not simply constituted, but it is actively *'exposed, seized upon, clarified, extended, coded, compared, measured, and subjected to mathematical operations'*.[44] In Chapter 8 I provide an in-depth description of how this work is achieved in the context of an ecological research project concerned with the construction of facts about a particular lizard species.

My review of the descriptions of work in the sciences suggests that graphing involves a lot of what Martin Heidegger called handwork ('Handwerk') or what Pierre Bourdieu refers to as practices. Practices do not require explicit coding, but constitute embodied forms of knowing available only to the 'logic of practice', a practical sense ('sens pratique').[45] I designed the investigations reported in this book to get a better understanding as to the nature of graphing. Particularly, I wanted to better understand the nature of the practices and as to the extent of which these practices are related to general skills, or whether these are in fact closely tied to the very contexts in which graphs have evolved.

1.6. GRAPHS AS CONSCRIPTION DEVICES

As a core scientific practice, graphs have other functions besides being inscriptions (semiotic objects) that are used for rhetorical purposes: As conscription devices, graphs bring together and engage collectivities to construct and interpret them. As conscription devices, graphs serve as the material grounds over and about which sustained interactions occur, and which serve in part to coordinate these interactions. In this, graphs function in the same way as other visual re-presentations that enlist the participation of those who employ them in the laboratory or in scientific publications, since users must engage in generating, editing, and correcting graphs during their construction. Here, then, graphs are central to interactions among scientists, a feature that will be particularly salient in Chapter 8. Graphs constitute a shared interactional space that facilitates communication because of their calibrating effect on what can be taken as shared, and what has to be negotiated when it becomes obvious that it cannot be taken as shared. Graphs can thereby become symbolic places where researchers meet with their phenomena. Working together over and about graphs allows scientists to 'blur the boundaries between themselves as subjects and physical systems as objects by using a type of indeterminate construction that blends properties of both animate and inanimate, subject and object'.[46] Such blurring can occur because the graphic representations constitute in a referential way both the scientists and the physical entities as simultaneous and co-existing participants in the ongoing event. Thus, graphs are not only tasks to be accomplished through talk, but they also make talk meaningful. In this, talk and graphs are in a reflexive relationship such that one draws on the other to be meaningful.

Another study of the role of graphs and the activity of graphing in scientific laboratories was concerned with the connections between specific research documents and the realities that are assumed to underlie these documents. Despite the philosophical and methodological arguments that question these connections—and even characterize them as indefensible, implausible, and even impossible—scientists succeed in characterizing them as good enough or practically adequate within the contextures of their collective work. Steve Woolgar shows that the presumably underlying phenomena are constituted in and through descriptive work, and that in this work scientists assign alternative versions, invoke mediating circumstances, and so on. The result of these sessions is therefore best described in terms of emergent and collectively achieved interpretations. For example, there is coordination work involved to make two instruments exchangeable:

> The graph cannot be seen as an obvious manifestation of a series of effects because strong uncertainties surround its interpretation. Only certain details can be indisputably ascertained and, very often, it is necessary to negotiate in order to establish the significance of a peak, or of a slope. Neither 'inductivity' nor 'deductivity' correctly describes the analytical strategy displayed by scientists: the coincidence of two signs, one on the inside and one on the outside of the graph, reinforces their mutual reality.[47]

Thus, how a graph is to be interpreted and what it 'means' depends on collective efforts and goes beyond the dichotomy of inductive and deductive styles of research. (Foreshadowing the framework I present in the next chapter, the semiotic approach offers a third process for mapping signs and their referents: abduction. Abduction is a tentative and hazardous tracing of a system of signification that allows the sign to become an articulation of understanding.) The claims in the quote are also consistent with my own research that suggested a concurrent stabilization of signs (features of a graph) and their referents (the natural objects they are said to refer to).

As a final example, let us look at the role of graphs in the ongoing work on a research vessel operating in the estuary of the Amazon River.[48] In this work, graphs and graphical displays plays a central role in coordinating the activities of individuals from different research disciplines. We already know that the same document, when used in a different discipline leads to different practices, and characteristically different language and discourse. However, there is also ample research to suggest that the same document in the same discipline but in different settings leads to different practices. However, when people from different disciplines work in the same site, they assume that the documents (e.g., graphs) are the same. Thus, graphical displays serve to coordinate the activities and the discourse. Goodwin showed how the surface of graphical displays becomes a work area, a site where cognition is enacted as a distributed phenomenon and where perception is enacted qua social process. Perception is therefore instantiated not primarily in the brain but rather in the social practices enacted over and about the unfolding graphical displays.

There is evidence that for some learners, graphs can serve as important anchors for the students' activities, for the negotiation of meaning, and for the topical cohesion of the conversations over and about the tasks they are engaged in. At various points I provided analyses of students' collective sense-making activities over and about graphs.[49] These analyses show that graphs are not only the objects of students' talk, but also provide students with additional communicative resources. Students point directly to data points, lines, and axes, use gestures to indicate trends, or invent indexical labels to describe aspects of the graph (e.g., 'question mark graph' to distinguish a graph resembling the curved shape of a question mark from a straight line graph). The physical presence of the graph supports the topical cohesion of the emerging conversation. Through their interactions over and about graphing tasks, students become competent users of graphing practices. This competence has a double nature in that students become proficient in using graphs to constitute and re-represent phenomena of their interest, but also to understand the relevance of graphing. Students no longer see graphs as unequivocal representations, but see graphing as a goal-dependent practice of re-representing.

1.7. CONCLUSION AND OUTLOOK

In this chapter, I make an argument for taking a cultural anthropological approach to graphing that allows us to articulate the practice in terms of the contingencies of everyday activity, cultural conditions, situated nature of cognition, etc. In particular, I focus on the nature of sign as part of cultural activity. In the remainder of this book, I will take such a cultural anthropological perspective on scientists' (and one technician's) graph-related interpretative actions in two different contexts. In Part I, *Graphing 'In Captivity'*, I report on the results from a study in which 16 scientists, mostly ecologists, were asked (a) to interpret with three graphing tasks culled from an introductory course and textbook in ecology and (b) to explain a graph from their own work.[50] (See Appendix for the tasks and their cognitive analysis.) I show that the phenomenon of graphing expertise is much more complex than theorized in current models. In particular, it turns out that there are different levels of graph use, ranging from blatant rejection of a particular graph as useful (Chapter 4) to their transparent use (Chapter 5). Somewhere in between we find that an interpretation unfolds in a way that shares similarity with my detective analogy. The results of this first part suggest that graph-related expertise is highly dependent on context and familiarity with the referent of a graph as well as details of its production. In Part II, *Graphing 'In the Wild'*, I report on graph-use in three different everyday context of the participants—a water technician, a Ph.D. student in field ecology, and two researchers from an advanced lab doing vision research in salmonid fishes.

Ultimately, my studies among scientists also provide mathematics and science educators with different images of what it means to know graphs and graphing in a community that is often held up as a standard that we should be working towards in schools.

PART ONE

GRAPHING IN CAPTIVITY

TWO

From 'Expertise' to Situated Reason

The Role of Experience, Familiarity, and Usefulness

A considerable number of experienced scientists were not successful on different aspects of three graphing tasks, despite their provenance from and frequency in undergraduate textbooks of their domain. The results suggest that only parts of some session protocols are consistent with existing cognitive models of graph interpretations. The chapter first unfolds a traditional analysis of interpretation and then provides details from one protocol to show why and in which way such an analysis has shortcomings. Even if not made thematic in a protocol, successful performance occurs against a background of a meaningful totality, a dense web of signification, in which graphs are used for the sake and in the service of particular purposes. The analyses show that the less successful non-university scientists also considered the graphs as unrelated to real ecological systems and therefore as useless.

2.1. HOW COMPETENT ARE 'EXPERT' SCIENTISTS?

Research on graphing often operates on the distinction between expert and novices. From the reasoning of 'experts', so goes the story, we learn a lot about the special ways of thinking that distinguish them from others, less experienced in the field. I maintain, though, that if the tasks are from a rather familiar domain, so that over-learned procedures, rationalizations, and tacit understandings can be brought to the task we learn little about how experts work when the going gets tough. That is, it is of little value to ask an economist to 'interpret' a supply-demand graph, which has become a stereotypical over-rationalized case in the field. Such tasks do not elicit the 'ethnomethods' that characterize the work of the experts in their daily work but personal and cultural rationalizations. Much less frequently are 'experts' asked to work on tasks in unfamiliar domains. When expert writers and expert historians are in situations where they cannot draw on rationalizations of their activity, the image of smooth and efficient expert that had come from the research on chess and physics problem solving, becomes more qualified.[1] For example, one study described the initial interpretative effort of a university-based historian as consisting of 'much cognitive flailing'.[2] In the face of lacking detailed content knowledge, the historian regained intellectual footing, worked through confusion, and resisted the urge to simplify. As it turned out, the scientists in my study had even more problems than the historians have had, although I had culled the three graphing tasks from an introductory course in ecology (see Appendix). But before presenting scientists' levels of performance, here a brief general description of the scientists. (Specific information on particular scientists is provided in the context of detailed protocol analyses. Pseudonyms are used throughout.)

2.1.1. Scientist Participants

The sample for this study was constituted by 16 practicing scientists.[3] The scientists were generally highly successful, some having received doctoral and postdoctoral awards, were recipients of national and international awards for their publications, and had sizeable research projects in terms of the funding attracted from private and public sources. They had been recruited at three universities and several federal and provincial governmental research branches. All had obtained either a M.Sc. (7) or a Ph.D. degree (9); among the former, three were currently working on their Ph.D. degrees. All had five or more years of experience in conducting research for the purposes of publishing the results in reports and scientific presentations. All but three scientists (two had completed physics degrees, one obtained a degree as a forest engineer) can be classified as ecologists.

The interview sessions lasted between one and two hours, which amounted to between 4,500 and 10,000 words per session. In several instances, because of repeated interviewing, we obtained between four and ten hours of materials from the same person; in such cases, more protocol materials were available. The interviews were conducted to accommodate the scientists. Eight scientists chose to conduct the sessions in their own offices, the others were recorded in my office or laboratory.

2.1.2. Survey of Performance

The tasks consisted of three graphs and associated captions (see the Appendix, which also includes analyses of each task). All three graphs were of types that appear with high frequency in introductory university ecology textbooks. When the scientists' levels of performance are compared with those of the teachers,

Table 2.1. Frequency of correct 'standard' answers

	Frequency (%)	
Task	Scientists (N = 16)	Teachers (N = 14)
Distribution		
Adaptation	56	44
Population Graph		
Unstable Equilibrium	75	44
Stable Equilibrium	63	22
Largest increase in N	6	0
Isoclines		
Essentiality	50	0
Substitutability	50	0
Complementarity	50	0

their characterization as 'experts' appears justified (Table 2.1); on all aspects, the frequency of correct answers was higher among the scientists than among the teachers. Yet one can also see from the table that the 16 scientists were far from perfect when it came to provide more than a literal reading and to arrive at the standard interpretations of the graphs. This was the case despite the fact that the graphs are similar to those encountered by students in an introductory ecology course and despite the fact that these types of graphs are standard for introductory textbooks on the subjects. The data also showed an interesting difference between university-based scientists and those working in the public sector. Using the number of correct interpretations as criterion variable, scientists who are based at the university or college level tended to be more successful than their non-teaching colleagues ($t(14) = 3.88$, $p = .002$). Clearly, the university-based scientists, involved in teaching or serving as teaching assistants were more familiar with doing such interpretive tasks (because they teach their students to do them) or with the materials than the non-university public sector scientists. Surprisingly, and contrary to the assumption that 'graphing' is a core scientific skill, a considerable number of scientists expressed difficulties reading the graphs (Table 2.2). Thus, with varying frequency, scientists suggested that a graph was 'a challenge to interpret,' 'not something I am dealing with,' 'a bad graph,' 'Christ almighty, confusing,' or 'Why do people make graphs like this?'[4]

In the case of the Distribution graph (see Appendix), 56% of the scientists causally linked the different positions of the distributions along the elevation gradient to the different photosynthetic mechanisms (C_3, C_4, CAM) or explicitly specified differential adaptation as the cause for the data as represented. The

Table 2.2. Frequency scientists expressed "difficulties" or lack of meaning

Task	Frequency (%)
Distribution	31
Population Graph	13
Isoclines	50

scientists did somewhat better on identifying the intersections on the Population graphs as stable equilibrium (75%) and unstable equilibrium (63%); only one scientist identified the largest increase of the population size where the function $(b[N] - d[N]) \cdot N$ was maximized; all others, in about equal numbers, suggested those abscissa values where $b = b_{max}$ and $(b - d) = (b - d)_{max}$. Finally, half of the scientists (50%) provided readings of the isoclines that are consistent with the concepts of 'essential,' 'substitutable,' and 'complementary resources.' In general, those scientists who taught at a university or college tended to be more successful than their non-teaching colleagues. Table 2.1 also shows that the science teachers provided fewer standard answers than the scientists did. Most frequently, the teachers did not perceptually isolate and therefore interpret those elements of the graph that are required by a standard reading (similar problems in the scientists' performances are analyzed in Chapter 4).

Scientists' readings of their own graphs differed strikingly from those related to my tasks. These graphs provided transparent access to, and representation of, real-world situations. These situations were rich in textures that included conceptual and methodological aspects, and, furthermore, also historical, economic, and socio-political details of the context in which the data were collected.

2.1.3. 'So What?' and 'What Next?'

My starting point were these disparities between the different readings and the fact that scientists at some time had been at the same point as the students and teachers whom had contributed to my database. The challenge therefore consists in developing a model of graph reading that accounts for the different types of readings and provides a mechanism for explaining learning (change). In this book, I develop a descriptive model for the kind of readings that we get in different situations. First, when scientists are familiar with a graph and the world (aspects thereof) represented, the two fuse and are used indistinguishably (see Chapters 5 and 6). In this, my data are consistent with other research that reports the transparent and skillful use of graphs in reasoning about instances. Second, the process of reading changes when scientists are no longer familiar with graphs, situations these refer to, or translations between the two. In many situations, such as university scientists who also teach courses, my basic semiotic model (Chapter 1) is sufficient to account for the process. The basic model assumes that graph (and caption) and their parts are recognized by the reader as cultural units (signs) referring to something that the graph is about (referent, content). The task of reading then concerns finding out what the graph is about by drawing on all the available signs as resources in this construction that constitutes the reverse of normal scientific activity that develops graphs given the particulars of the researched system. Typical examples of such readings are analyzed in Chapter 3.

Considerable parts of my database present situations where it is not clear to the reader (scientist) what the relevant signs are. In this case, much of the effort was spent in finding out how to structure the graph/text at hand and in constructing possible referents. This expanded model therefore undertakes to rebuild, from the beginning, the conditions necessary for the reading of graphical representations. Among these conditions is the perceptual structuring of graphs into signifying elements, which semiotics and information processing theories instead accept as data because they are the basic elements in theories of communication and mental representation. The expanded model therefore refers perception back to a stage where representations are no longer confronted as explicit messages. Rather, graphical representations and their captions are then treated as extremely ambiguous texts, akin to aesthetic ones, which are inherently open to different structuring through the experience-dependent interpretive horizon of the individual reader. This is a decidedly phenomenological project. This openness and underdetermination of graphs is analyzed in Chapter 4.

In this book, I show that reasoning with these paradigm cases by a small number of scientists highly familiar with them is not what makes expertise. In fact, it may be interesting to understand how 'experts' can be modeled when they talk over and about over-learned paradigmatic sign forms (e.g., particular graphs, graphical models, etc.). But these cases are far from general, do not model scientists' graphing and graph-reading practices when it comes to the novel situations they often face in cutting edge research, etc. What we get from 'expert' research, then, is a description of a particular case, after the fact rationalizations of over-learned graphs. Such research therefore describes only a fraction of what 'expertise' might mean. Often these cases are not interesting to me because scientists hardly ever use overlearned graphs in their daily work; I am interested in what and how scientists do when they face graphs in their real-time laboratory work. As the following analyses show, the kind of reasoning typical

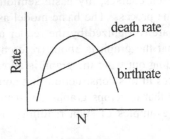

In the derivation of the logistic model, we assume that, as N increased, birthrates declined linearly and death rates increased linearly. Now, let's assume that the birthrates follow a quadratic function (e.g., $b = b_0 + (k_b)N - (k_c)N^2$), such that the birth- and death rates look like the figure. Such a function is biologically realistic if, for example, individuals have trouble finding mates when they are at very low density. Discuss the implication of the birth- and death rates in the figure, as regards conservation of such a species. Focus on the birth and death rates at the two intersection points of the lines, and on what happens to population sizes in the zones of population size below, between, and above the intersection points.

Figure 2.1. In this task, competent ecologists infer that there are stable and unstable equilibrium points for the size of a population. (For an analysis and discussion of the task see the Appendix.)

for expert-novice studies was found only among a small number of university professors whose teaching brings them into contact with these graphs. On the other hand, most researchers in the field from which the graphs were culled, were either unfamiliar or had difficulties reading the graph. They were clearly not expert, though they had many years of research experience in the field, had received many research grants, fellowships, or international awards for their research publications.

In this chapter, I home in on the notion of expertise and show that traditional models are concerned only with a limiting case of expertise, which does not explain the data that I collected for the present book. I focus particularly on the Population graph task (*Fig. 2.1*; see also Appendix) for at least four reasons. First, the task is of the type often used in expert-novice studies, which allows me to cross reference to these studies and articulate their limitations. Second, the graph belongs to the lore of ecology and can be found in this or similar version in any ecology textbook. Third, there are parallels to the graph used by others in an expert study that used a supply-demand graph.[5] Finally, there are interesting parallels with other domains, such as economy (physics, mathematics, and chemistry), that frequently were the source of the models that were subsequently taken over by ecologists. One of my theoretical ecologists described the origins of models in ecology in these words:

Richard: Well a lot of this stuff, the earlier stuff is from economically important areas, like fisheries. A lot of population models, I believe, are derived from economic models that came to us through fisheries. We inherit a lot of our modeling techniques directly from classical economic theories. Especially in population and evolutionary ecology...

Discussing an example of 'expert' reading of the population graph therefore constitutes a paradigm case both for showing the traditional approach, and beginning a critique of it.

2.2. DATA AND MODEL OF ONE 'EXPERT' READING

In this section, I show that under certain conditions we do in fact see scientists who reason in the way some classical cognitive theories describe it. I therefore analyze transcripts in the way we would expect it within a classical paradigm. However, I extend the analysis and show some of the background practices and experiences that make such reasoning possible. Furthermore, I show that even in the case of the 'experts' studied, one can find problematic elements. In the subsequent chapters, I then show that the present 'expert' readings constitute highly specific situations rather than what might be regarded as the general expertise of reading graphs.

2.2.1. Analysis of One 'Expert' Reading

'Expert' readings are not the norm in my database—at least if we follow the description Robert Glaser gave of the quintessential expert who possesses rich networks of highly elaborated knowledge and many problem-solving templates that allow fluid and rapid processing at relative ease.[6] In the present database there are only four scientists who provided readings of the population graph task (*Fig. 2.1*) that can be collapsed onto a traditional expert model; all of them teach ecology at the college or university level. However, the data are much too rich to be collapsed into the model, as I show in the unfolding analysis of this section.

The following analysis concerns the reading of one of the university professors in ecology (Brandon). Brandon had ten years of experience after his doctoral degree (at which time he had already five years of research experience), more than 30 published journal articles, and, according to his departmental colleagues is a very good and successful experimenter with a vigorous research program. His research interests pertain to the population and community consequences of adaptive behavior; his teaching interests lie in the areas of population and community ecology and ecological methodology. He had previously published research articles on ecological consequences of growth and mortality rates

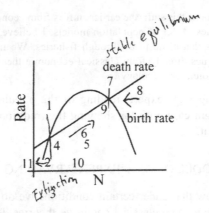

Figure 2.2. Population graph marked up by Brandon during his reading. Numbers in square brackets indicate locations to which pointing and gesturing were directed as indicated in the text.

(mediated by foraging activity or adaptive behavior) and density-dependent effects in populations. These projects are therefore directly related to one or another aspect in the population graph, such as density effect or mortality rates. With respect to the population graph, Brandon is therefore as close to being an expert as one can get it as both his research and teaching interests pertain to the specific disciplinary area from which the graph was taken. That is, the conditions for expecting an 'expert reading' on an a-priori basis were right.

Here I focus on that aspect of Brandon's reading of the population graph that fits a standard explanation of expertise, which I use to model the interpretation in the subsequent section. I deliberately leave out some of his readings that followed the present excerpts, but will return to these aspects in the discussion and critique of the standard expert model. As Brandon read the population graph, he added other signs such as lines, arrows, and words (*Fig. 2.2*).[7]

> Yeah, it only makes sense that way given this- (*Reads from sheet*) 'in the figure conservation of such a species' right. (*Reads from sheet*) 'Focus on the birth and death rates at the two intersection points of the lines, and on what happens to population sizes in the zones of population size below, between, and above the intersection points'. Well. OK. So, anything below this line (*Draws line [1]*), death rate is greater than the birthrate so that means the population size has to go this way (*Draws arrow [2]*) which leads to extinction (*Writes 'extinction' [3]*).

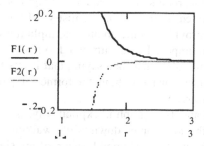 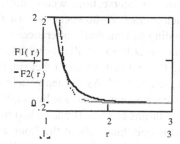

Figure 2.3. a. Forces between two atoms that are not chemically bonded as presented in physics and chemistry book; attractive and repulsive forces are plotted on positive and negative side of y-axis. A stable equilibrium exists where the two forces are equal in magnitude but opposite in sign. b. Positive and negative forces if they were plotted according to the same rules that were used in Fig. 2.1.

Brandon focused on the left intersection (*N[4]*), drew a vertical line through it, then posited a population smaller than the identified point (*N < N[4]*). He used the values of birthrate and death rate (rather than slope or curvature) as signs to be interpreted and identified the former as being smaller in absolute value, then drew the conclusion that the population size would get smaller and, ultimately would lead to extinction. In fact, he did not use '*N*' in his argument but immediately used the interpretant /population size/ in his reading. Although this reading of the event appears straightforward, several issues are important to my alternative approach to reading graphs. Thus, I already elaborated in my reading what Brandon actually said by adding that up on the abscissa means larger and left on the ordinate is coextensive with 'smaller'. This is not self-evident; for example, second-year university students repeatedly felt compelled to ask if they could make the assumption that *N* decreased toward the left of the graph or, alternatively, if *N* increased as one goes to the right.

How the geometric feature of a line is read depends on the constituent signs such that birthrate, though plotted on the positive axis, has to be treated mathematically as a quantity with the opposite effect on a population. In Brandon's approach to the graph, this understanding was implicit. He did not talk or wonder about this fact but treats it as an everyday occurrence. He therefore applied a convention that is not consistent across the sciences. For example, physicists and chemists plot opposite forces between two atoms on the opposite sides of the y-axis (*Fig. 2.3.a*). A plot of the forces equivalent to the rules Brandon used would show an intersection, where attraction and repulsion are equal in magnitude; such a graph is, however, not usually plotted in physics or chemistry. My own reading of Brandon's utterances already provides a clue about this implicitness that arises from familiarity with Cartesian graphs. Thus, I attributed to his reading the fact that N was in fact smaller than $N[4]$ rather than saying that it was to the left of $N[4]$. As the analyses in subsequent chapters show, the former may be a wrong assumption made by graphing researchers.

In this episode Brandon had to think about the graph in explicit ways as little as one thinks about the floor where the foot comes down while walking or about stopping when the traffic lights are (turn) red. Comprehension of the sign is not indicated by the action of staring at it and by ascertaining that it is a referring that occurs. Reading does not comprehend the sign that is at hand but orients to the surrounding world; it leaps beyond signs to the things they are said to be about. We do not interpret red lights and then stop but, taking it as a state in the world, we just stop. Such 'knowledge' of signs does not have to be modeled as explicit knowledge but rather in the way one would model the implicit knowledge that characterizes upright walking or knowing that things fall down when they are let go above ground. Some of this knowledge clearly comes from the metaphorical extension of bodily experience such as in the 'up = more, greater' relation that underlies the ordinate dimension of a Cartesian graph or the 'left to right = increase' that underlies conventional (Western) writing, number-line use, and graphical representation of time. Although these processes are likely to be implicit, we can analyze Brandon's rationalizations in terms of a model.

In the immediately following sequence, Brandon analyzed the graph to the right of the left intersection. Again, the changes in population size did not seem to be derived but were immediately given in the same way that the two lines themselves were given.

> Anything above this point (*Points to [4]*) means that you have birth greater than death, population growth to the point where birth and death equal each other again. So, in this range (*Points to [5]*) we've got population going this way (*Draws arrow [6]*) and above here (*Draws line [7]*) again death are greater than birthrates so the population goes this way (*Draws arrow [8]*).

The use of the descriptor 'anything above...' indicates a greater familiarity such that he knew the referent of the graph without going through the comparisons between birthrate and death rate for a number of points. Thus, he may have picked up directly from the graph the entire referent situation. Here, Brandon's focus shifted to values of N larger than that corresponding to the first intersection (*N[4]*). At these values and up to the second intersection (*[9]*), which he marked with a vertical line, the value of the birthrate is larger than that of the death rate, which implies a net increase of the referent population. Brandon marked this increase with an arrow pointing from left to right (*[6]*), accompanied by the utterance /population going this way [right]/. Finally, for population sizes *N* greater than *N[9]*, he described the death rates as being larger than the birthrates so that the population tends to decrease ('goes this way' *[8]*').

Again, we might easily conclude that Brandon drew an inference that took the result of the comparison of birthrate and death rate as the starting point. There was no doubt that /population decrease/ meant going to the left on the graph. Brandon stated the comparison between birthrate and death rate and then immediately drew the arrow pointing to the left. This becomes a crucial element in understanding the errors a significant number of scientists (and even more those with less scientific training) made in elaborating the references and implications of the population dynamic for *N > N[9]*. By way of foreshadowing the analyses in later chapters, a number of scientists perceived death rate as larger than birthrate in that section of the graph (*N > N[9]*) and elaborated it in terms of /population crash/. That is, they inferred not only decrease and therefore return to the equilibrium but in fact continuous decrease to the point of extinction.

Brandon then returned to make another pass through the analysis and then elaborated the core reference of this graph (in the teaching of ecology): The intersections index stable and unstable equilibria. That is, each physical instantiation of an intersection is a sign that has as its reference an «equilibrium situation». In the process of elaborating this referent through the production of interpretant signs, a process semioticians refer to as *semiosis*, Brandon articulated and repeated previous descriptions of the population dynamic.

> This (*Points to [9]*) is a stable equilibrium, meaning that if I displace the population to lower values then the higher birthrate than death rate means the population will grow until birth and death equal each other. If I increase the population by whatever means, say some random event whatever, so that the population above, is above this and the death rate is higher than the birthrate and it goes back to here (*Points to [9]*). However at this point (*Points to [4]*), imagine that we're sitting on this point, right here, exactly on that point, birth and death equal each other, so the population should be stable, no change. But all you have to do is add a single individual then the population is greater than the equilibrium point and it necessarily goes in this direction (*Gestures [10]*). And similarly we subtract just a single individual then it's below the equilibrium and it goes this direction (*Gestures [11]*). So this is an unstable equilibrium (*Points to [4]*)...

This episode begins with the identification of the intersection *as* a stable equilibrium. Brandon explained why the intersection denoted an equilibrium point and why this equilibrium should be a stable one. This articulation in fact reiterates the procedure: Posit higher/lower population size, compare birthrate and death rate, and draw inference as to tendency of population change. He then repeated the procedure for the left intersection. He first identified it as an equilibrium point (/should be stable/) and then hypothesized the direction of population change to the left and right of the intersection. Because these trends were different at the two sides of the intersection, the equilibrium had to be unstable. We can also see evidence that in the familiar use of signs, there is no longer a distinction between a sign and the phenomenon it signifies. Thus, rather than stating that the intersection signifies «stable population», a situation in which the population size does not vary unless disturbed by some random event, the intersection *is* the stable equilibrium.

This reading is structurally identical to that used by an economist in the Tabachneck-Schijf *et al.* study to rationalize a supply-demand graph, where the intersection also identifies a stable equilibrium. There was, however, no unstable equilibrium. I briefly develop a classical model of 'expertise' and subsequently show that even for the present case, the model is limited and only describes the (for me) less interesting aspects. Subsequent chapters show that this form of expertise is not general. That is, even highly successful ecologists and other scientists do not necessarily read the population graph in the way described by the model and the IF-THEN form of reasoning that lies at its core.

2.2.2. Standard Model of Expertise

The standard model of expertise on this graph type, as recently developed by Tabachneck-Schijf and her colleagues, is based on classical syllogistic reasoning. In the context of the population graph, the core of the reasoning pattern is constituted by:

1. IF the birthrate is greater than the death rate, THEN a population grows
2. The birthrate is greater than the death rate
3. THEREFORE, the population grows

Conditional syllogisms of this form are called production rules, defined as sets of rules of the general form IF... THEN... These production rules are combined with rule interpreters (control modules that determine which rule to activate) and working memory to form a production system.[8] As data driven control structures, production systems are easily implemented in some computer program. For a population graph problem, a production system for drawing arrows in the

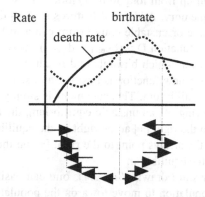

Figure 2.4. Results from an implementation of the production system for drawing arrows indicating the direction in which the population size will move to.

way some of the scientist had done could be structured as follows. First, the system is made to scan the graph along the abscissa and pick up the ordinate values of each of the two curves.

P1   IF   birthrate is greater than death rate
      THEN   set as subgoals
               draw line CONST * (birthrate minus death rate)
               draw arrow pointing right
P2   IF   the birthrate is less than the death rate
      THEN   set as subgoal
               draw line CONST * (death rate minus birthrate)
               draw arrow pointing left
P3   IF   birthrate minus death rate equal zero
      THEN   no action

Implementing this production system with specific values for CONST and INC yields the diagram displayed in Figure 2.4. Here, performance is modeled by assuming that information relevant to the problem is processed according to a set of given entities (i.e., the domain ontology), deterministic rules, and classical logic. For example, birthrate, death rate, and population size are declarative knowledge; the comparison 'greater than' and the instruction 'draw vector' are procedural knowledge. The IF... THEN... structure is a formal operation according to classical logic. Whereas the information (declarative and procedural knowledge) is called up from long term memory, the solution process is assembled in short-term memory. Additional features such as determining whether an intersection is a stable or unstable equilibrium can also be constructed. We can then construct a new function $D = (b_{Neq-\varepsilon} - d_{Neq-\varepsilon}) - (b_{Neq+\varepsilon} - d_{Neq+\varepsilon})$, where N_{eq} is the population size at which birthrate and death rate are equal and ε is some very small value. This new function is positive at a stable equilibrium and negative at an unstable equilibrium. This statement is simply a structural equivalent for saying the following. At an unstable equilibrium, the arrows point to the left [-] on the left and to the right [+] at the right of the equilibrium position; and at a stable equilibrium, the arrows point to the right [+] on the left and to the left [-] on the right side of the equilibrium.

Once the arrows are drawn (*Fig. 2.4*), one can easily 'see' that there is a tendency for the population to move towards the population size corresponding to the left-most or right-most intersection of birthrate and death rate curves. These are the points of stable equilibrium for the population size. There is also a tendency to move away to the left or right from the population size corresponding to the middle intersection between death rate and birthrate curves. This is a point of unstable equilibrium for the population size.

The model developed so far works as long as we pre-specify the domain ontology, that is, the sign elements that can be taken from the graph and accompanying text and the conceptual background that is taken for granted such as the directionality of the values along the two coordinate axes. I already foreshadowed, however, my contention that unnecessary assumptions are often made about what is going on. Certain information simply does not need to be represented. Subsequent chapters show that the assumption of a domain ontology (set of perceptual features) that is inherently fixed does not hold. Even in this select group of scientists, there were individuals who focused on the absolute and relative values of the slopes in the two graphs, or on perceptual patterns that could not easily be predicted from previous research and preceding interviews. In the next section, I outline limitations of the standard model of graph reading.

2.3. WHAT IS MISSING FROM THE STANDARD MODEL?

The information processing analysis that I provided in the previous section suggests that graph interpretation is a rational exercise in which the interpreter moves step by step from problem statement to solution. It also appears to suggest that something new, such as the temporal behavior of the population size, can be inferred from a graph. In a sense, I only provided a part of a full production system. To have a complete problem solving system, our production system would be embedded in a larger production system in which solutions are constructed using one of several problem-solving heuristics. The most important aspect, however, remains the context-independent problem solution process and product. Information-processing approaches generally do not differentiate problem solvers according to their ordinary practices other than classifying them along a singular dimension ranging from novice to expert. What's wrong with the picture that I have developed thus far about expertise in reading and thinking with graphs? The following sections show that 'expertise' does not stand on its own. Experts such as Brandon draw on rich relevance structures, including embodied, material, and conceptual matters; these networks of familiar segmentations of the world serve as the (tacit) ground on which successful expertise is founded. Insufficient attention has been paid to this ground in the past.

2.3.1. Semiotic Networks, Networks of Significance

The analysis provided so far is not different from many traditional analyses of graph reading (interpretation). The major problem lies with abstraction (Lat. *ab-*, away + *trahere*, to draw) of interpretation from the familiar practices that scientists normally bring to their work with graphs. Such abstraction leads to a simplification because it reduces the complexity of life for purposes of economy and narrative. Chapters 6–8 show that much of what is lost in the process of abstraction is actually the fundamentally necessary background against which any representation makes sense. Thus, Brandon's graph reading does not stand on its own, as an abstract skill that can be applied across contexts. Rather, it is couched in his familiarity with the 'standard model', many different ecosystems and relationships including laboratory cultures, and many networks of semiotic relations which connect to and therefore embed each sign he uses. As his additional readings unfolded, he provided a glimpse of the vast and multiple sign relations each of the terms is participating in and the signifying networks they are connected to. Furthermore, Brandon is not only familiar with the graph *per se* but also with the way this graph is normally introduced to college students. Much of his discourse is similar to the explanatory remarks made by another professor in an introductory ecology course that we had recorded in its entirety.

Each of the following excerpts provides additional evidence for the connectedness of 'expert' reading in a much larger and heterogeneous signifying network that can be thought of as being constituted by smaller networks all interconnected and accessible from other parts. These visible (audible) parts of those networks are woven into an unfolding stream of interpretants. Other signifying representations are less material, some have little to do with language (sketches, gestures), and some are unspeakable, cannot be said or drawn or otherwise be made explicit.[9] The following excerpt constituted the end of the first part of Brandon's reading of the population graph. Here, he elaborated the unstable equilibrium in terms of an analogy. That is, the interpretant translated the sign /unstable equilibrium/ into another sub-network of significance and thereby was further connected to and elaborated by other networks of signs.

> ... sometimes called the saddle point, imagine a marble sitting on top of a saddle, it's stable until the horse moves, then it rolls off. I mean this (*Points to [9]*) is like a bowl, a marble at the bottom of a bowl as it goes away from the edges, it still rolls actively. And obviously, rather than, extinction is in fact guaranteed in any point below the second intersection which is somewhat different than the standard models.

The instability is elaborated with the analogy of a /horse saddle/ on which someone has placed a marble. In the ideal situation, when the horse does not move and other external influences do not exist, the marble remains in equilibrium. The stable equilibrium is elaborated in terms of the interpretant /bowl/. Here, /stable equilibrium/ and /unstable equilibrium/ are connected to other signs, /marble on saddle/ and /marble in bowl/ respectively. They are also related to experiences that almost every child has with attempting to balance a pencil on its tip or broom on the end of the handle, very embodied experiences indeed. Given Brandon's preparation and ongoing research concerns, it may not be incidental at all that 'saddle points' are central to the solution of the mathematical treatment of predator-prey (Lotka-Volterra multi-species) models, which are expressed and solved in terms of general autonomous second-order ordinary differential equations.[10]

These additional signs and less material representations, and the imagery they evoke, translate the original signs into new relationships. Multiple representations, and the web of signification associated with translations between nodes, lead to deep understanding. That is, in these situations, understanding is elaborated in the emergent construction and unfolding chain of differential reference. The fact that Brandon read the graph in an 'expert' manner rests on his familiarity with using each of these signs in multiple networks of relations, which signify multiple situations. These networks constitute the *con*text, that is, the additional text that situates the primary text, the graph, and the initial barebones elaboration provided by the 'expert' reader. Evidence provided in subsequent chapters further support these contentions. Thus, when (for one reason or another) scientists cannot perceptually isolate and relate the signs that constitute a graph to referents, troublesome readings arise. With increasing familiarity, however, the signs become more and more transparent so that in the end, scientists appear to have direct access to their phenomena.

2.3.2. Drawing on the Standard Model

In the following episode, even more of the *con*text that grounds Brandon's 'expert' reading becomes apparent. In this episode, we learn that there is a standard model on which the current task is simply a further elaboration; this in fact is the same standard model on which the work of Tabachneck-Schijf *et al.* was based.

> Well, the standard models assume that basically population- That the standard model would look like this (*Draws axes*) where birthrate declines (*Draws birthrate line*) like so and death rate increases (*Draws death rate line*) like this. So this is (*Writes 'death'*) death, this is (*Writes 'birth'*) birth, this is (*Writes 'N'*) population size, this is (*Writes 'rate'*) rate. So that at very low values of *N* (*Points to [12], Fig. 2.2*), the population is increasing maximally. And this (*Points to [12], Fig. 2.2*) says that that's not true. And here the argument they're using is just that if you can't encounter a mate when the population is extremely low then there'll be lots of unmated females and the birthrate will be less than the death rate. So, that's, known as the Allee effect. But the simple models make this (*Gestures along birthrate and death rate*) assumption, simply that death rates are a linear function of population size and birthrates are linear function but negative of population size. And so that gives you a single (*Points to intersection*) stable equilibrium point.

Brandon constructed relations between the model of the task and that which he knows as the /standard model/. At first, he produced a sketch. This sketch can be thought of as the referent of /standard model/ (involving a translation between the typological sign and graphical expression). At the same time, the new graph can be viewed as an elaboration of the sign that the interviewer had not understood. Upon completion of the sketch, Brandon then constructed a relation between the birthrates in the original graph (*Fig. 2.2*) and his graph (*Fig. 2.5*) by comparing the behavior of the birthrate curve for small population sizes ($N \approx 0$). He pointed out that the large difference between birthrates and death rates for small population sizes cause maximum increases, which is, as he said, contradicted by our more realistic model of the task. He immediately elaborated this distinction. That is, the improved model is more reasonable given that with low population sizes (really densities), many females do not find mates so that the birthrates have to be low (as in *Fig. 2.1*) rather than high (*Fig. 2.5*). Because each rate is a linear function of population size, there is only one intersection that signifies a stable population.

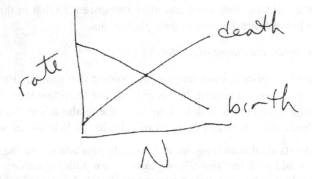

Figure 2.5. The 'standard model' is an interpretant used by Brandon to elaborate the population graph.

Brandon is not only familiar with the standard model, and the standard argument for modifying it, but he associated the modification with a particular effect associated with the name /Allee/. He elaborated:

> I think Frank Allee suggested it for data that he had on, I wanted to say, triboli.
> I think that if you have really low densities, if you've got a huge container and only a few triboli, they go extinct. Basically males and females never encounter each other and they're not so fertilizing. I think that's what it is from originally.
> So, there have been suggestions of Allee effects of that type.

The original interpretant /Allee effect/ was further elaborated in another chain of differential reference. Brandon suggested that some scientist 'Frank Allee' had conducted research ('data *he* had on...') with one organism (*tribolium*, flour beetle) and had observed that it would go extinct at low population densities. This extinction was explained in terms of low female-male encounter rates and mating probabilities. Thus, the shape of the birthrate function in our task graph 'made sense' in terms of these additional relations to research and theoretical statements that Brandon was familiar with. However, it is also noteworthy that some of the details of the elaboration are incorrect although the gist is consistent with textbook presentations of the subject (low population densities lead to low mating rates).

Frank and Allee were actually two researchers rather than one ('data *he* had on'). Warder Clyde Allee was a sociologist who had become interested in ecology and published an important textbook on animal aggregations in 1931, and, in collaboration with others, another one in 1949.[11] P. W. Frank and colleagues

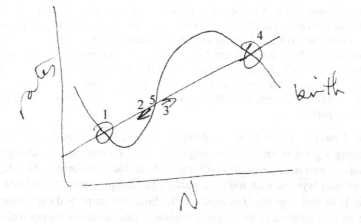

Figure 2.6. Graph constructed by Brandon in order to elaborate the original graph in the context of the Atlantic cod population that had recently crashed leading to a moratorium on fisheries.

had conducted research on cultured water fleas (*Daphnia pulex*) in small beakers noting the fecundity of females as a function of density.[12] The data of Frank and colleagues relate birth- and death rates to population density, but does not elucidate the mechanism of density dependence.[13] Despite these inaccuracies, his elaboration worked because it could account for the low birthrates at low population densities.

Further details about the different networks of significance that *contextual-ize* Brandon's 'expert' reading unfolded after the interviewer had asked Brandon whether the graph would illuminate the crash of the Atlantic cod stocks.[14] For many ecologists, Atlantic cod is a classic case for discussing the relationship between fisheries policies, particularly catch quotas, and scientific models of ecological systems. Despite the quotas determined by fisheries scientists, the Atlantic cod population crashed in the mid-1980s leading to a complete moratorium for fishing the species within Canadian waters. Here, Brandon further elaborated the original graph by tying it into another network of significance that deals with the particulars of Atlantic cod. In the process of elaborating, he produced a new graph (*Fig. 2.6*).

> The cod is not a good example, because that actually requires more than just two species. It requires that the cod are at a point, what's known as a multiple stable state. So, this is an unstable equilibrium, right. What you want is a third

equilibrium that is in fact stable and that requires something that looks like this (*Draws axes*) so this (*Draws death rate curve*) is the death rate and this is the birthrate here (*Draws birthrate curve*). So assuming that the death rate remains linear with *N* (*Writes 'N'*), and these are rates (*Writes 'rates'*), and this is birth (*Writes 'birth'*). Or some equivalent from the death rate. OK. So if you have this well you've got a decline, so that this is a stable equilibrium (*Draws circle [1]*), this again is unstable (*Draws arrows [2] [3]*) and this is stable (*Draws circle [4]*).

Without having to reflect on the Atlantic cod situation, Brandon described it as not being a good referent for the original graph. To describe the changes in cod population one needs a system with multiple stable equilibrium. That is, 'Atlantic cod' and 'system with multiple stable equilibrium states' are related as a red traffic light and stopping. His rapid production of a graph and explanation needs to be seen in contrast to the more problematic production of an equivalent graph for the wolf and moose populations below. Here we have again an articulation of a familiar situation that requires as little (conscious) representational activity as stopping upon seeing a red light. He then proceeded to draw the population graph that would serve as a signifier for the cod population. In this new graph, there were three intersections, two of which have a «stable equilibrium» as their referent. In Brandon's use, such a graph only comes about when there more than two populations.

2.3.3. Drawing on an Inappropriate Rule?

Brandon then attempted to provide a rationale for rapidly deciding when an intersection represents a stable or unstable equilibrium; he stumbled between making a statement about second and first derivatives of the two curves at the point of an intersection.

So what you need is that the birthrate and the death rate are, have different, that the second derivative- No let's put it this way, that the first derivative is, are of opposite sign, so that the slopes are of opposite sign. (*Points to intersection [5]*). So this [*birthrate*] is positive and this [*death rate*] is positive so they are unstable. (*Points to intersection [1].*) This [*death rate*] is positive, this [*birthrate*] is negative so it's stable. (*Points to intersection [4].*) This [*death rate*] is positive, this [*birthrate*] is negative, so it's stable.

Brandon remembered that it was possible to predict whether an intersection signifies a stable or an unstable equilibrium by comparing the derivatives of the two curves. At first, he mentioned the second derivative, then the first derivative (slopes). He posited that an intersection refers to a stable equilibrium when the slopes are of different signs. As he framed this hypothesis, he gestured over the center intersection. He immediately followed this conjecture by conducting (for the audience) a test of the relationships between slopes at the unstable equilibrium. His comment 'so its unstable' confirmed what he had earlier identified as a property of the intersection. There was a convergence between the initial procedure (as I modeled it in the previous section) and a new procedure, which identified the type of equilibrium on the basis of the relationship between slopes. He repeated the procedure for the other two intersections, each time finding his hypothesis confirmed.

At this point, my data do not contain any information as to the roots of Brandon's hunch that the slopes could be used for determining the stability of the system. Furthermore, he did not simply remember the strategy, only that it had something to do with some form of derivative (first, second). From this, and given the example at hand, he (re-) constructed a rule—which in this case was valid across the different graphs in front of him. It is notable, however, that in this situation, the slopes, which heretofore had not been salient to the interpretation, were presented as signs that provided a shortcut and therefore a more transparent access to the referent of an intersection. It was noted earlier that an intersection of birthrate and death rate curves has a possible referent in «equilibrium»; now, Brandon stated a rule of thumb that permits rapid perceptual distinction of «stable equilibrium» and «unstable equilibrium».

It is interesting that despite what we had identified as expertise in the earlier parts of this chapter, the procedure outlined here is not general. In fact, it is easy to show a counter example where the rules he established ad hoc provide the reader with an incorrect referent. Based on the general model outlined by Brandon (*Fig. 2.5*) and his graph, which elaborated the case of Atlantic cod (*Fig. 2.6*) where birthrates begin above, I constructed a new graph (*Fig. 2.7*). Readers can easily verify for themselves that we are dealing with a stable equilibrium despite the fact that both slopes are positive.

My intent is not to critique the scientist for making an error. Real-time everyday practical problem solving by scientists and other practitioners alike, is characterized by mumbles, stumbles, malapropisms, mid-word stoppages, self-evident errors, and so forth to which collaborators and audiences have to adapt their passing interpretations about what the speaker is saying. I am not interested in using a normative model for understanding what people do and therefore offer the following ideas for consideration. First, for all practical purposes, that is, all the cases Brandon is familiar with and that make sense from an ecological perspective, the rule may hold. It does not really matter in this case that this does not hold as a general rule to cover all mathematically possible cases. We know

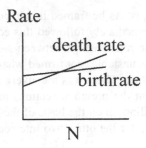

Figure 2.7. Example of a population graph where both curves have a positive slope, but the resulting equilibrium is stable.

that scientific practices are often such that they exploit statements that are not truly general but hold for all cases that are reasonable in practice. Second, such a rule can be constructed but in the context of population graphs used to show predator-prey interactions. In that type of graph, intersections also signify (stable, unstable) equilibrium states of the predator-prey system and their nature can be quickly determined with a simple rule such as that Brandon provided (*Fig. 2.8*). It therefore appears that Brandon vaguely remembered the rule but brought it to the inappropriate context; he used it to analyze a single-species population graph rather than a graph that shows the interaction between the population sizes of a predator and its prey. That is, Brandon was guided by a vague image and drew implications from it, which, though they sounded reasonable, were correct only incidentally. We will encounter other such situations in subsequent chapters.

2.3.4. Back to Familiar Predator-Prey Relations

After having constructed and tested his rule of thumb for determining the nature of the referent of an intersection, Brandon returned to the content of the graphs themselves. In the following episode, he returned to the question whether the graph in the task could be used to model the well-known collapse of cod stocks in Canadian waters off Newfoundland.

> So, it's not quite the case here (*Points to Fig. 2.2*), this is only two thirds of the required story. And probably it's not a birthrate function but a death rate function (*Waves at Fig. 2.6*) that has to do with that. And the reason you get it is that the things that are eating, the cod are being subsidized by a third species. So you maintain a high predator density because it's eating something else. But

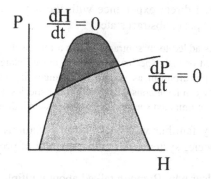

$$P \quad \frac{dH}{dt} = 0$$

$$\frac{dP}{dt} = 0$$

$$H$$

Figure 2.8. Example of a population graph which models predator-prey relationships and in which the relative slopes of the curves predict the relative stability of the predator-prey system under study.

because there's such a high predator density they have a very high predation rate on the cod.

In the first comment, Brandon distinguished the task's two-intersection model with his own three-intersection model required for understanding a situation like Atlantic cod, which is thought to have collapsed from one stable equilibrium state with large population size to another stable equilibrium state with very small numbers. In the second statement, Brandon qualified his own drawing, where he used a straight line to model death rates, by hypothesizing that the death rate function (in the cod case) might be different from the one he had drawn.

To understand the significance of 'more than two populations', one has to be familiar with yet another context (i.e., network of significance) and the experimental practices of laboratory ecology. In modeling the population dynamics of a single species, such as the flower beetle (*Tribolium confusum*) or water fleas (*Daphnia pulex*), ecologists culture organisms in beakers or aquaria ('tanks') that contain a certain amount of food supply (another organism). One can model density-dependent population dynamics of one species by using a second as a food source. Many ecologists are familiar with the general approach to studying individual or pairs of organisms in the laboratory. But Brandon is also very familiar with this approach as his own research deals with population dynamics of different frog larvae as they depend on the presence of predator. Furthermore—something I learned only two years after the interview had been conducted—Brandon had direct experience with the topic at hand. Thus, the abstract of a conference paper abstract stated:

> When a species is added to or subtracted from a food web, predator-prey linkages will be added or lost. Species involved in those linkages will be directly affected by the perturbation as shown by changes in their density. These changes in density can have consequences to other species not directly affected by the introduced/ eliminated species (density-mediated indirect effects).

That is, he was very familiar with the kind of situations that pertained to the stability of multi-species systems and the interaction of predator and prey densities.

It is not quite clear why Brandon talked about multiple species in the face of my population graph and why he might have talked about high predator densities in the context of the last third of the story. His familiarity with such systems may be one explanation. But there is also a possibility that he took the graph before him as another one, in the way I suggested in the previous section. Let us take a look at this possibility. Brandon elaborated a scenario that would give rise to a third intersection. He suggested that rather than being a function of a changed birthrate, the intersection should come about because of a change in death rate as the species preying on the cod shifts to another organism when the density of cod is too low. But with higher population density of the cod, the predator shifts food preferences when the cod populations are sufficiently high.

I begin with the 'standard' predator-prey model (*Fig. 2.9.a*). Here, the two predator isoclines (population change = 0) intersect with the H (prey) axis, which means that predators exist only when there is a sufficient prey population. The two predator isoclines intersect with the prey isocline (constant population) at S and S'.[15] Now, the mathematical solution for the standard model is such that S represents a stable equilibrium and S' has, depending on the specific shape of the curves, either a stable equilibrium or a limit cycle periodic situation (oscillating predator and prey populations) as referent. I drew curves with the same shape as those Brandon had drawn but in the coordinate system used for predator-prey systems (*Fig. 2.9.b*). In this case, S and S" have stable equilibria as their referents, whereas S' corresponds to an unstable state that does not have a limit cycle as its referent. My drawing therefore represents a system with two stable states S and S". Furthermore, perturbations can 'switch' the systems between S and S". That is, my diagram constitutes a model that could explain the 'collapse' of cod from one stable state with high population numbers into another one with very low numbers. More importantly, in this model, the predator isocline intersects with the P (predator) axis, which means that predators exist

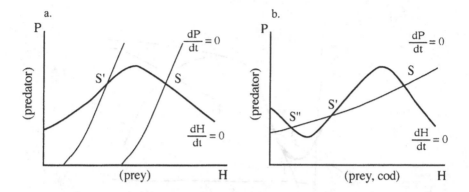

Figure 2.9. Predator-prey models. a. Standard models where the predator population does not come into play until there is a sufficient number of prey available (dP/dt intersects with H-axis). b. There are predators even for low prey numbers (dP/dt intersects with P-axis), which is only possible if they also feed on another species.

even if there are no cod as prey and for low numbers of prey. This, then, is only possible if the predators also feed on another prey. Such a situation was what Brandon talked about.

Brandon's talk therefore makes sense if the ordinate was labeled 'P' and the two curves were relabeled to represent the zero growth in the predator (dP/dt = 0) and prey (dH/dt = 0) populations. In this way, at least the abscissa continues to represent population size and both curves signify rates ('clines'). One final clue suggests that Brandon was talking about the graphs I suggested (*Fig. 2.8, 2.9*). The slopes of the predator and prey curves in these figures with respect to the H and P axes can indeed be used to classify the type of steady state an intersection represents. But the evaluation is much more complex than can be established by a quick look. It is therefore quite likely that Brandon talked about the population graph as if it had been another one, which has some similarities, but pertains to a substantially different mathematical framework. A one-species model involves a simple differential equation whereas a two-species model involves a second order differential equation. What looked like clean and smooth expertise at the surface begins to unravel when the deep structure of mathematical biology is brought to the situation.

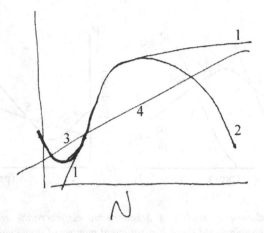

Figure 2.10. Graph drawn by Brandon as part of his elaboration of the relationship between moose-wolf populations, and as an extension of the web of meaning that embeds his 'expert' episode.

2.3.5. Trouble with Less Familiar Predator-Prey Relations

The interviewer than attempted to elicit further comments from Brandon concerning the usefulness of graphical models such as the population graph within the discipline. Brandon was asked whether he knew of any natural population that would be a reasonable referent for the population graph in the task. As he answered this question, Brandon drew two additional graphs (*Fig. 2.10, 2.11*).

> There's been some suggestions, not this one (*Points toward Fig. 2.2*), but this one (*Points toward Fig. 2.5*), some suggestions for instance of moose and wolves have been suggested. For instance, one of the things is that you can get, how does it go, I'll make sure I get this right (*Draws axes of Fig. 2.10*), because the wolves will have a (*11-second pause*), it looks like this (*Draws curve [1]*). OK, so they increase to the point where, oh no, it has to decline in fact, has to decline (*Draws branch [2]*). The rate of mortality (*Gestures an inverse parabola along [2]*) due to wolves goes up as the number of prey. (*Writes 'N'*) so this is N, meaning prey, so this is the mortality rate due to wolves (*Gestures an inverse parabola along [1] and [2]*), it goes up and then goes down because the wolves are territorial and there's a limit and there's a limit on how many they take. So, the rate of death declines with population size eventually. Now birthrate... (*9-seconds pause*) Oh yeah, and then you get this (*Draws branch [3]*)

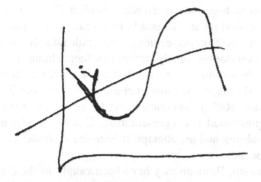

Figure 2.11. Sketch drawn by Brandon to further elaborate on the wolf-moose example.

because at some point the population size is too small to support wolves at all. So the death rate declines, again. Oh God, I can't remember the details but there is a way of doing this with the birth and death rates that gives you this (*Draws straight line [4]*). Again, you use the multiple stable states. I mean I probably got the figure inverted somehow (pause). Yeah, somewhat it's actually, it does, it goes like this (*Draws Fig. 2.11*) and so this is increasing. (*6-second pause*)

Brandon's talk had become more stocking, long pauses sometimes accompanied by requests to say what he thought, marked the transcript. Although he was somewhat familiar with the relationship of moose and wolf populations, he could not generate a graph that expressed the relation described. As he talked about a referent for the original graph, Brandon drew on his experience and familiar populations that both elaborated the original graph, folding yet another network of significance onto it, and also served as a referent. He began with the hunch that wolf and moose make such a system. Graph and familiar (common sense) ecological knowledge were used in a mutually constitutive manner. Curve [1], as the next sentence indicates, signified the death rate of the moose due to wolf predation. However, as he began to explain he added branch [2] to make the curve consistent with his understanding that the predation rate comes back down 'because wolves are territorial' and only take so much prey. Then, after a long pause, Brandon added another branch thereby changing the death rates for small prey densities because 'the population size is too small to support wolves'.

At this point he began to experience a conflict. On the one hand, branch [3] moved upward indicating an increase in death rate, but in his experience-driven description of the wolf-moose ecology, the death rate should decline. He then abandoned his explanation, simply stating that there should be a way of drawing and elaborating the example so that in fact there are three intersections with the birthrate graph. He made one more attempt suggesting that he had inverted the graphical situation. Rather than questioning his sense about the 'real' ecological situation, he questioned his representation. Despite rendering another graph (*Fig. 2.11*), he abandoned his attempt in bringing consistency to his verbal and graphical representations.

In this situation, Brandon may have been caught in the dilemma that he remembered the existence of a simple model for the population, and, perhaps even a graph that goes with it. For example, the moose-wolf relation can be explained with an even simpler birthrate and death rate model than the one he had at hand under the condition that he described (a constant territorial, non-migrating wolf pack) and some maximum number of prey they take ('limit on what they can take'). The growth rate of such a moose population leads to the logistic growth curve,[16] with multiple stable equilibria (*Fig. 2.12.a*); in this situation, and as Brandon suggested, the predator-induced mortality increases up to a certain point with increasing prey density and then decreases.[17] The shape of this curve is like the one that Brandon had initially drawn, that is line [2] (*Fig. 2.11*). His 'death rate due to wolves' maps onto the downward shift of the parabola rather than onto the curve itself. But then Brandon became tangled up in his own statements. He added a new branch to his parabolic curve moving upward, that is, to larger values of mortality; this graph contrasted his words that suggested that death rate should be going down. His description would be consistent with a graph similar in shape but with growth rate as the ordinate (*Fig. 2.12.b*). At the point where the growth rate increases (left of the broken line), his statement 'death rate decreases' is consistent with the increase (in growth rate) plotted. Finally, it is possible that Brandon vaguely remembered such diagrams (*Fig. 2.12.b*), which are used to represent multiple equilibrium models for a single species that is being preyed upon. One can easily see that the elements drawn by Brandon are present in my diagram (*Fig. 2.12.c*), the horizontal curve for number of prey killed by wolves, the wiggle of this curve at low densities, and the overall parabolic shape of the logistic growth curve.

2.3.6. Fizzling Out

After a very smooth start, Brandon had become uncertain. In the absence of a familiar situation, his talk became stocking and there were an increasing number

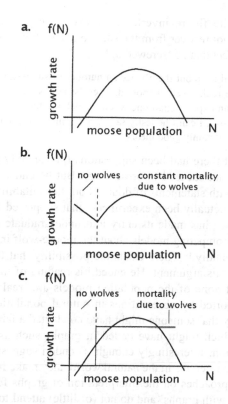

Figure 2.12. a. Single species model for a population in which a constant number of individuals are removed. There are two equilibrium points, one unstable and the other stable corresponding to the left and right intercept, respectively. b. The same model but assuming that a below a certain moose population size (broken line), there is no predation by wolves. c. Single species model as in (a) including the mortality rate due to the predator.

of discrepancies between the diagrams and his talk. Because of the iconic nature of the diagram, the discrepancies might be similar to those that are often observed between gestures and utterances when people talk about topics relatively unfamiliar at the moment.[18] Near the end of the previous episode, Brandon suggested that he could no longer remember the details of the wolf-moose model but he remembered that there is a way to combine the death rates and birthrates (he may have meant recruitment rates) to yield a model with two steady states. In the penultimate sentence of the previous episode, Brandon intimated that he might have 'got the figure inverted'. However, despite attempting another drawing, he could not recover from the situation. This final episode began with a statement to the effect that he 'screwed up'.

> No, I've screwed it up but there's been a number of suggestions- I don't think any of them are really well supported. There's been some, I'd have to think, there might be an experimental one where somebody have done it with proto-zoans but it doesn't, I can't honestly say. Yeah, I mean, I'm not gonna say that it is real but it is certainly good theory.

He maintained that there had been suggestions (presumably of the type that he wanted to but did not succeed in presenting). But he ended the session on the population graph with statements of doubt about the suitability of the models or whether there had actually been experiments that supported the model. A well-known article, which has made its entry into undergraduate courses, in fact argues that multiple competing models about the moose-wolf interaction needed to be tested experimentally.[19] There exists the possibility that Brandon was therefore familiar with this argument. He ended his reading of the population graph by suggesting that none of the population models are 'really well supported'. That is, it is supported by observations in natural population. Though he left open the possibility that someone might have conducted a laboratory experiment the outcome of which might have ended in graphs such as those he had discussed in the session. Interestingly enough, Brandon suggested that even if the graphs did not have referents in the natural world, they make for 'good theory'.

Normative approaches on the interpretation of graphs focus on a small aspect of 'reasoning with graphs' and do not (or little) attend to the enormous web of signification that makes these overt forms of didactically exemplary reasoning possible. The problems that arise from such an approach are the following. First, when the same scientists face novel graphs, even from their own domain, their reasoning becomes much less expert-like. We saw in the present case study, that as soon as the 'expert' was asked whether there is a natural population that fits the graphical model at hand, he began to flounder despite the fact that the domain is part of his interests and regular teaching. Graphing and reasoning with models are a function of the familiarity with particular graphs rather than general skills. This was quite clear when Brandon suggested that he could not 'remember the details'. How familiarity interacts with the reading of graph is the topic of Chapter 5, and how this familiarity comes about as scientists engage in research that ultimately leads to graphs, is also described in the chapters of the second part of this book.

Second, if 'expertise' is so much a function of familiarity, we need to understand much better the background assumptions, practices, and so forth that allow the kind of overt and explicit forms of reasoning to function. In terms of my semiotic model, we need a better understanding of the networks of significance that allow scientists to enact and develop rich and cross experience-related interpretants, unfold networks of other signs, and networks of text that make the very *con*text that ground the interpretation. This context then provides the existential condition for other forms of reasoning. I do not suggest that 'experts' need to know everything in their domain; there is no reason to believe that Brandon would not be able to piece graphs and narratives about the wolf-moose population together were he to prepare himself for his next course or topic in population ecology. Rather, I suggest that graphing expertise is highly contextual and a function of the purposes for which graphs are used in a familiar world. Reading graphs requires substantial familiarity with the natural phenomena that are signified (represented) by the graph and with the particular representational form.

In the end, Brandon commented on the use of population graphs such as that at hand and those he had drawn. It is clear that in his (very rich) experience, there are no natural populations that behave in the way the graph expresses—other than a protozoan culture in a test tube or Petri dish. The graph, in his words, does not describe anything that is 'real', that is, does not have a referent in the phenomenal world. But, he added, 'it is certainly good theory'. In his case, playing with this theory and the representation has some interesting aspects, though the results of this activity do not have applications in describing the ordinary phenomena an ecologist is dealing with. Brandon and other university scientists were relatively successful on the graphing tasks because these and similar graphs belong to their lifeworld of teaching even if these graphs do not have referents in the natural world; these graphs are part of the lifeworld of a biology professor who regularly teaches. Non-university scientists were relatively unsuccessful because they did not have referents; these graphs did not belong to their lifeworlds. This points toward a possible dialectical relation between graph and what it refers to, that is, a context within which the graph is employed for the sake of some goal, which itself is part of the scientists' unarticulated understanding of the situation. This in turn suggests a dialectical relation between the graph and a person's experience-derived understanding of what the world is like. Thus, it is not surprising that the university scientists expressed attitudes similar to Brandon whereas non-university scientists generally found the graphs useless. They found them useless exactly for the reason that it did not have a referent in the phenomenal world and because it 'misrepresented' important relationships. They therefore abandoned any reading of the graph or, provided readings that were inconsistent with those provided by the university scientists (the high culture of the domain).

2.4. READING VERSUS INTERPRETING

Much of the research on graphing appears to be guided by the assumption that 'experts' use some explicit process in which they take information available in the graph and draw inferences that they make available to the researcher in the protocol. The expert models also make assumptions about the degree to which external representations are also represented internally. Such assumptions have come under considerable critique in a number of domains concerned with cognition. Such research shows that internal representation would make such high demands on the presumed mental apparatus that it would be overloaded. Rather, we are better off thinking in terms of cognition as distributed across the environment, thereby avoiding the informational bottleneck that would otherwise exist. Something similar occurs in the reading of graphs. In cases that can be classified as 'expertise' one observes a-posteriori rationalizations. Sociological studies have shown that such rationalizations are no better than what an outside observer may perceive and therefore bring into the order of discourse.

Already Martin Heidegger and Ludwig Wittgenstein warned us not to think about everyday competence in terms of representation (e.g., language) and represented (e.g., world). Both encouraged us instead to think about cognition in familiar situation in terms of use, useful things, and purposes. When we hear words, see a stop sign, or apperceive a familiar graph, we do not so much *comprehend* what is at hand but acquire an orientation to a familiar world. Signs are useful things that explicitly bring a totality of mutually interrelated things to circumspect attention. Simultaneously, the worldly character of the signs itself is known. When scientists gazed at familiar graphs, such as Brandon in the above example, then it is not just the sign in itself that is available to his circumspect attention but a totality of familiar things. Brandon did not have to think about what 'N' might refer to but he used it from the beginning synonymously with 'population size' and 'population density'. Similarly, the 'standard model' and the fact that my population graph was inappropriate for modeling the Atlantic cod population were not facts to be inductively derived from the graph. Rather, they were immediately available to Brandon together with the graph itself.

The details of the totality (network of signification) are not there present, as we think about 'there' in representational terms. But when asked, Brandon articulated aspects of this totality, such as when he talked about the standard model, which 'would look like this (draws graph) where birthrate declines like so and death rate increases like so'. That is, he articulated a familiar world rather than inferring something new. Interpretation merely articulated an understanding already present and presupposing the interpretation. In the same way, when asked whether the graph at hand could be used to explain the recent developments of Atlantic cod populations, he suggested without hesitation and without the need to reflect 'No, that's not a good one because that actually requires more than just two species. It requires that the cod are at a point what's known as a multiple stable state'. If this is so, then the traditional explanations of graphing expertise have it the wrong way around. The explanation provided and recorded in the verbal protocol is an after-the-fact rationalization rather than an inference-generating interpretation. The totality of things is already apprehended, immediately available to the scientist, who can subsequently and ad lib talk about and elaborate the situation. Rather than reasoning about the differences between two graphs, the graphs and their relationships become available in their entirety. The 'expert' scientists then unfolded some of the interpretants that they have at their disposal for the benefit of the listening researcher rather than for their need of interpretants required for a subsequent understanding. The entire problem is reduced to recognition. For example, the university-based scientists who also taught undergraduate courses did not need to think about the implications of the intersections. The intersections in population graphs are immediately associated with 'equilibrium' states. It is part of the background that comes with the familiarity of having worked with or taught population graphs. The details can then be worked out in the specific context of a problem, for example, the question whether an equilibrium is a stable or unstable one.

To get to closer to the heart of the matter, let us consider yet another situation. When Brandon was presented with the isograph task (*Fig. 2.13*, see also Appendix), he immediately attributed the graphs to particular ecological concepts of resource limitation and substitutability. He reacted in the knowledgeable way that he would had he seen a red light and stopped. The kind of reasoning traditionally associated with expertise was provided only after the initial attribution.

It's a question about resource limitation basically and this (*Fig. 2.13.a*) is absolute limitation, this (*Fig. 2.13.b*) is substitutability and this (*Fig. 2.13.c*) is partial substitutability. So, basically it (*Fig. 2.13.a*) says that, that, if you are anywhere in this region, that you have a growth rate of 20. Which means that no matter how much R_1 you have, if you have less than this much R_2, then you can't grow more than 20.

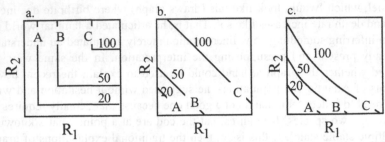

The amount a plant grows depends on a number of factors, for instance, the availability of nutrients (R). A shortage of any single nutrient can limit plant growth. Sometimes scientists study the effect of pairs of nutrients. The graphs depict three different biologically realistic scenarios of how two nutrients (R_1 and R_2) might combine to affect plant growth. Discuss the effects of different levels of the two nutrients on each amount of plant growth (20, 50, 100) in each scenario (i, ii, iii).

Figure 2.13. In the isographs, two independent variables are related to a third, dependent variable, which is represented only in the form of isoclines, lines of equal value.

In the first sentence Brandon already indexed the entire totality of resource limitations and substitutability. His explanation, which is very similar to that provided by another ecology professor, is an after-the-fact elaboration of this totality. We may also think about this in the following terms of getting around in our familiar environment.[20] When we decide to shop for groceries, we do not need to make a plan how to get to the supermarket; we do not need to represent the road to get there, or even how to drive the car. Rather, know that we can trust ourselves getting there, taking the right turns, adapting to the traffic situations that we find, and stopping at the red light. If we are asked, we can rationalize our actions, provide reasons for turning or stopping.[21] The three isographs, I propose, made present to Brandon the totality of useful conceptual things relating to the ecology of resources, their limitations and substitutability. The everydayness of isographs became quite clear when Brandon was asked about the use of such graphs in his work.

> I do actually use this kind of isographs in my own work. I mean, when you've got three-dimensional data then you got to use three dimensions in your presentation. And this in fact is four-dimensional because we've also, the fourth dimension being the model right, so we got three panels each of which has three dimensions in it. So, yeah- But my general rule is you have as many dimensions in the graph as you do in the data. Unlike three-dimensional graphs in

Excel, which basically has a bogus third dimension that has nothing to do with the data.

For an adequate explanation of graphing, we do not require processes in which 'experts' construct the meaning of a graph by making inferences from its components. Rather, we can model the entire process as one in which a graph and the situations in which it is used and those to which it refers become available at once, as a totality related to specific intentions for the sake of which graphs are used as tools or objects. It is this totality that is meaningful and in the context of which the graph becomes a useful tool and intelligible entity that stands for something else. It might be recognized as a particular type, namely a model for equilibrium states (stable, unstable) in a population of organisms. Without enacting any comparisons, any hypotheses or the like, the entire graph has become a sign with «equilibrium states» as a referent. I found evidence for this kind of recognition among theoretical ecologists who, at a glance, recognized the graph and seemingly without any further processing, identified equilibrium states and other typical features.

When we ask the scientists, they elaborate rationales why some graph and the text that accompanies it has «adaptation of photosynthetic mechanisms» as a referent, but the fact is that they do not need to engage in this reasoning. They pick this up immediately from the environment and recognize the text as an instance of a sign from a particular context, itself existing as a totality. Such recognition requires considerable familiarity with the various sign objects and the habitual referents they signify. In this case, the remainder of the interview session concerned with one scientist's understanding of the various aspects he talked about provided ample evidence of the various ways in which significance can be elaborated into webs of signs.

This consideration leads to the following situation. Scientists may know a specific graph but do not have an explicit structure that can be said to underlie it. Through use, they have come to be familiar with such graphs, and only later learn to frame understanding in terms of more abstract descriptions. When these scientists see a graph that can be said to have the same structure, they may not recognize the structural similarity. This is just what I observed. There are situations in which scientists do not transfer readings to structurally identical representations or experience difficulties with reading a particular representation despite having used them previously in their own work.

2.5. CRITIQUE OF THE TRADITIONAL EXPERT MODEL

Based on these observations I would assume that the case of the economist analyzed by Tabachneck-Schijf *et al.* is similar. What these researchers obtained as 'expert reasoning' was really a post-hoc rationalization of a recognized sign. That is, the 'expert' does not even need the kind of descriptions provided here by Brandon, other cases of expertise in my database, or by the mentioned economist. Rather, «supply», «demand», and «equilibrium» are so much part of an economist's familiar world that the graph is not even necessary for communicating or thinking about the equilibrium model. Whether the 'expert' actually uses these structural processes in difficult problems of his everyday work is another question. It makes little sense to bracket a particular aspect of the practices and to call it 'expert reasoning' without also considering all the networks of signification that embed the different aspects, and which often go without processing explicit representations. Without the meaningful totality into which each sign can be connected through the production of interpretants, the kind of reasoning with population graphs or supply-demand curves is impossible.

Other models poorly handle the kind of readings that I describe in the second part of this and in subsequent chapters. Brandon remembered familiar images that related to the domain evoked by my population graph. But neither could he articulate and elaborate them nor did he find an appropriate situation description that corresponded to the graphs he drew. Take another example. The Tabachneck-Schijf *et al.* model specifies /supply/ in terms of a 'proportional relation' between price and quantity. However, in my database, such an explicit framing of relationships (other than in the case of one physicist) does not exist. Furthermore, some of the erroneous readings provided by scientists in the case of the population graph arise exactly because the mathematized relationship was not explicit. These scientists did not make salient the mutual dependencies of birth and death rates on the one side and population size on the other. Thus, it is *not* the case that my scientists regard the relationships as mathematical ones. If the Tabachneck-Schijf *et al.* model has some validity, we still have to go a long way before arriving at a genetic model that allow us not only to model instances of extreme familiarity with the objects of interpretation but also model graphing competencies from the very beginning. Such a genetic model needs to account for the changing ensembles of signs salient to some reader as she becomes familiar with some domain and even within the confines of a single session. In Chapter 7, I provide the kind of representational work that preceded and therefore provided a background to the graphical expertise of one doctoral student in ecology pertaining to her graphs.

So far I have put into relief traditional conceptions of expertise in problem solving related to graphs that take excerpts such as those at the beginning of this chapter as evidence for expertise. Such a move is questionable for at least two reasons. First, the claim is particular to this scientist's reasoning with this graph. Second, although there appear to be multiple representations, the question even in traditional models has to be how much is actually represented in short term memory. Here, some recent research on the indexical functions of short term memory and visual pointers of other forms shows that much of the world that is being reasoned about does not have to be represented in the mind. Furthermore, we also know that much of the expertise in tool use lies in the fact that the tool does not have to be represented in explicit forms. For example, rather than representing a hammer in normative space with all its coordinates, speed, or acceleration, a skilled carpenter engages in hammering. Rather than representing a keyboard as I am writing these sentences, searching for those keys to be pressed, or finding the correct sequences, I focus on the contents of the sentences that unfold from my typing (as well as on potential typographic errors and possibly some other things). In both examples, the tools become transparent to the activity in which the person is engaged. There is no additional load on the reasoning processes, memory, or the like. In a similar way, I have begun in this chapter (and do so in more detail in Chapter 6) to articulate graphs as signs (tools) that become, under certain conditions, transparent. I also show that the processes enacted by the agents are not general such as to transfer from one graph to the next. There are other, highly contextual factors that mediate any reading, making inferences, or interpreting activity (in the presence) of graphs.

2.6. DISCIPLINARY CRITIQUE OF THE POPULATION GRAPH

In searching an understanding of graph reading as we can observe it in the everyday work of scientists, we also need to know more about the context of the work, the contexts of use of the graphs as tools in mathematizing, producing rhetorically strong arguments, and so on. For, knowing how these graphs are employed by a small number of scientists, or rather, how they are rationalized, may tell us little about the graph-related practices in the everyday work of scientists. Thus, models of graph-related competencies are severely limited if it turns out that (a) the tasks have little to do with real-world tasks and (b) graphing and graph reading is a highly contextual activity. Here, I provide evidence that these models find little use in field ecology and that these models are in fact misrepresenting the understandings of real-world ecosystems. Such graphs are mainly used for didactic purposes. In the discussion of the networks of significance that grounded Brandon's interpretations of the population graph, we have already seen that 'expert reasoning' does not stand on its own. In fact, extensive webs of signification, that can be articulated when needed, buttress this expert reasoning; and scientists unfold these webs in chains of differential references when asked about the signs they use. In subsequent chapters, I show that the relationship between graph reading and contextual specifics arises from familiarity with signifying representation and the «natural settings» thus signified.

At the end of his session about the population graph, Brandon suggested that even if these 'graphs are not real', that is, describe natural systems, they make for good theory. Among university (especially theoretical) ecologists and physicists, I could observe an interest in the models, and engaging in the 'Gedanken' experiments irrespective of the question whether there are natural ecological systems that they signify. In some instances, field ecologists described such models as 'the bread and butter of theoretical ecologists'. The university scientists were in general agreement with Brandon who suggested that 'I would expect students at the end of a second-year ecology course to be able to look at that and do this little analysis here'.

> Tom: We don't worry about the fact that it may not be an accurate representation of what really goes on. But the graph provides a theoretically fertile sort of way of talking about things. For an introductory class we just throw in the curve that as the density increases we expect the birthrate to go up only to a point and then as the density goes even further beyond this inflection point up here it gets worse. One reason for that could be that increase in disease, that the number of miscarriages that occur goes up because of bumping and the stresses of finding food in a dense population. Whether or not that's true I have no idea but- it would be uninteresting if these lines never intersected.

Tom described the graph as providing a 'theoretically fertile' activity to talk about 'things', although these graphs may have no referent in natural ecological situations and although the shape of the curves are not signifying those processes in a population which he told students about. This is interesting given that so much of Brandon's understanding of the population graph was grounded in his familiarity with natural populations and other representations that elaborate the graph he was presented with. Furthermore, missing referents in the natural world was an important aspect in the rejection of some ecologists of graphical models such as the population-graph and isograph tasks.

Among the ecologists who work for various governmental and other non-university agencies, the attitudes toward the graphs in the tasks were different. Rather than seeing in these graphs opportunities for engaging in modeling practices, they described the graphs as 'useless' and, sometimes, as detrimental to the development of new ecologists. They regard these graphs as an aspect of the institutional requirements to become an ecologist but that have little to do with the everyday practices in the field. The practitioners in the field suggested that such models tend to enculturate young ecologists into reductionist ways of thinking about the environment, which may have disastrous consequences to ecosystems because of inappropriate model-based management practices.

> Rick: I kind of get a bit bitter about behavioral ecology by assuming graphs like this because they're really idealized, abstractions. Well, if you want to be realistic about it, I'd say that this model is probably complete crap and that you'd have to represent N as go through time as some sort of stable limit cycle thing. So you know that, given an N between some range, you can predict the N between some other range. I mean this type of modeling is a type of modeling that I believe contributed to the complete destruction of the economy in Eastern Canada. I won't blame the fisheries scientists too harshly, because the politicians are always bastards misrepresenting everything the scientists tell them because they have a time horizon on a scale of four years and fish often have a time horizon on a scale of much longer than four years.

Others suggested that these models only serve didactic purposes but in this, actually may have a tremendous effect on the enculturation of new people to ecology, particularly to modeling phenomena in unrealistic and detrimental ways. Thus, one scientist said, 'I have become a little bit more skeptical over the years of how applicable these models really are- this is a model to teach ecology, to help you get certified'. Let us take a closer look at the evaluation of the population graph by one scientist.

At the time of my research, Drew headed a research unit, which in turn administered a research consortium with a long-term plan for understanding a particular marine ecosystem in the Pacific Northwest. He generates between $300,000 and $500,000 of research funding per annum, and has a considerable publication- and report-related scholarly output. (In the four-year period from 1996–99, there are 15 refereed articles, 6 reports.) At the time of the interview, he had conducted research on his core species for nearly 18 years, and had 8 years of experience after his Ph.D.

In his research, Drew directly works with issues concerning population dynamics (population biology and bioenergetics), which arise from changes in birthrate and death rate, changes in food availability, and other factors that describe complex marine ecosystems. In particular, Drew focused on one species that, as some of its relatives, has experienced significant decline in population size. In one of his research articles, he argued that the decline represented a change from on stable equilibrium to another, that is, from one carrying capacity to another. Although his research unit is attached to a university, Drew is not involved in teaching. Despite his research experience directly related to population dynamics and despite a substantial record, Drew had difficulties with the population graph. He interpreted, for example, the intersection at lower values of N as a stable equilibrium and elaborated the birthrate-death rate relations above the second intersection as a case that leads to a population crash.

> You're never gonna find a data set that looks like this. This is a theoretical model, it's based on, you know, nice mathematics and equations, and it's the way we think the world probably works. But I don't know any data set where you ever find this and you can ever point out there are probably two steady states. It's just, in the real world it is a constant fluctuation. There are two ways to approach the problem. One is, I mean, you collect real data then look where the patterns are, and you might- I mean, it's theoretically possible to create something like this from real data. The other approach is to start at the other end, and just to start with pure mathematics and conceptualize how these populations must change, use equations to describe it, and then go on for data that might fit this to try to validate your hypothesized model. So you can approach it in two different ways but I've never seen in this case. I don't know any example where this fits.

Drew suggested that graphical models only serve a sensible purpose when they describe natural populations. As he was not familiar with any such population, he considered the population graph (and even more so the isocline graphs) as having no use in ecology research. In his own work, Drew attempted to understand the population changes in his species of interest as the result of interactions and energy transfer between 24 different species. While he recognized that the population graph could be used to teach people. (Drew used 'people' not necessarily to signify «students» but also the stakeholders in the nature and activist organizations that fund his research.) Although he judged the population model as unrealistic, he felt that it could be used to help non-specialists understand the population dynamics and the idea of 'steady states':

> What is useful about it is making people understand that you can have, for example, two different steadies, two different points of these populations would move to. Because over time, you can change the population size and, but this model here is going to move to one of the steady states but you could perturb it and see it move away perhaps to another one. So, it's useful to make people understand how numbers change over time and how they can move to different points but at the same time. It isn't necessary realistic, you know, but you are taking people conceptually thinking in different ways and so that when they do consider a real population they are at least aware of its possibilities.

Associated with his rejection of the population graph as useless in the ecologists' work and of little use in the discipline, he described the two intersections as steady states. He, as several other field ecologists, did not distinguish the equilibria in terms of their stability but simply noted that at these points, the population should stay the same. In the following episode, Drew further debunked the usefulness of population graphs because they are unrealistic for there are no equilibrium states, and policies based on these models have caused havoc (wiped out entire species) in natural ecosystems.

> There's a danger of that and so the case is where people apply these sorts of models in real-life situations and started, for example, thinking, well, 'What if I am increasing the death rate?' And so we'll go and we'll kill more because we should see a response from the birthrates. It never happened, but they almost wiped out entire species in their pursuit of something that theoretically should have happened. Which usually shows you how it is more complicated, the world really isn't so nice and clean and simple. I don't think it would be an equilibrium. It's a sort of myth that there is a balance of nature and there is a magic fixed point. I mean, it could be here (*intersection*) it could be there (*intersection*). The reality is that the world is in constant flux.

Drew described the referent «balance of nature» as a myth. That is, in his use, the referents of the intersections, the equilibrium points do not exist and therefore are myths maintained by some. He suggested that there are population fluctuations that drastically change even with minor changes of other aspects in the system. Even the removal of just one species from a complex system or a slight change of ocean conditions will shift species distributions and relative abundance. I have little reason to doubt this successful researcher.

> In many ways, it's a mental exercise, just using differential equations and, the beauty of it, you know, for many mathematicians. What they love about this is that it is pure, that is exact, and there is no error in it. But that isn't the way the real world works. The real world has so much randomness to it and sorting out the randomness from the signal or the trend that's there within is a real challenge.

Drew distinguished between the domains of mathematics, interested in simple and beautiful, exact models and the domain of ecology, concerned with real-worldly, messy ecological systems. Other ecologists echoed Drew's concerns with the population graph and therefore threw its usefulness even more in relief. Furthermore, many field ecologists suggested that when the models actually were used, they often led to the destruction of habitat and animal populations.

Field ecologists in particular highlighted the tension between the abstractions that are associated with representations and natural situations. Rick and Drew not only related the problems of the population models to disastrous effect on the animal populations per se, but also *con*textualized the graphs in terms of the effects crashing populations have on the economy of entire regions. For other field ecologists, the population graph not only had a low fit but also contained conceptual errors. One scientist suggested that in natural settings, animals do not just stay in one place but move and react to their circumstances. In his experience, if the density were to decrease to precarious levels, animals tended to move and congregate into higher densities. To him, the Allee effect signified in the birthrate curve for low densities is conceptually flawed.

These excerpts from the scientists' talk about the use and usefulness of models embodied by the population graphs or isographs are clear. Most of them neither used population graphs in their work nor were they familiar with natural populations that were suitably described by the graphs. On the contrary, these scientists seem to suggest that if the graphs or some related model is actually used it has disastrous effects not only on the animal populations, but also on the economy of entire regions.

From the reactions of the scientists to the graphs I presented, particularly the population graph and the isographs, it is clear that graphical models were not unanimously accepted as useful tools for doing ecology. Theoretical ecologists and the two physicists regarded these models as tools that deserve inquiry in their own right, which allowed a researcher to understand the dynamics of a system, even if it did not have a natural equivalent. The general attitude was that these models allowed researchers to get a feel for the dynamics of simple systems, providing them with a basis to investigate more complex systems. This is very much the approach that others have reported for the way physicists go about developing models for understanding gravitational waves and their sources. In this approach, it does not really matter to the scientists whether there are natural referents that fit the model. On the other hand, the field ecologists were generally concerned with the models both because they were simplistic leading to disastrous outcomes in population management and because they enculturate students to reductionist ways of doing ecology.

In a seminal paper, Jill Larkin and Herbert Simon provided their answer to the question why diagrams are (sometimes) worth ten thousand words.[22] Wisely, they qualified their answer by adding 'sometimes'. The present study provides even more qualifications to the usefulness of diagrams. Similar to the research of mathematical representations in a variety of contexts my research shows that scientists are not inherently experts, nor are graphs inherently useful. Rather, the usefulness of graphs to particular scientists seems to be substantially mediated by familiarity with the graph as tool and with the setting in which the tool is used, and for the purposes to which it is employed.

THREE

THREE

Unfolding Interpretations

Graph Interpretation as Abduction

To better understand 'competent' reading of unfamiliar graphs, two detailed case studies are presented. 'Competent' reading of unfamiliar graph, which involves the process of abduction, shares features with the work of successful detectives, who engage in a dual process: they structure the situation to isolate hypothetical signs and construct hypothetical referents. Through the production of interpretants, which produce a mutual elaboration of sign and referent, the graph reader seeks convergence so that the sign-referent relation is intelligible and plausible. This referent situation is an articulation of the reader's understanding of the world rather than something that the author of the graph attempted to transmit. All interpretation is therefore already grounded in understanding, which is further articulated and elaborated in the interpretive process. It is shown that, even though the scientists form a relatively homogeneous group, perceptual structures such as intersections or intercepts are not inherently signs and that a sign on a graph such as 'N' may have different interpretants.

3.1. ABDUCTION

In ordinary everyday life, words and signs are better understood as useful tools used to attend to the concern at hand rather than to represent the world or an aspect of it. Words and signs are part of the way in which we are attuned to and understand the world, and in this way they are transparent; they are a form of direct relation rather than a (mental) representation of something else. Graphs and graphical features, too, can be transparent, such as when Karen, a water technician, points to a spike in a graph and says, 'This is a clogged pipe'. She *interprets* the spike as little as she *interprets* the sound that is, according to the International Phonetic Association, phonetically transcribed as '[paip]'; rather, the spike is an aspect of a world in which a water pipe on a farm is clogged. (This transparency is the topic of Chapters 5 and 6.) Familiar graphs occasion the making present the totality in which they normally take their part. One could also say that in such cases, graph reading leaps beyond the material basis of the text to an intimately familiar world. When scientists face an unfamiliar graph, such occasioning cannot occur in principle. But what they may do is find some aspect of their familiar world that could stand in a mutually constitutive and transparent relation with the graph at hand. This requires a two-part process. First, the reader has to perceptually analyze the graph and caption and isolate some features as signs. Second, for the signs and their structural relations, the reader has to hypothesize a situation in the world that could be the equivalent of the graph. As a constraint, the internal structure and dynamic of the graph and caption has to coincide with the structure and dynamic of the intelligible world: Interpretation is a search for coherence. In this case, reading does not leap beyond the material basis of the text but constitutes a protracted process in which the reader moves back and forth between structural analysis and grounding the signs in the familiar world. The analogy of Detective Bertrand developed in Chapter 1 may be helpful in thinking about this interpretive process.

When Detective Bertrand entered the room where Professor Ashmore was found dead, he did not know which aspect would allow him to find the murderer, that is, which material object or constellation constitutes the sign relevant to reconstruct the events. But, by moving between various hypothetical event scenarios and hypothetical signs—the placement of objects in the situation and perhaps various testimonies he already collected—he can construct a scenario and a set of signs that stand in a mutually consistent and constitutive relation. Although he may arrive at some such scenario, there are no guarantees that it corresponds to what 'really' happened and he may inculpate the wrong person for the murder. However, his reading is said to be competent when he arrives at an intelligible and reasonable scenario and corresponding set of signs. The situation is not transparent, for Detective Bertrand cannot say what happened at the instant that he walks into a situation. Rather, his process of reading involves a protracted process in which he constructs tentative signs, links these signs into a coherent, self-stabilizing network of significance. He also analyzes signs, discards some perceptual aspects as relevant signs, and seeks new ways of looking at the situation to provide him with clues (signs) to the original event. It is possible that the process and the results of another detective's work would be different, involving different scenarios and culprits. That is, the situation is like an *open* text (work) read differently by different people. In contrast to the closed text (work), built on known codes and set cause and effect chains, open texts (or works) force readers to make their own interpretation and draw their own conclusions. This process is referred to as creative abduction.

Creative abduction is a synthetic inference whereby a specific sign is explained to be the result of a general rule applied to a specific case. Take the case of the spike in Karen's graph. A scientist unfamiliar with the graph may nevertheless hypothesize that rapid changes in water levels are due to problems in the water level measuring device (rule). One kind of problem is a clogged outflow pipe that makes the water rise rapidly in the measurement cylinder (case), which would directly entail a spike in the strip chart (result). Seeing the spike as a spike (artifact, non-natural event) rather than as a sign of a natural event then leads to the particular scenario constructed. However, it is clear that other scenarios could be found that are also consistent with the reading of the trace as 'spike'. Although less likely, the rule may be inappropriate in this situation if the water could rapidly rise for some other natural reason.

In more formal terms, creative abduction is the process of finding the in Chapter 1 discussed set of parameters $\{R, S, c, r\}$ that satisfies the *relation R =* $f_r(S,c)$. Although previous cultural experiences make scientists parse the graph into particular features that serve as signs S, they never exhaust the realm of possible signs nor do they elaborate the signs identified in the same way. The possible contents R are functions of the scientists' experiences and understanding of the world.

The 'competent' readings in my study follow the patterns of creative abduction. When scientists read graphs with which they are not familiar (though these might represent standard fare in the undergraduate training of their own field), their task of interpretation is twofold. First, they have to reconstruct the internal dynamics of the multimodal text. Second, they have to restore to the text the ability to project itself outside in the representation of a world that we can inhabit. This double task constitutes the structuring and grounding components of my semiotic model. Despite the unfamiliarity of a graph, scientists succeed in establishing a web of signification that is intelligible and corresponds to a standard (and therefore 'correct') interpretation. The following case studies show, however, that the outcomes of interpretative process can be quite different. That is, depending on their background and experiences, scientists arrived at different sets $\{R, S, c, r\}$ and interpretant signs that elaborated the relation $R = f_r(S,c)$.

Ultimately, then, confronted with the graphs and captions, scientists asked to make sense of this compound 'text' and to elaborate it in terms of their fundamental and existential understanding that they always and already have available, though frequently not in articulated form. The task occasions this understanding to be articulated in the face of the graph. That is, rather than decoding information, readers unfold their understanding of an intelligible world. At best, therefore, there will be agreement between the author and reader of a graph when the worlds they inhabit are similar to a certain degree.

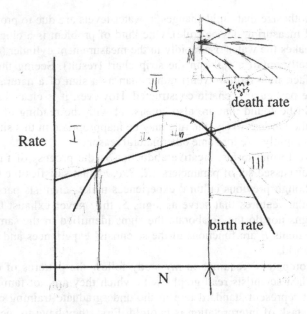

Figure 3.1. In the process of his interpretation, Ted elaborated the original population graph by drawing another graph that displays population change over time (top right). This graph can be thought of as a referent of the original graph (sign). As part of his interpretive effort, Ted produced additional signs to demarcate and name features.

3.2. BETWEEN ECOLOGY AND REPRESENTATION

The scientist in my first case study is Todd. Todd had studied physics (Ph.D.) and, for the past 15 years, has been an active researcher in a government research unit. Todd regularly publishes research articles (15 alone for the 1996–99 period) that include Cartesian graphs. Todd was interviewed in his own office and talked for nearly two hours about the three graphs presented to him and about the graph from his own work.

In the course of his reading, Todd marked up the population graph and added his rendering of what the population dynamic would be over time given different population sizes as starting point (*Fig. 3.1*). That is, he constructed a new graph that itself can be regarded as one referent of the original graph. He also drew on many mathematical resources such as the concepts of 'family of curves', 'convergence', or 'logistic curve' as part of his elaborations of the target graph. Compared to the field ecologists in my database, his interpretants include more references to mathematical processes and descriptions and fewer references to actual animal populations. In this way, his reading was more like that of the theoretical ecologists and the other physicist (see below) in my database.

3.2.1. 'They are obviously plotting some population...'

As his interpretation unfolded, Todd engaged in the processes outlined in my semiotic model—including perceptual structuring to construct signs and the process of abduction that integrates these signs into an intelligible world. When Todd encountered trouble, he sometimes displayed difficulties that are analogous to those identified among high school students. Unlike most of his peers, Todd did not begin with the caption, but engaged in a reading of the graphical representation and the textual information within it before reading the caption. (Roman numerals are used to number utterances.)

> [i] N would be the number like the individual in a population, rate would be a number differentiated by time, so this would be a measure of change. [ii] They are obviously plotting some population where we take- find the number of individuals and we see that as the number of individuals goes up the rate, the death rate increases- [iii] There are more individuals dying per unit time as the number goes up and birthrate increases as the- [iv] It probably should go to zero if there are no, well it should go to two probably; if there are no parents, there will be no births. [v] The birthrate increases to a maximum, at some optimum number and then the birthrate falls of as the number of individuals increases, [vi] probably because of limits in the environment or competition or disease or overcrowding or social problems within that population. [vii] Now, the logistic model is a mathematical formulation and it's very common I know in the ecological literature and this particular curve (*Points to birthrate curve*), logistic curve, has been shown, it has been shown or has gained a lot of notoriety because it demonstrates chaotic behavior in non-linear...

In this opening reading, and without reference to the caption, Todd immediately elaborated the signs /N/ and /rate/ with the interpretants /number of individuals in a population/ and /number differentiated by time/, respectively [i]. Both interpretants were not directly available from the task, but arose from common conventions, *r*, with which more theoretically oriented scientists are normally familiar. Todd then enacted a literal reading (i.e., describes the graph in words) and produced an interpretant that translated the trajectory of lines with respect to the coordinate grid into a verbal representation [ii]. This he further elaborated in [iii] with a second interpretant. All of a sudden, Todd engaged in a reverse movement: he took everyday understandings of reproduction (no organisms no births, it takes two parents to procreate) to project where the birthrate curve should intersect with the abscissa [iv]. In this, we observe a movement from the world of experience (i.e., possible content of the sign) to the sign. Todd continued to provide interpretants that translated the shape of the birthrate curve [v]; then again, he used familiar understanding of ecology (a referent) that legitimized the dropping of the birthrate (S) with increasing N [vi]. In this situation, the situation descriptions (i.e., 'limits in the environment', 'competition', 'overcrowding', and 'social problem') were not available in the graph or in the caption. Yet here, Todd understood ecology in a way that was consistent with his interpretant of the birthrate curve for larger N. Therefore, the ecological understanding (and prior experiences) and his reading of the graph reified each other (double arrow in the bottom right triangle). Todd then turned to the text and read the term 'logistic model', leading him to relate it to mathematical formulations and chaotic behavior in non-linear systems [vii].

3.2.2. 'At intersection points you are at equilibrium...'

Up to this point, Todd engaged in the internal processes of structuring the graph by producing interpretants (e.g., birthrate drops), which then became signs that projected beyond themselves to a possible world. We also observe the reverse movement where he took descriptions of the world as it was taken to be (e.g., competition, overcrowding) to reify and legitimize the shape of the birthrate curve.

> [viii] At intersection points, presumably, you are at equilibrium, the number of individuals in the population, they are being replaced as quickly as they're being removed and the same will hold at the other equilibrium point. [ix] In between, the population is increasing in size and if you're concerned with harvesting this population for some reasons, hunting or trapping or fishing, you would want to make sure that you're keeping your population numbers in this range. [x] The most increase you have is here (*points to maximum difference between b and d*) because you calculate the population, say the P is equal to the initial P plus birthrate plus death rate.

Todd made salient both intersection points ([viii]). This in itself is significant given that in a different graph, he (as other scientists) did not highlight the existing intersections. Thus, the intersections became salient (or are made salient by the caption) not in themselves but in the light of other signs that his reading individuated (which therefore constitute the context c of the intersection sign), 'birthrate' and 'death rate'. Furthermore, for the intersection to be significant, the graph relies on the reader's adherence to the conventions regulating the use of the two signs 'birthrate' and 'death rate' as opposite trends that tend to increase and decrease an existing population. This information is not provided in the graph itself. It therefore constitutes background information that is always and already presumed to exist in the way that a movement along the ordinate away from the reader is usually presumed to be 'up'. This background, as the previous chapter showed, depends on the culture of the discipline (see *Fig. 2.3*, p. 33).

An intersection therefore acquires significance in the context of the signs /birthrate/ and /death rate/ rather than being significant in and by itself. This is even more evident when we compare this graph with its copy where 'birthrate of deer' and 'birthrate of moose' replace the two original labels. In this case, the implications of the intersections are of a very different nature. This example is equivalent to the task involving the distribution of CAM, C_3 and C_4 plants (see Apppendix). In that task, Todd (as his peers) did not make salient the intersections of the distributions of different plant types. This is a significant observation that contradicts traditional assumptions about salience in graphs (see also Table 3.1, p. 94). Rather than being composed of features that are salient a priori, what graphical features are salient depends on the context of the interpretation; that is, it depends on the background and experience brought by the reading individual and by the contingent associations and history of the interpretative session.

Todd's reading of the intersections as points of equilibrium therefore presupposed complex interactions between extant (socially shared) understandings of how the world operates and the graph at hand. His interpretant /equilibrium/ for the sign constituted by the intersecting curves was legitimized by a further interpretant /birthrate equals death rate/ in the form /they are being replaced as quickly as they are being removed/ [viii]. Todd's next statement [ix] showed again a complex interaction between knowing the world and the topology of the signifying graph. Here, in the context of those conventions that regulate the use of 'birthrate' and 'death rate', the relation between the two curves suggested an increase in number of individuals in the referent population. Conversely, the concepts 'hunting', 'trapping', 'fishing', and 'harvesting' are from the domain of the experienced lived-in world. His ecological common sense suggested that to engage in these activities, sufficient population sizes are required unless the species is to become extinct. Todd used such a common understanding to suggest that activities leading to the density-independent reduction of population size should be encouraged when the population tends to increase its size. This interpretation is only partially correct, however, as it does not include the region for population sizes above the second intersection. But, from Todd's perspective correctness relative to some normative model did not matter to the unfolding interpretation. He continued drawing on his resources, which were always taken (at least temporarily) as the way things *really* are. On the basis of what he had done so far, Todd then elaborated this non-standard interpretation even further in subsequent statements.

The last interpretant in this section related to the maximum difference between birth and death rate curves [x]. As he pointed to that part of the graph, he extended the already constructed meaning relations in terms of the maximum increase in the population size and provided an algebra-based rationale. This, however, is a non-standard interpretation for the maximum increase in individuals occurs when the function $f(N) = (b(N) - d(N))\cdot N$ rather than $f(N) = (b(N) - d(N))$ is maximized. Here, Todd did not attend to or observe the convention of interpreting birthrate and death rate as the increase and decrease in individuals relative to the population size. This non-standard interpretation was not unlike the slope/height difficulty observed among students reading a distance-time graph, for in both cases, the interpretant is based on some maximum in the graph currently available rather than in a transformed graph. In both instances, situated reasoning (inappropriately so from within the normative model) associates the utterance of 'most' (or faster) with the differences visibly at hand rather than with the maximum differences in the associated f(N) or velocity-time graphs.

3.2.3. 'So you have a family of curves...'

In the subsequent episode, Todd continued by shifting his attention to the relationship between birth- and death rates to the right of the second intersection. In the process, he began to draw his own graph in which the population size was represented as a function of time (top right of *Fig. 3.1*, *Fig. 3.2.a-d*). In this episode, the drawing progressed to the point where a family of downward sloped curves intersected the abscissa at various points.

[xi] If we're working out in this range, the birthrate is much lower than the death rate, so the number in the population versus time which would be the other curve that's behind this one will show a declining population. [xii] I believe that if we looked at the population versus time, in year say (*Draws graph axes*), if we were in, I call this Domain I, Domain II, Domain III, we are in Domain III (*Labels sections, see Fig. 3.1*) because the initial population sitting up here is dying at a rate greater than it's been born, then over time you would have a decline in the population. [xiii] It wouldn't be strictly linear because this is a power curve and this is linear. So you would have a convexity or a concavity to it, I'm not sure. I think it will be like this, down, because if we took a tangent at this point, this would be a linear response and as the numbers get higher, you're actually getting. So, you have a family of curves that all do this (*Draws the family of curves*).

Todd began this episode by producing the interpretant 'declining population' for the sign constituted by a birthrate curve lying below the death rate curve [xi]. At this point, it was not clear how far the population might decline. However, after sectioning the graph into three domains (*Fig. 3.1*, [xii]), Todd drew a 'family of curves' representing different developments. Beginning with different popula-

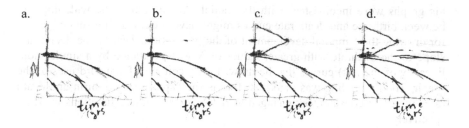

Figure 3.2. Ted's graph unfolds as another sign that elaborates the original graph. a. Family of curves that incorrectly shows the population-time graph for Section III. b. Ted adds the population dynamic for the upper equilibrium. c. This line constrains Ted to revise the earlier population dynamics, now moving towards the equilibrium. d. Ted's final graph exhibits two lines for the equilibrium states and the correct dynamics for populations below, between, and above the two states.

tion sizes from Domain III (right of the second intersection), he sketched six curves that resembled the arms of inverted parabola (*Fig. 3.2.a*). That is, he drew population-time curves that showed, incorrectly, the extinction of the species. The curves in this family intersect with the abscissa and therefore signify crashing populations. This interpretant would have been standard if birthrates and death rates were not functions of population density (or size) but constant. About 37% of the scientists and 78% of all non-scientists arrived at this non-standard interpretation (see Table 2.1, p. 27).

At this point, it is interesting to note that Todd made his populations collapse for different starting values of the population size. That is, his reading here was consistent with the simplest models taught in introductory ecology courses where birthrate (b) and death rate (d) are constants leading to the differential equation (3.1).

$$\frac{d\,N}{d\,t} = (b - d)N \qquad (3.1)$$

But his graphs are not consistent with the solution of this differential equation, which has a solution whereby population changes follow an exponential curve

$$N(t) = N_0\, e^{(b-d)t} \qquad (3.2)$$

where $N_0 = N(t = 0)$ is the initial value of the population size and $N(t)$ represents the size of the population at time t. Todd's reading would be closer to a standard reading if the variable N on the abscissa corresponded to N_0 in (3.2). However, his graphs were inconsistent with the model. At this point, the widening gap between birthrate and death rate curves might have provided a perceptual model for drawing the temporal development of the population. That is, the decreasing slope of the birthrate with greater values of N was paralleled by a similar non-linearity of the $N(t)$ curves that he drew. In the context of standard uses of the graph, Todd's reading was incorrect, for, as the caption stated, birth and death rates were not constant, but changed as quadratic and linear functions in N, respectively.

3.2.4. '... that's when you get in this chaotic thing...'

When Todd elaborated on his earlier interpretants, there arose the opportunity for the problem to become salient.

> [xiv] If you were at these points here, birth and death rates, you would have a family of, we have, this in here, I call this, I should call this N, whatever this value is. [xv] You have a straight line across here (*Draws horizontal line in Fig. 3.3.b*), versus time, and this value, which is a little higher, you'd also have a straight line versus time, it's an equilibrium. [xvi] And if you get an N that's higher than that, then it would appear that that is going to drop, to drop off (*Draws descending curve in Fig. 3.3.c*). [xvii] But then you- that's when you get in this chaotic thing, it'd fall back down in this zone and it would be an equilibrium again I guess.

As he returned to include Domain II in his own graph, he drew a line horizontal to the abscissa representing one of the two equilibrium states [xiv] (*Fig. 3.2.b*). However, conventional wisdom in mathematical analysis holds that these lines cannot intersect with the family of curves so that he drew them directly above. When he then drew a new curve beginning with an N above the equilibrium [xvi], his line stopped at the upper horizontal line (*Fig. 3.2.c*). Apparently without conflict, he continued in his interpretation adding the population dynamic for a size of N between the two equilibrium states (*Fig. 3.2.d*). Such a conflict might have been expected given that he earlier had drawn a family of curves starting for Ns above the equilibrium point but which terminated at the abscissa. The final interpretant [xvii] was somewhat cryptic at this point, but became clearer subsequently when he explained that from his understanding of natural populations, he expected oscillations. We also find an allusion to the 'chaotic thing', which he associated earlier with the behavior of non-linear systems and the 'logistic curve'; the latter was also an interpretant of the quadratic birthrate function [vii].

3.2.5. '... and talk about slopes of convergence'

In the next episode, Todd then elaborated the 'chaotic thing'. As his subsequent interpretants [xx] indicate, he searched for something in the original graph [xix, xxi] that would lead to an oscillation, an instability, of his own curve around the equilibrium. In ecology, it is a truism that population numbers normally oscillate around some value. This value is the point of equilibrium in population models. Oscillating population numbers have been reported for hare-lynx populations in Newfoundland and other pairs of species; they are standard fare in ecology books. Models that include oscillations are either discrete delay models or two-species predator-prey models that operate under limit-cycle conditions; even moderately informed individuals know about such oscillations.

> [xviii] I never worked with a logistic curve- But if you were up, depending whether you're inside that zone or outside that zone and whether you're on this part of. And we can call that Zone IIa and IIb (*Labels the zones*), because you're diverging from the death rate and here you're converging toward it, you probably, you could probably divide this one into an upper and lower portion and talk about slopes of convergence toward the equilibrium within that. [xix] I should be able to come up with some way that would go unstable but I don't see it right now. [xx] By unstable I would mean something that would, an oscillation that would put you sometimes above this curve and sometimes below it so that you didn't converge to this equilibrium. [xxi] That would be to me an unstable population and I don't see how these two curves, I'm not seeing it right now anyway.

Todd suggested that he never has worked with the logistic curve. In fact, however, the population dynamic near the second (right) equilibrium point is approximately equal to the linear logistic model. For such a model, the population size asymptotically approaches an equilibrium value (*Fig. 3.3.a*) or, depending on the recruitment parameter r_j, there may be multiple stable states or chaotic solutions to the logistic equation (*Fig. 3.3.b*). Thus, although Todd's own model did not predict asymptotic behavior it incorporated a dynamic in which any population size above the first, unstable equilibrium would eventually end up in the second equilibrium. However, Todd appeared to associate the logistic model with oscillating population numbers (*Fig. 3.3.b*) for higher values of r_j, which is a discrete population model for a single species. However, oscillations can be modeled in very different ways such as was the case in the two species model that Brandon attempted to elaborate in the previous chapter. We cannot know what the system would have looked like had Todd continued to elaborate on the idea of the chaotic model. It makes sense to think of this situation in a way similar to what had gone on in Brandon's attempt to model the wolf-moose system (see Chapter 2). The problem solvers make some associations that are often

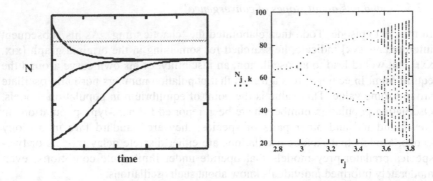

Figure 3.3. a. Population dynamics for logistic population growth with different starting populations. Whatever the starting value of the population size, it will end up in the equilibrium state. b. Changing the recruitment parameter r_j causes the emergence of multiple stable states for logistic growth, and eventually chaotic solutions t the logistic equation.

quite vague and indeterminate so that what they might lead to, and therefore tell us about the person's understanding, is undetermined. What we do know is that Todd began to question his own structuring of the graph and to look for a feature (sign) that would make his model oscillate.

Despite engaging in a focused perceptual and structural analysis of the original graph, Todd did not find a feature S that would be consistent with an oscillating (unstable) population, R, to satisfy the relation $R = f_r(S,c)$. Here, we observe a process where the unsatisfied expectations of a scientist drive him to question his original reading. That is, from the expectations of population oscillations based in his understanding of the world, Todd not only returned to the level of the signs, but to the structural analysis of the graph/text which no longer stood in a mutually constitutive relationship with the phenomenon. Graphs can therefore be said to contain neither a fixed set of signs nor a set that is immediately apparent. A real-time analysis may initially turn up signs only to abandon some of them and isolate new ones that previously were not taken into account or did not even exist (see Table 3.1, p. 96). Here, the graph turned out to be a highly ambiguous text requiring an analysis not unlike those of esthetic texts (e.g., paintings, music). In fact, Todd's comment 'I don't see it' indicates a reverse movement from text to structural analysis questioning the process of perception itself, which lies at the origin of the graphical text as a structured object. Much like Detective Bertrand, when at a seeming impasse, Todd attempted to take a fresh look at the graph and perhaps discover signs that he had not attended to until now.

As he searched, Todd perceived differences in the relative slopes of the two curves in the middle zone, which he sectioned off into two sub-zones IIa and IIb. These differences became salient as he was looking for features that would produce oscillations. These differences therefore are related to a specific purpose, *for the sake of which* he needed certain evidence. Here then, the relative slopes heretofore absent in his analysis became part of the resources to his interpretation. Because of their absence, there is no reason for us to believe that slopes actually existed in his perception. We can say that his ontology, the ensemble of the signs in the task world had changed. Without elaborating the point, the interpretant [xxviii] raised the possibility that the instability comes from the different 'slopes of convergence', the fact that the two lines converge and diverge in the two subsections, respectively. As we saw in the analysis of Brandon's interpretative work, there are slopes of curves that one can use to decide whether a two-population predator-prey model has a stable equilibrium, a limit cycle solution, or is unstable. However, the slopes and their relations that need to be considered are more complex.[1] This approach bears similarity with iconic errors, as Todd treated convergent and divergent lines in one graph as isomorphic with the expected convergence to and divergence from the equilibrium so that it could not be reached. This becomes evident in the final episode discussed here.

3.2.6. '... that's what I mean by looking for some kind of instability'

Todd sought a feature in the graph such that the curves in his own graphs never reach the lines representing the equilibrium state, but oscillate about them [xxii].

> [xxii] If you use it in a management sense, you're aiming for a particular population estimate. And what you find, certainly what is always there- All of the case examples of population management by man occurs that there's always overshoot, undershoot and feedback loops that prevent that population from ever reaching these theoretical equilibrium points. [xxiii] And that's what I mean by looking for some kind of instability in this relationship that would- And it could be that it has to do with the- [xxiv] If you change these coefficients k_b and k_c you change the shape of this curve- death rate is linear, and if it were not, it may be, it's also, has a logistic model shaped to it, maybe consider this.

Todd did not pursue the implications of the convergent and divergent slopes. He moved on to make salient other aspects of the graph, which heretofore were in the background, or did not exist at all from his perspective. His ecological common sense provided a strong suggestion that the representation (sign) should oscillate. As this was not the case, Todd returned to the issue and explored ways of perceiving the graph. As part of the process of perceptual restructuring, he turned up new entities that had the potential for becoming signs that denoted a meaningful entity or situation [xxiv]. In the light of an earlier interpretant that the logistic model had 'gained a lot of notoriety because it demonstrates chaotic behavior in non linear' may introduce the instability that Todd attributed to natural systems. He ultimately abandoned his efforts without further elaborating on the issues raised so far.

In summary, the episodes provide glimpses of a scientist attempting to read the graph with the purpose to leap beyond to relate it to the world he knows. In the process, it was not that graphical signs were provided with meaning; rather, these signs accrued to a world already understood, that is, to existing webs of signification. But this reading was not effortless and in some instances nonstandard for three reasons. First, there were instances where Todd did not observe standard conventions. Second, his structural analysis produced individual and ensembles of signs that he did not always use to constrain their respective interpretants. Third, he used some visual features of the graph to make inappropriate structural arguments. We also see an interaction between structural analysis and understanding of how the world works, and dialectic relations between sign and referent dimensions in each of the two processes at work. The analysis of the protocols reveals that (a) 6 of 16 experienced scientists experienced even more trouble than Todd with this graph and (b) scientists generally connected their reading of this graph much more than non-scientists to their understanding of the world or mathematics. The difficulties faced by scientists starkly contrasted with their reading and use of graphs from their own work.

The case study of Todd shows that even scientists do not read unfamiliar graphs transparently, that is, in the way that expert graph interpretations are often portrayed. However, it is not only that the particular graph and the elements from which it is constituted (expressive domain) but also the referent domain that was unfamiliar. In this case, despite Todd's extensive experience in plotting data related to intimately familiar natural phenomena, his reading involved considerable efforts, which were not always successful within a normative frame. In his reading, Todd moved back and forth between the structuring process (identifying aspects that signify) and the process of making the signs converge with the world he understands so well. Dialectic movements between expressive (S) and content (R) domains characterize both processes. In these dialectic movements, both domains are reified as Todd found consistency between situation descriptions and interpretants. An interpretation emerged as the reader sought a translation from the expressive domain, and by finding appropriate contextual constraints that decrease the number of solutions in the relation $R = C_r(S,c)$. In the process, what is a sign is not always certain nor whether a sign required by the understanding of the situation is actually available in the graphical representation. We also observe how familiar practices allowed Todd to read some aspects of the graph in a transparent way such that it became difficult to assess whether 'birthrates' or 'death rates' referred to some state in the world or to the graphs at hand.

Previous research showed how people, for a variety of reasons, 'misread' graphs. One source of problems arises when the relevant signs are not perceived and another when the context constituted by other signs is articulated so that the relevant constraints supporting a canonical reading do not exist. Despite their experience and training, scientists engaged in similar misreading. For example, Todd also 'misread' the population graph (as did a number of his colleagues). Thus, he initially read 'birthrate less than death rate' for the right-most section of the graph (*Fig. 3.1*) and inferred from it a population collapse rather than a dynamic situation in which N, $b(N)$, and $d(N)$ are continuously adjusted until an equilibrium state is reached. Todd recovered from this inference after constructing his own graph that made his readings of the different sections and intersections incompatible with each other.

Slope/height errors constitute another common form of 'misreading'. This error comes about when people are asked to read something from a graph that is designed to exhibit another relationship. For example, it is common practice to provide research participants with a set of position-time graphs but to ask questions about speed. Thus, the information sought for is only implicitly available and, in the case of velocity from position-time graphs, only because of the coincidence of derivative and the definition of velocity. Rather than comparing relative slopes, people unfamiliar with such graphs make height comparisons and therefore provide incorrect answers to the speed question—not one of the studies I am familiar with actually ascertained what the real question was that these participants answered. From a perspective of graphs as rhetorical constructions—consistent with the general tenor of my respondents—using one graph when the core issue can be made more salient in another type of graph is 'bad practice'. The point here is that scientists committed the equivalent error to slope/height confusion when asked where the maximum number of individuals would be added to an existing population. Todd and all but one of his colleagues pointed to b_{max} or $(b - d)_{max}$ rather than to the appropriate $[(b - d) \cdot N]_{max}$. Similar to 'novices', scientists also used a salient feature of the unfamiliar graph (an existing maximum) in answer to a question that more easily could have been read from another graph. In fact, slope/height errors may be classified as iconic. It may turn out that the question 'Which is more...?' biases the participants' attention and perception processes to a graphical representation of 'more' in the same way that a representation of, for example, a curve on race track biases attention and perception to curves in a Cartesian velocity-time graph.

3.3. PROLIFERATION OF INSCRIPTIONS

In the previous example, we have seen how one scientist elaborated the original graph by drawing his own graph that displayed the population dynamics rather than the functional relation between birth and death rates and population size. Another university-based scientist, Nelson, trained as a physicist but working as a statistician, with 20 years of experience in research, constructed even more mathematical interpretants. He publishes regularly and has had continued external funding to conduct his research. Having agreed to be interviewed in his familiar environment, he asked that the interview be stretched over several meetings, which all took place in his office and amounted to a total of about 4 hours. In his office, he asked to make use of some of his familiar computer modeling resources. This gave the sessions quite a different feel as the interpretations arose from the use of familiar tools. He agreed to being interviewed in his office repeatedly.

So far, we have seen that scientists articulated and elaborated networks of signification, where they drew on a variety of experiences and diffuse images. The interpretants produced were not so much inferences from the graphs to the facts communicated by the authors of the graphs, but more articulations of understanding of personally familiar worlds. The interpretants, in fact, then provided a link between structures that the scientists perceptually isolated in the display and the structures of the world known to them. Before taking a peek at the physicist-statistician, let us reflect on what we would expect him to produce based on what we know so far. If interpretation is the structuring and development of (existing) understanding, if interpretation is based on and presupposes understanding of that which it interprets, then we might expect very different interpretants from someone who spends his life creating models of a variety of phenomena. We would probably expect to hear and see more from the world of modeling in general, a variety of numerical and graphical practices, then from the natural world concerned with organisms in general or specific populations in particular, as we have seen in the case of Brandon. We might also expect similarities with the interpretation sessions among theoretical ecologists and biologists more concerned with the exploration of complex mathematical and graphical models than with the description of specific populations. Yet, because these models are tested using specific animal populations or ecosystems, the theoretical ecologists usually also bring exemplary or paradigmatic cases to bear on their discussion of the graphs. Concern for mathematical and graphical models is much less likely among the field ecologists, who, as the fieldwork in my research group showed, are wrapped up in long and detailed studies of particular organisms and who, as my ethnographic work among them has shown, find the use of mathematics far too abstract and irrelevant. But let us turn to the sessions with the physicist Nelson.

3.3.1. Population Graph and In/Stability

After reading the problem and looking at the graph for a few seconds, Nelson elaborated the graph:

> There are two points where death rate and birthrate are the same. So there is no change in the population at these points. So, a little to the left of this left point, the death rate is larger than the birthrate, so the population will disappear; a little to the right of the point, and the population will grow. An unstable equilibrium point. Near the other point, the population increases on the left, but decreases on the right side. A stable equilibrium point. This is evident. But if there are stable and unstable equilibria, one should be able to find a better representation.

Nelson said that in physics and physical chemistry, atomic and nuclear potentials are usually represented so that the valleys correspond to distances where the systems are stable. Peaks correspond to distances where there are unstable equilibria. He quickly drew by hand a series of curves including the potential energies of inter-atomic distances in binary molecules, atomic nuclei, deformed atomic nuclei (*Fig. 3.4. top row*). In addition, Nelson indicated that the phenomena such as the ammonium molecule and the two-dimensional magnetic pendulum have topologically similar energy profiles, though the causes were considerably different, each system featuring two stable equilibria and one unstable equilibrium (*Fig. 3.4.d, e*). In his final interpretant, he developed a new chain of signification. He sketched the last of his figures (*Fig. 3.4.f*) and continued:

He suggested that it should be possible to model this [ecology] situation in a similar way. He noted

$$\frac{dN}{N} = b(N) - d(N) = F$$

and proposed

> The birth- and death rate are 'tugging' on the system N. So they are like forces, F, which in our case is d-N-over N, plays the role of the gradient that pulls the population. U is the potential in which the population finds itself. But I have potential energy functions here (*Gestures toward graphs in Fig. 3.4*), so I should be getting the same.

Nelson then took a mathematical modeling program and created b, d similar to the problem and graphed them (*Fig. 3.5.top*) and then integrated F (*Fig.*

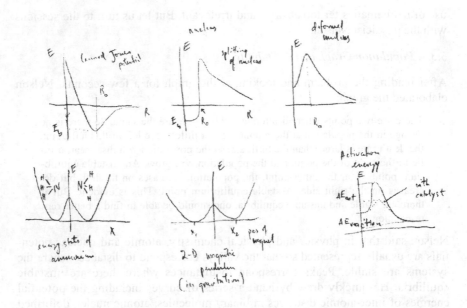

Figure 3.4. The physicist generates multiple graphs, all of which related to cases with stable and/or unstable equilibrium states. Energy of a binary molecule (a), nuclear particle around the nucleus (b), nuclear particle in deformed nucleus (c), ammonium molecule (d), two-dimensional magnetic pendulum (e), and activation energy of chemical reactions (f).

3.5.center). But the curve had a hump where there should have been a valley, and vice versa. He reverses the sign and notes:

> That's right (*Writes,* $F = \dfrac{-\partial U}{\partial x}$) F equal minus d-u-d-x. Force is the negative derivative of the potential energy function.

Running the new model, he obtained a new graph, which had the hump where he had expected it, at the value of N where the equilibrium is unstable (*Fig. 3.5*, bottom right). He then pointed to the earlier figures (*Fig. 3.4*) and said, 'It's just like in these cases, the hump where the system is unstable, the valley where it is stable'. That is, drawing on mathematical resources—representation in terms of differentials and integrals—and familiarity with mathematical modeling programs, he created several new inscriptions within a matter of minutes. It is notable that his concerns were not whether the particular example described

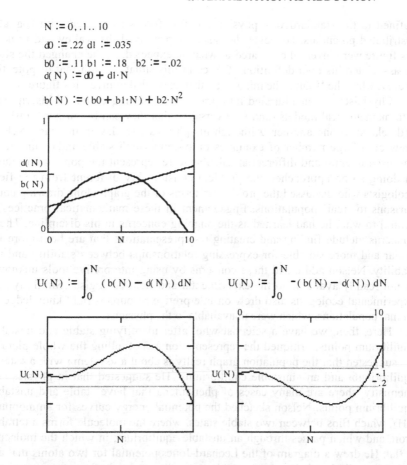

$N := 0, .1 .. 10$

$d0 := .22 \quad d1 := .035$

$b0 := .11 \quad b1 := .18 \quad b2 := -.02$

$d(N) := d0 + d1 \cdot N$

$b(N) := (b0 + b1 \cdot N) + b2 \cdot N^2$

$$U(N) := \int_0^N (b(N) - d(N)) \, dN \qquad U(N) := \int_0^N -(b(N) - d(N)) \, dN$$

Figure 3.5. Translation of the population graph into another representation supposed to feature stable unstable equilibria.

some actual population, but with finding a re-representation that suggests stability and instability in a more intuitive way, based on everyday experience. It is also notable that he talked about the stable and unstable equilibrium in terms of the interpretants /hump/ and /valley/, which are structurally identical to the /saddle/ and /bowl/ Brandon had used as interpretants.

Nelson's performance was not error free, but working with the mathematical modeling program, he 'caught' the problem with the sign. Beginning with the notions of 'tugging' and 'force', he did not immediately link the force he defined to the standard way physicists define forces associated with spatially distributed potentials. However, he instantly perceived that valley and peak in his figure were inverted compared to what he expected. As he changed the sign, he associated his own definition of force in this situation with that of potential energy, where he 'found' the minus sign that he needed to invert his figure.

Physicists are enculturated into practices of turning graphical inscriptions into mathematical models (and vice versa) so that they can be further translated and related to one another. Although his physics days lay in the past, Nelson drew on a large number of examples of systems (both stable and/or unstable) and used integral and differential calculus to re-represent the population graph. In doing so he approached the problem considerably different from the field ecologists who discussed the problem in terms of the graph alone drawing comparisons to 'real' populations. Engagement in these mathematical practices is central to what he had learned as the ongoing concerns in his discipline. These concerns include finding and creating re-representations that are both more familiar and more suitable for expressing relationships between stability and instability. Nelson addressed these concerns by using interpretive tools unknown or not readily available to the experimental and field ecologists. Similarly, the experimental ecologists also drew on interpretive resources, field knowledge of animal populations, which were unavailable to the physicist.

Here, then, we have a scientist who, after identifying stable and unstable equilibrium points, critiqued the representation as not telling the whole picture. He suggested that the population graph really is about a «system» with a «stable equilibrium» and an «unstable equilibrium». He suggested that in physics and chemistry, there are many cases of phenomena that have stable and unstable equilibrium points. Nelson sketched the potential energy curves for ammonium, NH_3, which flips between two stable states, where the molecule forms a tetrahedron, and which passes through an unstable equilibrium, in which the molecule is flat. He drew a diagram of the Leonard-Jones potential for two atoms that are not chemically bonded with one stable state, the potential of an atomic nucleus, and the potential of a chaotic pendulum. In all these cases, the unstable equilibrium is associated with a local maximum in the potential energy and with (local) minima for the stable states. He pursed the hunch that birthrates and death rates function similar to forces in mechanical systems, and the fact that force is defined as the negative derivative of the potential energy.[2] He then constructed an equation that he integrated to yield a 'potential energy' curve for the population situation that resembled the curves he had sketched earlier for the physical situations. In his view, this new curve was a better representation of «stable equilibrium» and «unstable equilibrium» than the graph in the task.

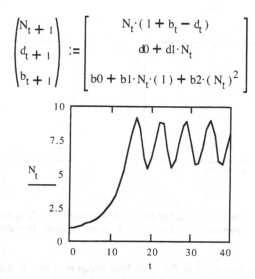

$$\begin{pmatrix} N_{t+1} \\ d_{t+1} \\ b_{t+1} \end{pmatrix} := \begin{bmatrix} N_t \cdot (1 + b_t - d_t) \\ d0 + d1 \cdot N_t \\ b0 + b1 \cdot N_t \cdot (1) + b2 \cdot (N_t)^2 \end{bmatrix}$$

Figure 3.6. The first mathematical model and its corresponding graphical implementation produced by Nelson as an interpretant of the original population graph.

3.3.2. Modeling the Population Dynamics

Nelson then suggested that another referent might be the «temporal evolution of a population». Again, this dynamical nature of the situation is not expressed in the original graph. He sketched out the problem as an iterative discrete one in which population size, death rate, and birthrate are updated according to the sets of equation for population size, death rate, and birthrate (*Fig. 3.6*). He commented on the resulting graph 'This is just like the population model I have seen in *Scientific American* published sometime in the 1980s'.

At first, he appeared contended, but then, upon inspecting the system of equations he noted that there was an inconsistency in his notation in that birth- and death rates at t + 1 were inappropriately functions of N_t rather than N_{t+1}. He changed his notation to

$$\begin{bmatrix} N_{t+1} \\ d_{t+1} \\ b_{t+1} \end{bmatrix} = \begin{bmatrix} N_t(1 + b_t - d_t \\ d0 + d1N_{t+1} \\ b0 + b1N_{t+1} + b2N_{t+1}^2 \end{bmatrix}$$

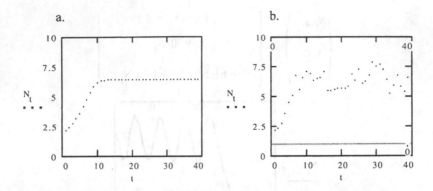

Figure 3.7. a. Deterministic model of population dynamic. b. Random variations in birth and death rates will provide a stochastic development of population dynamic.

which, when implemented in the modeling program (which requires a re-writing of N_{t+1} on the right hand sides), yields a deterministic graph (*Fig. 3.7*, top). Because he expected oscillations, he therefore suggested that the present model was deterministic and that, to get the expected oscillations, small random fluctuations to birthrate and death rate had to be added. These random variations then provided a graph that he commented upon:

> These are not the oscillations that I have seen in the *Scientific American* article about the lynx and hare in Newfoundland. It's OK, but the reason for the difference may lie in the systems themselves. I would have to check in the ecology literature to find out more about the nature of these systems. I think that the oscillations in the graph from *Scientific American* come about because we have two interacting populations, a predator and a prey. In the population model we have here, all we know is that the birth- and death rates are density dependent, which might be independent of a second population. Relating to physics again, it matters to the modeling whether the supply to the system can be regarded infinite, or constant, or whether the supply to the system, or forces on a system, changes during the time span modeled.

It is evident that his unfamiliarity with existing ecological systems constrained any further modeling without additional information. Without knowing, for example, that population graphs such as those presented here are in fact tested on one species samples, he did not have the necessary background understanding to expand the analyses he had conducted so far. In the early part of the century such experiments in which the evolution of a population to the stable equilibrium was tested had indeed been conducted. These experiments were conducted

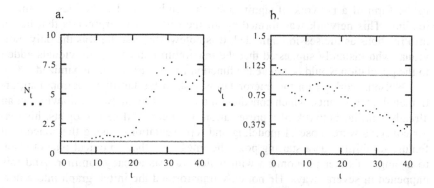

Figure 3.8. Population dynamic around the unstable equilibrium as modeled by Nelson. Due to the random influences built into the mathematical model, and beginning with the same population value, the population may eventually oscillate around the stable equilibrium (top) or move to extinction (bottom).

in Petri dishes, where the cultures had plenty of food supply such as cultured yeast cells, the gut bacterium *Escherichia coli* cultured in artificial medium, Drosophila, flour beetles, water fleas, and the protozoan *Paramecium*.

Nelson ended this session by showing that the model he had obtained was 'pretty good' in predicting a range of 'plausible' situations. By changing the initial value of *N*, he showed how the population tended to oscillate around the equilibrium value (*Fig. 3.8.b*), but close to the unstable equilibrium, could both collapse as well as eventually stabilize at the upper equilibrium (*Fig. 3.8.a*). He suggested that additional refinements could be included if birthrate and death rate were modeled and updated on a monthly level.

Here again, using the mathematical modeling software enabled him to move from quick sketches to representations that 'make sense' in terms of his own (more limited) understanding of biological systems. When he made an error, unexpected results provided him immediately with feedback, which encouraged him to seek further. Evidently he did not hesitate in implementing series of equations, even if it turned out that he had to revise them. In the course of his previous experiences, he had learned to interactively model phenomena using this modeling tool. However, the situation where he faced an unfamiliar graph occasioned new uses and new explorations that in this way he had not done before. His reading therefore made apparent how someone develops the possibilities that are already projected in his understanding rather than constituting an acknowledgement of what he had understood. These possibilities are articulated in the form of a network of significance that embeds and elaborates his initial reading. This network was formed by all the different interpretants that he set into relations as the session unfolded. It is noteworthy that he was the only individual who instantly suggested that the maximum number of individuals added to the population would be where the function $f = (b - d) \cdot N$ was maximized.

Nelson, working on the interpretation task in his familiar environment, articulated interpretants, which constitute the nodes of the network, drawing on an already familiar network of representation practices. Unlike ecologists, his major concerns were those of modeling and representations. Using the conceptual feature «stability» as a starting point, he integrated the biological phenomenon to a range of other phenomena with which *he* was already familiar. And this happened in several ways. He not only transformed the initial graph into a new graph representing the stability and instability, but also compared the structural feature of the 'hump' associated with the instability as one that prevented the system to move to another equilibrium state. Such movements might occur either from higher value to extinction or from low population sizes to the stable equilibrium. In this way, the hump was associated with another signification, another interpretant 'that makes intuitive sense'.

3.4. PERCEPTUAL STRUCTURES AND INTERPRETANTS

The case studies showed that particular interpretations depend on (a) the specific features perceptually isolated and (b) the interpretants used to elaborate the feature. Thus, Todd initially compared birthrate and death rate; when the model resulting from his deliberations turned out to be deterministically approaching the equilibrium in an asymptotic manner, he began to parse the visual display to find a feature that would bring about the desired oscillations. In the process, the slopes of birthrate and death rate curves became salient. That is, he became aware of the differences in slope while attuned to the goal of bringing about oscillations. Nelson, on the other hand, produced the interpretants /stable equilibrium/ and /unstable equilibrium/ and proceeded in producing a chain of other interpretants concerned with equilibrium states of physical and chemical systems. Here, the initial interpretant set of a chain of associations and significations that led to a very different result. If these two results generalize then it would not come as a surprise that scientists enact very different readings of the graphs that they were presented with. Furthermore, to arrive at the standard interpretation, an individual has to isolate the right set of features, produce the right referent situations, and use the right interpretants of those features selected. How did the scientists in my group fare in this regard?

In Table 2.1, I presented the frequencies of scientists who provided, during their session, a correct interpretation of the graph (that provided by the professor in an introductory course on ecology in the lectures and seminar activities). For the different graphs, the frequencies of the standard interpretation varied from 50 to 75%. In Chapters 2 and 3, we saw scientists who either recognized the graphs as familiar objects or occasioned extended readings. These scientists ended their reading having a sense of success, despite some errors and uncertainties that they discovered subsequently. In Chapter 4, we encounter three scientists whose graph readings not only failed to produce the normative standard but also who ended particular graph sessions with frustration, confusion, and a sense that they had dealt with graphs that violated good practices. Some of the interpretations, which a normative approach picks out as problematic, shared similarities with errors that middle and high school students are said to commit. That is, when the graphs were unfamiliar to the scientists, I frequently found them making inferences based on perceptual cues, picking out aspects of the graph (signs) that are not usually interpreted in the standard interpretation (such as slope in the population graph). Two issues are salient from a consideration of all interpretations as finding viable (correct) sets $\{R, S, r, c\}$. For a standard reading to be achieved an individual has to perceptually isolate some feature as a relevant sign (which pertains to S and c) and the relevant feature has to be seen in a culturally appropriate way (which pertains to r). Neither the identification of structural features in the graph nor their reading can be taken *a priori*.

3.4.1. Perceptual Isolation of Structures that Becomes Signs

Correct interpretations assume that the experts perceptually isolate the appropriate sign, S, to enter the relation $R = f_r(S, c)$. It has already been noted that scientists constructed different contextual constituents for the signs they isolated. I was even more surprised by the fact that scientists differed in their isolation of features from a graph. I even observed within-scientist variations, which occurred when the initially isolated features did not lead to a result satisfying to the expert. For example, when Daniel (forest scientist) asked himself where the population would be constant, he began to focus (incorrectly so) on the point where the slopes of birthrate and death rate graphs were the same.

> Well, you know, you reach a point of course where the slopes are the same (*Points to graphs where slopes of birthrate and death rate are the same*). You got sort of a constant, you know, birth and death rates, and population sizes maintain I guess whereas over here (*Points to right-most side of graph*) it's decreasing.

Such episodes show that the context of the scientist's current inquiry shaped which aspect of the graph would be salient.

Features of a graph are usually accepted by graphing researchers as given a priori. Thus, features such as intercepts, intersections, (relative) slopes, (relative)

Table 3.1. Frequency of particular features salient in interpretation across three tasks

Features	Frequency (count)		
	Distribution	Population	Isocline
Intercepts	0	5	1
Intersections	0	16	11
Slope, rel. slope	0	8	4
Maxima, rel. minima	8	7	n/a

maxima, and (relative) minima are taken as a priori salient elements. However, my analysis of all transcripts shows that such features are not attended to by default. There is a considerable variation within and across individuals for making salient and attending to possible referents of such features (Table 3.1). (During the analysis, all features made salient by the participants were noted and later compared, by type, across tasks and sessions.) For example, in the population graph task, all scientists attended to the intersections; this is not surprising given that the task asked them explicitly to 'focus on the birth and death rates at the intersection points'. On the other hand, the intersections in the distribution graph task were addressed by none of the scientists. Finally, 11 of the scientists (69%) made explicit reference to the absence of intersections in the isocline graph.

In the same vein, a pertinent feature of the population dynamics graph, the maximum of the birthrate curve, $/b_{max}/$, was a salient sign in only two interpretations (S01, S02). On the other hand, all but the physicist Nelson pointed to either $/b_{max}/$ or $/(b - d)_{max}/$ when asked where they would expect the maximum number of individuals added to the population between two years.

> Drew: You got some optimum population size then we're gonna get the maximum return (*Points to* b_{max}) and [...] there is a trade off between death rates and birthrates but we're looking here at the point where you're getting the maximum number born (*Points to* b_{max}) to what, compared to the number that are being lost (*Points to death rate at* $N=N(b_{max})$).

For Drew, the highest point of the birthrate curve and the greatest distance between birthrate and death rate stood out perceptually and led him to his answers. However, these differences became salient not as a matter of course but in the context of a question. Furthermore, both answers are inappropriate because the largest increase in individuals (over a year) is given by the maximum of (b - d)·N, the answer suggested by Nelson.

> Nelson: Because the curves represent rates rather than actual numbers, the absolute numbers of dying and born animals will be b times N and d times N. So, the change in actual animals in the present case will be b minus d times N.

Thus, the nature of salient geometric features in a graph cannot be taken as given but has to be established empirically, especially when graphs are unfamiliar.

It is evident that the nature of the response depends on which geometric features are salient and therefore become a signifying element /S/. For example, Brandon tracked all three isographs at a constant value R_2 from left to right and then suggested that the third isograph (*Fig. 2.13.c*, p. 58; see also Appendix) referred to 'absolute limits above which additional quantities in Resource 1 do not have any effects'.

> Brandon: So, basically it says that (*Fig. 2.13.c*), that, if you are anywhere in this region (*Moves finger along horizontal of Line A, 20*), that you have a growth rate of twenty. Which means that no matter how much R-one you have, if you have less than this much R-two, then you can't grow more than twenty.

Figure 2.12.c meant to him 'two resources are perfectly substitutable'. Finally, in Figure 2.12.c, the resources signified to him 'partially substitutable resources', for it took much more of R_1 to substitute an equal amount of R_2. Barry (university-based field ecologist) tracked all three types of curves, each one along an entire isocline. He then suggested that the content of Figure 2.12.a was 'essential resources'. He reasoned that 'you had to have a certain amount of R_1 and R_2, indicated by the values at the corners, for certain growth rates to occur'. He further elaborated Figure 2.12.b in terms of /two resources of which one could substitute the other/; and in Figure 2.12.c, he made salient the elbow where 'the total amount of R-one plus R-two was a minimum suggesting that the curve represented 'complementary resources'.

Here, we have two different interpretations, both acceptable, but based on different geometric features (instantiated in gestures and pencil lines added), and making salient different concepts. In the first interpretation, amounts R_1 along some horizontal line (R_2 = constant) are salient as are the distances between adjacent curves. In the second interpretation, the amount of R_1 and R_2 at the elbows (*Fig. 2.12.a, c*), and the non-existence of an elbow in case (*Fig. 2.12.b*) were salient and formed the starting point for a chain of interpretants, that is, for the interpretation as it unfolded. Furthermore, for the isocline graph there was no within- and between subject consistency across university-based and non-university-based scientists for focusing on elbows, shapes, or distance between curves. Some interpretations made salient the elbows in one curve, but the overall shape in a second curve; interpretations highlighted the corners in Figure 2.12.a, but centered on the location of the intercepts in the other two figures (*Fig. 2.12.b, c*). The geometrically salient features of the isographs differed considerably between individuals. Two scientists (Brandon, Rodney) perceived the isographs holistically, ordered the contours (Fig 2.12) and suggested that Figure

2.12.c lay between the other two (*Fig. 2.12.a, b*), which constituted extreme cases. Such results, of course, challenge all those models that are built on the assumption that certain geometric features of a graph are inherently part of the interpretive domain.

3.4.2. Interpretants of Given Signs

The model embodied in $R = C_r(S, c)$ shows that arriving at some normative interpretation requires the individuation of all relevant sign elements S^i, the identification of appropriate constraints c operating on them, the observation of conventions of sign use, c, and an understanding of a relevant content domain (R). Interpreting then means to satisfy a series mutually constraining influences of individual signs $/S_i/$ within a matrix of signs. Many 'errors' (in the normative frame) made by scientists who did not appropriately interpret the population graph were attributable either to the fact that a sign element was not salient, or to the lack of a constraint between those sign elements actually perceived. In the context of the population graph, for example, the most common problem was that individuals did not construct a functional dependency of birthrate and death rate on the population density. Thus, they read the sign /b - d < 0/ on the right end of the graph as /population decreases/ (interpretant). They did not articulate /N/ as constituent of /b - d/. These scientists did not contextualize their reading of /d[N] - b[N]/ in terms of /N/. They therefore concluded that /d - b < 0/ signified «population crash».

Very common among non-scientists, but less common among scientists were identifications of referents for /N/. In some cases, scientists noted that /N/ 'always' signifies «population (organisms, atoms) size», «number of individuals», or «population density». This use of 'always' suggests a typical case of conventional constraint, r, such that $C_r(/N/, c)$ = «population density». These constraints are consistent within science such that /N/ is used in the same way across disciplines (physics, chemistry, statistics). Other scientists did not use conventional constraint, but other signs as constituents to triangulate «population density» as the referent of /N/. Thus, /birth rate/, $/b = b_0 + k_b \cdot N + k_c \cdot N^2/$ and /such a function is reasonable... at very low density/ are articulated constituents of /N/ that allowed participants to recover the appropriate referent «population density» even when the conventional use of /N/ was not salient at the moment. However, scientists and more frequently non-scientists did not make maximum use of the nomological and typological signs they identified to mutually constrain their referents. As an example, Table 3.2 features the different referential attributions made.

Table 3.2. Frequencies of referents for the label /N/ on the abscissa

Label	Frequency (count)
«Number of individuals»	3
«Population size»	8
«Population density»	5
«Unspecified variable»	2
«Time [years]»	1

Each aspect that can be constructed as a separate element of a Cartesian graph (e.g., axis labels, scales, units, and so on) can be understood as an articulated constituent that is a potential resource for the reader in the recovery of meaning. When they are salient (which is in part a function of the degree to which the reader has been 'disciplined'), these elements constitute informational resources and therefore reduce the interpretive flexibility of the primary line or series of data points. To guide readers' perception in a desired direction, authors have available a wide range of resources. These resources are particularly important when the inscriptions serve demonstrative purposes in scholarly texts and textbooks. Some of these resources include axis labels; nature of scales, which can be log-log, linear-log, log-linear, linear-linear; units; legends; error bars or confidence ellipses; titles; statistics in the figure; caption or text; color, shape, and size of data points; color and nature of lines (broken, fat, dotted); labels to curves, arrows, and other identifiers such as numbers and words; or juxtaposition of lines within the same graphs and series of graphs.

These resources assist readers in a process that allows them to imagine 'real' situations from which these graphs can be thought to have been abstracted. Scientific authors usually draw on these resources to limit the interpretations of the reader to those they intended. However, even with these resources that limit interpretation, the reading process which reconstitutes referents through elaboration is inherently open. Such interpretive flexibility is unavoidable even in the case of a very simple diagram (*Fig. 3.9*). It may be seen as a wire frame, as a glass cube; an aquarium of cubical shape that Brandon used in his experiments; a box closed on all sides or open on one of them; a schematic cube viewed from the top or from the bottom.[3] In fact, if such a diagram appears in the context of an article or book, the caption accompanying the drawing will promote any particular way of seeing the diagram over all the others that are possible. How the diagram should be read depends on the context, in particular the interpretive resources made available by the author. The fewer the resources, the greater the interpretive flexibility, and therefore range of situations from which the graphs could have been derived. At the same time, the problematic outlined here suggests that research on graphing requires a much more phenomenological take.

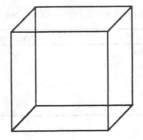

Figure 3.9. Simple diagram that can stand for different things in different contexts.

Rather than saying that an individual incorrectly interpreted a graph, it may be more fruitful to begin by noting first which features the individual attended to and second which interpretants are used to elaborate them. It may turn out that once the ensemble of signs (domain ontology) on which the individual operates, their interpretive process are not as flawed as it is made to appear in the literature. To gain a better understanding of 'incorrect' interpretations, I suggest that we attempt to understand the framework within which the individual's statements are perfectly reasonable.

3.5. REFERENCE, SENSE AND MEANING

From the two case studies considered in this chapter, we can take two basic messages. First, 'meaning' neither resides in nor is brought to a graph. Second, interpretation does not reveal something that the author of the graph somehow transferred. For example, we saw that Nelson did not just bring existing understanding to the task in the form of systems characterized by the co-presence of stable and unstable equilibria. There was some of it. But what was more important was that his understanding of such systems was further articulated by the addition of biological systems and has been further articulated in terms of the analogy between birthrates and death rates in biological systems and physical forces in chemical and physical systems. In a similar way, when another scientist first analyzed the internal dynamic of the distribution graph (see Appendix), he articulated the covariation of elevation and climate. This description brought forth a previous experience in which elevation and climate changed as he had hiked across Mount Kenya. That is, he made sense of the covariation *in terms of* the memory of an earlier experience, which acted as an interpretant of the sign (covariation) that he previously articulated. He also brought forth the changing fauna (semi-arid, plantations of coffee) that he experienced as being associated with the changing elevation and climate. Again, he articulated an experience that had the potential to become an interpretant of the changing fauna articulated in the graph. Here, then, the earlier experience became further elaborated in terms of a (novel) graph at the same time that the graph was bound in with the existing understanding. Readers of graphs do not make meaning; it already exists as an existential of being a person in this world. Graphs *accrue tò* meaning (webs of existing significations) rather than becoming or being constructed as meaningful. In this, we can see how a possible content of the graph is articulated as scientists produced interpretants. They translate individual aspects of the graph into interpretants that stood for (make present again) their experiences.

The intelligibility of a graph therefore lies in its possibility to engage the individual in articulating the relations that constitute 'meaning'. In this, meaning has to be considered as an existential of human beings rather than as a property that is attached, lies behind, or floats between signs, texts, or objects. Rather, 'meaning' circumscribes the formal structure of those things that interpretation articulates, which is, as we have said, understanding. Meaning therefore is that which can be articulated, the network of interpretant signifiers produced by the individual in the task. It is an existential of simply being human and being with others, that is, being part of and sharing a physical and social world. Meaning is not a property that somehow is or becomes attached to things. During interpretation, meaning is disclosed *in front of* the graph. In the present situation, the graphs are therefore sites where this disclosure occurs. Interpretations arise from the dialectical tension between graph and elements thereof as these are salient in the perception of the reader and the already existing understanding of the individual. When the reader of a graph perceived a word, line, or text as meaningless in the way a substantial number of scientists did (see Table 2.2, p. 29), they in effect said that it had no relation to their present understanding. That is, that they could not accrue it into an existing network of significations.

Already in the early part of the twentieth century, Martin Heidegger articulated this close relationship between understanding and interpretation that was most recently further developed by Paul Ricœur[4] was perhaps Heidegger's unusual way of writing that made it difficult for other scholars to appreciate what he had to say with respect to interpretation and understanding. Heidegger foreshadowed this analysis by equating the articulation and elaboration of understanding with interpretation. But his term of 'understanding' cannot be equated with explicit factual or procedural knowledge. Rather, understanding is associated with the fundamental condition of being in this world that we always and already share with others. Understanding is always relational and has the character of in-order-to, what-for, what-in, and for-the-sake-of-which. These relations are interlocked, forming a network of signification, which Heidegger called *significance*, and constitutes the structure of the world that we always and already find ourselves.

In this view, then, interpretation, the production of interpretants, reveals the rhizomic nature of the network of signification that constitutes that aspect of understanding that can be articulated. It is necessarily specific of individual biography, the historical contingency of the enveloping culture and its language. In this way, graphs are never just graphs but are objects or tools that accrue to networks of signification that can be articulated in a variety of ways and to different extents when needed. The extent of this network is related to how familiar we find a graph to be and exactly what kind of understanding we elaborate as we work with it. Finally, these networks should be thought of as consisting of segmentations of matter, the entire range of material constellations that human beings may find significant rather than just texts (if these are thought in terms of words).

Nelson arrived at some content rather quickly, which he described in terms of a system with stable and unstable equilibrium. All subsequent work, then, was a further articulation of his understanding of models in the particular context of biological systems evoked in the present task. It is evident that the graph provided a site, a catalyst, where he worked out, developed, elaborated, and articulated always and already existing understanding. The interpretation of a graph (as that of any text) did not lead to new existential understanding but to a better articulation of that which is already understood. All interpretation therefore requires that the text to be interpreted be already understood in some deep sense. Again, even the process of interpretation presupposes understanding. The process of interpretation (structural analysis) allows a movement from surface understanding to deep understanding. Here, depth understanding (semantics) is what the text is about.

We also see that there are neither natural ends nor natural criteria that set limits to the interpretations by some individual. An interpretation ends, momentarily, when the individual has in this situation the impression that any new interpretant is not further elaborating what has been said before. That is, the network of enunciated relations that elaborate a possible content of the graph does not articulate further understandings of the sign or the natural object it is said to represent (refer to, stand for). The relationship between each individual interpretant and the original sign articulates the sense of the sign; that is, sense is provided by the relationship of synonymy with the original sign.

In the situations analyzed here, the person did not know the content (referent) of the graph (sign) beforehand. That is, a spontaneous understanding does not exist in the way it does when we read a sentence such as 'the mouse has three off-spring', 'five mice died', or 'mice give birth to 5–8 young each year'. (In Chapter 6, I provide a case study were such spontaneous understanding exists because the graph is part of the everyday work of the individual scientist [scientific professional].) In all of these cases, we know immediately what the text is about. When such spontaneous understanding does not exist, that is, when the content of the graphical expression is not immediately evident, readers have to engage in a process where possible contents are disclosed as the sign is elaborated in terms of existing understandings.

In this work, we see that even the nature of the sign is not self-evident, for there are different features that can be made salient in the same graph. The content of the graph, however, is a function what is considered as the relevant sign. Thus, Todd did not know a priori whether the slope of the graph is relevant, during much of his session it is not even salient. However, when he arrived at a juncture where his interpretant graph was not consistent with his experience, features of the graph became salient that might produce the kind of oscillatory behavior in the graph that he knew natural systems to exhibit. In this situation, the population dynamics graph was problematic because it did not describe the natural phenomenon Todd already knew about. As he evolved the population dynamics graph through a translation (i.e., an interpretant), he sought the problem in the original graph.

On the basis of the data and analyses presented in this chapter, I contend that classical models of graph interpretations are limited in its scope: they refer to situations where readers have a spontaneous understanding of the graph. What follows are articulations of understanding in terms of accepted schema rather than unfolding elaborations of initial and partial reading of novel graphs. On the other hand, readers of novel and other unfamiliar graphs do not know whether what they term to be the content of the graph is in fact the normatively accepted one. Their interpretation sessions constitute the elaboration of possible content models that stand in a reference relation to the original sign. In these elaborations, readers essentially segment two different portions of the material continuum, one that gives rise to the sign (inscription before them), the other a segment of the 'natural' world. These two segmentations correspond to the R and S in the relation $R = f_r(S,c)$. The elaborations therefore can be viewed as a process of reification in which the two segmentations are articulated and set in relation through reference. 'Correct' interpretations will only be observed when the perceptive processes isolate content R, sign S, context c, and convention r according to existing norms. That is, a likeness with a normative reading requires that relevant signs and their context be salient, conventions in the appropriate domain be followed, and readers' familiarity with the content (i.e., some 'natural' phenomena).

FOUR

Problematic Readings

Case Studies of Scientists Struggling with Graph Interpretation

Three case studies are provided of scientists who, despite normally being very successful in terms of publications and funding, struggled with making any sense of the graphs. I articulate in some detail the particular problems that these scientists had and the resources they drew on to cope with their difficulties. The chapter ends with a discussion of the notion of expertise and its relationship with familiarity in the domain. This discussion sheds light on other studies that use cognitive deficits to explain why students do not provide standard answers when asked to interpret graphs.

4.1. TOWARD AN ALTERNATIVE TO MENTAL DEFICIENCY

> Mental structures are needed in order to cognitively manipulate some forms of content and graphical representations. Without cognitive development, students are dependent upon their perceptions and low-level thinking. These deficiencies are not often apparent.[1]

Research on graphing has long been dominated by normative models, whereby people with little experience ('novices') are compared to others, often scientists or mathematicians, with lots of experience in a domain ('experts'). This research focuses on what sections of the population do *not* do. It attributes the differences between a normative population, 'experts', and some other population, 'novices', to mental or cognitive deficits (see opening quote). By implication, experts do not experience the difficulties that novices do, and therefore do not have the deficits diagnosed with an expert-novice model. In the opening quote, science educators, too, attributed the fact that students did not interpret graphs according to the standard that they had defined as evidence for lack of mental structures and cognitive development. As a consequence of lacking the cognitive development required for graphing, students are said to rely on perceptions and low level thinking. In the past, deficit models have become to be the norm among many researchers on graphing.[2]

Having been a scientist and worked among scientists as ethnographer and having conducted research on the emergence of mathematical representations from scientific fieldwork, I find the theoretical positions underlying the quote wanting. The previous two chapters showed that highly trained individuals proceed with their interpretation not in an error-free manner but their talk was full of the stumbles that characterizes real-time activity and there were errors, sometimes corrected and sometimes not. However, these cases presented individuals who eventually articulated a solution that was counted as correct. Not yet presented so far were examples of those readings that were far from smooth (see Table 2.1, p. 27), some of which ended with the indication that the tasks were difficult (see Table 2.2, p. 29). To understand these examples and to develop a better framework than the deficit model, I develop three case studies in this chapter, each of which focuses on a different one of the three graphs.

In each of the three cases, the participants ended feeling that they could not make sense of the graph. They ended with a sense of dissatisfaction, and, in the first two cases, critiqued the graphs as inappropriate, as 'evidently' not showing what their authors intended to show. Of course, one of the problems may lie in the way we define experts and novices. We can normatively define expertise in terms of particular competencies independent of the professional experiences of a person. In this case, we would have no trouble categorizing all those scientists in my studies who did not arrive at the normative model as non-experts. However, we then end up defining expertise in terms of models restricted to reading graphs 'in captivity', which often have very little relevance to the way graphs are used and interpreted 'in the wild'.

Throughout this book I indicate my interest in developing a way of approaching graphs and graphing that are useful for looking at ordinary, everyday reading and interpretation. As part of this overall goal I want to show that otherwise highly competent scientists—who have demonstrated their competence repeatedly as part of getting their degree, publishing their research, and in getting grants—are much less competent when solving unfamiliar graphing tasks in a traditional experimental paradigm. Such results are evidence in support of the contingent nature of graphing competency, which is a function of the familiarity with the particular graphs, content domain of the graph at hand, translations between content and graph, and the conventions in the particular discipline in which the graphs have high currency. Such competencies do not easily transfer.

In the academic culture, many think of texts and graphs in terms of information—university professors tell me nearly every day that they see their job as one of 'giving students information'. In a similar way, instructions are seen as information provided to the potential user to engage in a proper sequence of actions; when users are not successful, they blame the instructions. In both examples, the underlying epistemology is that of information transmission: some sender sends a signal that encodes the information, which is subsequently received and decoded. In such an account, a graph and its accompanying text constitutes compact information, which can be fully unpacked through appropriate interpretation. It is assumed that encoding and decoding are exactly the reverse processes. In human communication, the exact equivalence of sender and receiver is only a limiting case. That is, we have to abandon the notion of informational signs as unambiguous in all other cases.

In the previous chapter, I began to elaborate a view of interpretation as a process in which existential understanding is articulated in the form of an increasing (potentially rhizomic) network of interpretant signs. In this view, an unfamiliar graph constitutes an occasion for articulating an understanding in the form of a network of interpretant signs that provide connecting points for appropriating the graph in the task. As a consequence, if graphing researchers are attuned to the interpretants they are provided with glimpses into the universe of the participant's world. I encourage graphing researchers to look at what people reveal about their worlds rather than at the difference of utterances with respect to some normative frame. In other words, every graph interpretation that is supposed to contribute some understanding must already have understood. Particular interpretants produced during a session therefore tell us more about the existential understanding occasioned *by* the graph than about knowledge *of* the graph.

4.2. A GRAPH THAT DOES NOT CONVEY ANY INFORMATION

At the time of the graphing session, I had known Daniel for a period of five years prior to conducting the interviews over and about the graphs. During this period, I had informally interviewed him about his work in the context of another study, particularly his research and the kind of analyses he conducted. I also had asked him to complete tasks used in other research projects on the interpretation of data to ascertain face validity and as calibration for student responses to scientific test questions. Daniel holds a master's degree and works as a senior researcher for a private, non-profit research and development organization. Leading companies, the Government of Canada, and the provinces fund the organization. He had worked for over eight years in this organization as an engineer conducting independent research projects. He regularly attended and presented at applied research conferences, and writes and publishes reports and research articles. Whenever I talked to him about his work, he articulated a particular concern for helping clients to understand the different kind of inscriptions that he produced as a result of his research. These clients were both from the public and private sectors and, as Daniel frequently pointed out, had trouble understanding graphs 'that include too much information' or 'where the relationship is too complex'. 'They [clients] like it simple', he added.

As Daniel does all statistical analyses for the studies that he designs, he is very familiar with a variety of techniques and with plotting the results of regression analyses. It was evident from the readings that he provided of his own graphs that Daniel expected graphs to be transparent and that they readily tell the user the story that they are intended to tell, a story that he quickly provided upon request. Graphs had to be transparent in his work such 'that people in the industry would understand' and the kind of graphs he was asked to read are 'not charts that you give someone'. The graphs had to be such that they could instantly find the information they needed. Despite his educational background and experiences as a researcher, Daniel found interpreting my graphs a troublesome activity.

4.2.1. Understanding the Graph or Understanding Forestry?

Knowing that he expected graphs to be transparent, it came as no surprise that Daniel began his session with the distribution graph (*Fig. 4.1*, for an explanation of the task see the Appendix) questioning whether he should read the caption of the graph [i]. Despite an initial reluctance, he partially, and in episodes, read the entire paragraph.

> [i] You want me to read this down here? Or does it matter? I won't bother reading this stuff. So, anyway. [ii] Anyway, so they're different plants, relative importance, OK, relative importance? Given? OK, so the relative importance of the plant at a certain elevation? Importance in terms of...? Ah, what are we talking about here? [iii] (*Reads from text*) 'are maximally important under any...'. (*Points to the caption and reads*) 'Internal gas exchange for...'. OK, so it's their importance relating to gas exchange for water conservation. [iv] I think that importance here may be relating to how much, how successful they are in growing in that. [v] Yeah, I mean, you know, for ecologists, forest ecologists, that's something that they're dealing with, I mean, the ecosystem classification systems or problems, are looking at indicator plants given certain temperature and moisture gradients. [vi] It's not something I'm dealing with, as an engineer. But anyway, so, OK, what's- So exactly what are you asking here?

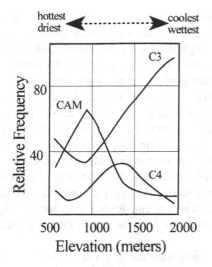

hottest ◄ - - - - - - - - - - - ► coolest
driest   wettest

Distribution of C_3, C_4 and CAM (succulent) plants in the desert and semi-desert vegetation of Big Bend National Park, Texas, along a moisture and temperature gradient due to differences in elevation. CAM plants with nocturnal gas exchange for water conservation predominate in the hottest, driest environment, C_4 plants are maximally important under intermediate temperature and moisture conditions, and C_3 plants predominate at the cooler, least dry end of the gradient. (After data of W.B. Eickmeier [1978], Photosynthetica, 12, 290–297). What implications can you draw from this graph?

Figure 4.1. In this task, the different distributions of three types of plants is used to make inferences about the differential adaptation to environmental conditions (here climate).

After noting that the three lines each has one «type of plant» as a possible referent, Daniel stumbled over the notion /relative importance/ [ii]. Not being familiar with it in the context of graphs, he checked the caption for further interpretants, and then produced the interpretant /importance to gas exchange for water conservation/ [iii]. This interpretant was further elaborated by another interpretant /importance relating to successful growth/ [iv]. At work, he often interacted with forest ecologists and was therefore familiar with the kind of studies they conduct, including ecosystem classifications using indicator plants for different climate measures ([v]). However, rather than continuing this chain of interpretants to build a network of significance, he constructed himself as a non-expert in this type of problem normally dealt with by forest ecologists [vi].

In this first excerpt, Daniel perceptually isolated some signs in the graph, including /relative importance/, /elevation/, and others from text (/internal gas exchange/ or /maximally important/). He also provided the interpretant /different plants/, which elaborated /C_3, C_4, and CAM (succulent) plants/. Although it is quite possible that the graphs were perceptually salient in some way, the transcript does not provide any indication about the nature or detail of this salience. However, what is more interesting than the literal reading are the interpretants that are indicative of the everyday understanding occasioned by the graph. These interpretants include /how successful they are in growing that/, /forest ecologists/, /ecosystem classifications systems/, and /indicator plants given certain temperature and moisture gradients/. Although these signs are merely bits, insufficient to piece together a detailed picture, they constitute sketches of a world in which there are people interested in growing plants, who might use graphs to show how successful they are in doing that. There are forest ecologists perhaps conducting ecosystem classifications, and, in this, drawing on 'indicator plants' that characterize particular temperature and moisture levels. Knowing that Daniel and his coworkers attempt to (a) improve forestry operations related to the harvesting and transportation of wood, (b) improve the growing of trees within a framework of sustainable development, and (c) increase the efficiency of current forestry equipment and reducing costs, we realize that his interpretants give us indeed a glimpse of his everyday work environment. Rather than just telling something about information that the author put *into* the graph, Daniel in fact told me about his work.

We may be led to think that Daniel interpreted the graph correctly. For example, his interpretant /importance here may be relating to how much, how successful they are in growing in that/ could be read as a way of expressing adaptation to 'that', which itself can be read as a reference to the abscissa variables. However, as the following excerpts show, this interpretant did not stabilize. It was fleeting, raised as a possibility, but not woven into a stable network of interpretants (consistent or not) that elaborated the referents of the graph, a background to which the graph could have accrued. He did not yet have a sense of what the graph might express, which we can gauge from his question to the interviewer what he was being asked to do.

4.2.2. Confusion: 'There's a bit too much happening here'

After being encouraged to tell how he understood the graph and what implications he could draw, he continued by indicating confusion.

> [vii] Yeah, it's fair a bit going on, I mean, they got this gradient up here (*Points to gradient on top part of graph*) hottest driest, coolest least dry. Well, you know, for me there's a bit too much happening here, I mean, you have got the elevation scale and, your independent variable down there, and you've got, hum. [viii] For me, you know, this relative importance, you know, you've got, OK the relative importance of this plant, relative to? I don't know, It's tough for me- to pick much, get a lot out of that. [ix] Hmm, I would like to see some scale in the y-axis that it's a little more quantitative. [x] And then, for me anyway (*Swiping gesture along temperature gradient*), I'd like to avoid having another scale on the same axis (*Gestures along abscissa*). It's just, you know, I

don't think it conveys any information- that really clouds the, what you're trying....

As he continued, Daniel made salient multiple scales along the horizontal axis
(above and below graph) [vi], although it was not certain whether he actually
perceived two or three scales. He then shifted to and questioned again the ordinate label /relative importance/ [viii], a term of which he apparently could not
make sense. Subsequently, he returns to the multiple scales along the abscissa
[x]. He suggested that the graph clouds rather than illuminates what the author
had attempted to convey, though he had no answer about what the author intended to say.

During this entire episode, Daniel engaged in meta-talk about the graph.
Rather than seeking what it could be said to express, he suggested that there was
'a fair bit going on' and 'a bit too much happening' which made it 'tough ... to
get a lot out of that'. He attributed the problems to the use of multiple scales
along the horizontal axis [x], the apparent lack of a referent for /relative impor-
tance/, and the confusion about the nature of the vertical scale [viii] and its
measure [ix]. The display was apparently too complex for Daniel to configure
the graph in a way that he could use to seek convergence with his understanding
of a particular aspect of the world. Although he attempted to engage in a structuring activity that is internal to the multimodal text he could not isolate one that
could be linked to other more familiar realms. In the end of the episode, he
faulted the author of the display for not having avoided multiple scales and
thereby clouding the information that was to be communicated.

Daniel continued his meta-discourse relating to the quality of the graph before him. He recognized that there are authors who do use multiple scales on the
same axis [xiv]. From his own work—that is, in his experience with the users of
his research, including business people and government officials—he knew that
increasing the structural complexity of graphical displays leads to confusion
among his clients.

> [xi] I really don't like to do that... You know, I don't think it's a good thing to
> do in charting a graph- I mean people do it but I think that if- The people who
> were trying to present information should really think hard about different ways
> of doing it. [xii] If- and there's usually some other way to present your infor-
> mation, I think, or your results or data, or observations what ever you are doing,
> I would think....

He suggested that the information really should have been presented differently
[xii]; elsewhere in the session, while talking about his own graphs, he contrasted
them with the ones I had brought to the session. He pointed out that the num-

bers, money and productivity, are what his clients could relate to. The graphs he and his colleagues produced had to make sense to people in the industry, who have to set up production sites under difficult conditions, and who use his graphs to maximize their productivity. Key to the graph always has to be that 'people can really relate to'. In the graph before him, this was not given.

I had explained my research to Daniel as an attempt to understand how students, on the one hand, and scientists and engineers like him, on the other hand, use, read, and interpret graphs. I had told him that I ultimately wanted to be able to provide recommendations for designing curricula that allow students to develop greater competence than they do at present. Throughout the session, Daniel was concerned about the graphs that I had presented to him, suggesting that they were confusing and violating conventions of good graph use. He was so concerned that a few weeks following our interview, he faxed me an article entitled 'Crafting Better Charts' with the suggestion that it might help me to construct better graphs for my research.[3]

4.2.3. Understanding: More about Forestry

From this excursion to the demands he and his clients have regarding graphs, he then returned to an analysis of the ordinate axis, questioning in particular the notion of 'relative importance' [xiii]. He began by producing a possible interpretant /abundance/ and /area covered/. He briefly elaborated /area covered/ by articulating conditions under which it might be a suitable interpretant for /relative importance/ [ivx]. He then returned to talking about situations from his lifeworld, which he understands not only explicitly but also intuitively, in the way we know gravity without ever having studied physics.

[xiii] What is the relative importance? The abundance of it? Or something? I mean, or the area covered or, I don't know, some area that was covered, say it was occupied by those plants or something on that scale, I mean, sure, I mean, you know. [xiv] Importance, well, I mean, you know, at a certain elevation, I mean, you're gonna, you're looking at, well it depends on the value, value of the project, I mean, you know, whatever the most abundant is obviously the target of them. [xv] Unless it's, you know, a very high value, if a species or whatever, you know, a couple of trees of this valuable species what's there, you know. It could be more, it could be more than more scarce. [xvi] If it's more scarce, it might be more valued, I don't know, I mean. [xvii] It is pretty rare, you don't log at those, I mean, at two thousand meters, six, seven thousand feet? That's, you gotta be, it's not commonly done, I mean, at those elevations of course, it's probably timber types. [xviii] I mean, because you've got, you're talking about relative importance, you've got three different species here or

whatever. Are you talking about the relative importance of this (*Points to C4*) one to these other ones (*Points to C3, CAM*) or to some other external influences, I don't know. (*At this point, Daniel abandons the interpretative work on the distribution graph.*)

Daniel continued attempting to produce possible interpretants for /relative importance/ or rather /importance/ alone. This led him to suggest that contexts such as /value of the project/ determine what /importance/ denotes [xiv]. He then specified that 'whatever the most important is obviously the target of them'. Although the graph was constructed by a plant ecologist interested in the relative distribution and adaptation of plants with particular photosynthetic mechanisms, Daniel clearly situated the problem in his own work as shown by the chain of signification he developed: /importance/ → /value of project/ → /abundance/. Implicit in this chain is the nature of the plants. As Daniel's own work is related to forestry, it is possible that he talked about trees. (In the same way, Newfoundland and Labrador fishermen used to talk about «fish» and everyone would know that they were talking about «cod», the species that people had fished in these waters for four centuries.)

Daniel then articulated interpretants from the domain of harvesting commercial species of plants—which is in fact the domain in which he has done research for many years. In this context, then, high value of the species may mediate /most abundant/ so that species of lesser abundance are actually also of interest and contribute to the contextual /value of the project/ [xv]. In this new context, the sense (interpretant) /abundance/ for /importance/ is no longer valid, for scarcity may increase the «commercial value of the project» and therefore its correlate «importance» [xvi]. Daniel tied his discussion of the commercial value back to one of the variables, elevation [xvii]. He made salient the abscissa value at the right end of the scale, 2000 meters (which he translated into six to seven thousand feet [another interpretant]) and suggested that at those elevations, logging is not commonly done. In the final throws of this session, Daniel first hypothesized that the three curves have different species as their referents, and then asked if /relative importance/ pertained to the relation between the three different curves. At this point, Daniel abandoned the session.

Beginning with the interpretants for /importance/, Daniel articulated an understanding of *his* world; that is, the graph provided an occasion to articulate a world that he is intimately familiar with. It is a world of projects, trees that have different values depending on their abundance, sometimes difficult conditions for logging them, and costs that have to be incurred to log in difficult sites. This is also a world where logging at an elevation of 2,000 meters is not commonly done. The world that is sketched in the talk over and about the distribution graph was nearly continuous with the talk about another graph that he presented from his own work.

> You don't have total flexibility [how to log] and you have to design [the pro-
> duction system]. Given what the terrain gives you and a number of other fac-
> tors, and when you got a certain particular set of conditions out there that you
> have to deal with, you can estimate, your cost, your productivity and then, in
> your costs.... They've got difficult situations where they have to get timber
> from the mountain down to the road, and there is no other way to do it other
> than a helicopter so...

In this excerpt he talked about a company that made use of his research in which he had provided regression analyses that identified the key indicators ('condi-tions') for the cost of logging from difficult sites. It is clear that although we were at a very different point in the interview session, it was the same world that he articulated to me. What we see, then, is Daniel as a specific reader who ar-ticulates his understanding in terms of the specific relevance, purposes, and knowledge at hand; and these terms are integral to any determination of sense that he makes.

4.2.4. Forestry, Graphing and Expertise

Daniel is successful in his work as a scientist, publishes research articles, writes reports, and produces graphs for clients in the public and private sectors. At the time of the interview, he has had experience in doing complex studies using multiple regression analyses and plots with at least three variables displayed in a single graph. Consider one of the graphs that he had brought to our session (*Fig. 4.2*). Daniel had spent many weeks with the operator of a certain logging sys-tem, following him into different sites and recording data on a variety of aspects involved in the operation of the system. Daniel then conducted his analysis of the data to build a model of the logging system, its productivity and cost effec-tiveness under different conditions. He conducted a multiple regression analysis, based on backward elimination of variables—those variables that do not con-tribute significant amounts of variance are dropped from the analysis. His graph represents the type of results he achieved, in which productivity is plotted as a function of two independent variables. As such, his graph is probably as com-plex as the distribution graph that he found confusing and difficult to interpret. In both instances, a dependent measure is plotted as a function of multiple inde-pendent measures. In contrast to his graph, the distribution functions are non-linear but the additional independent variables (temperature, moisture) are di-rectly correlated with the main variable (elevation).

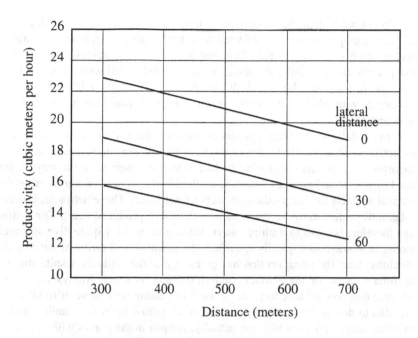

Figure 4.2. Graph involving three dimensions similar to the one presented by Daniel from his own work.

Despite similar complexity in the two graphs, Daniel did not arrive at the expected (normative, correct) interpretation of the distribution graph. He failed to establish coherence between the structures of the graph and his understanding of the world: The foreign remained foreign, and Daniel remained at the doorstep of the graph. He ended the session with some frustration about the poor nature of the graphs that I had brought. Within a normative frame, Daniel's reading of this and the other graphs were problematic; he failed to achieve the standard reading. Viewed differently, the distribution graph occasioned the highly competent articulation of a familiar world in which differential plant distributions would affect productivity and cost of harvesting. His own graphs were of similar structural complexity and yet his reading was tremendously competent. A likely conclusion therefore is that graph reading is not context independent. We therefore need to find out how levels of knowledge of situations, data collection, and processes of graph construction mediate the reading of graphs.

When we look at the session not in terms of what Daniel said or did not say *about* the graph but in terms of what he actually said, we gain considerable insights into his world, a world of logging and research on diverse aspects of forestry, particularly in difficult situations. It is an individual with elaborate understanding of a domain, but which did not provide sites that could be brought into coherence with what he saw in the graph. The graph could not accrue to his understanding.

In his daily work, Daniel produces graphs at the end of extended sequences of activities that constitute, for example, a production study. These graphs are therefore records that index what he had done and observed in the way. A standard of (social and natural) scientific method is the reconstruction of order from natural artifacts or the residual artifacts of activities. These artifacts, however, inherently underdetermine the situations that have produced them. The artifacts can therefore be read in multiple ways, leading to possibly quite diverse conclusions. Ethnomethodologically specified, the scientist will arrive at definite conclusions about the situation that has given rise to the artifacts despite the indeterminate nature of the artifacts, individually or as a set. Analysts can do so because they too are able to contrive ways of dealing with these difficulties, and are able to do this because they too are able to draw upon their understandings of what things may possibly, and actually, happen in the places with which they are familiar. That is, informed readers of artifacts do not encounter difficulties with the indeterminate nature of (natural or social) artifacts, or rather, they encounter difficulties as occasional rather than chronic ones. They read records under the auspices of their involvement in, and familiarity with, the circumstances in which the natural objects were found or under which the social records were created. In this, scientists can appeal to understanding how research normally occurs, how research in the specific instance occurs, what things actually versus conceivably happen. If there is trouble, they may even elaborate what the producer of the representation must have meant. In Chapter 7, I will provide a detailed study of how such knowledge about situation is built up by one scientist as she spent three summer seasons in one area, where she hunted lizards 'rain or shine', and developed an intimate and often unarticulated understanding that resurfaced when there were troublesome data. If the origin of a graph is known, we therefore expect readers to articulate and elaborate their understanding of how the place works and of procedures that might have given rise to the graph. If the nature and origin of a graph is unknown, and the reader is nevertheless complicit in a game of producing an interpretation, we then expect the articulation and elaboration of some other situation with which the reader is familiar. When Daniel talked about the distribution graph, this is just what we have seen. He articulated his understanding of doing forestry with different species in different (and difficult) mountainous situations. However, he never found the information that he thought was put into the graph by its author and which he should have been able to disclose but could not because the problems with its construction.

4.3. GRAPH DEMANDS KNOWLEDGE OF POPULATION ECOLOGY

In the previous section, we saw how a graph occasioned a researcher if not to articulate so at least to hint at his understanding of a familiar, daily-experienced world. We might ask the question what individuals say if the graph and text do not trigger interpretants related to their familiar world. What might an individual say about a population graph when he clearly identifies himself as not being a population ecologist and who suggests that reading the graph demands more knowledge of population ecology than he had? At first, we might be tempted to suggest that the individual would be referentially stuck, reading the graph literally and describing features, without however making suggestions what the graph and text are about. Of course, the words in the caption and on the graph presuppose understanding so that a graph accompanied by text probably never leads to complete referential isolation. We saw, on the other hand, that Nelson seemed to disregard concrete populations altogether and moved to investigating the graph qua model and its mathematical structures.

At the time of the interview, Rodney was in his seventh year after receiving a Ph.D. in environmental biology particularly on issues in entomology, taxonomy, and systematics. He was working for a government institute focusing on animal pests endangering plants produced for commercial purposes. This involves pest damage assessments in research, damage predictions, and pest management. Rodney is an active member in an entomology society and editor of a journal in the domain. Having averaged more than two articles per year over a ten-year period prior to the interviews, he continues to publish regularly at that rate.

4.3.1. It's climbing dramatically but it's still [below?]

After reading aloud the caption of the population graph, Rodney began his reading of the graph and verbal text. As he talked, he pointed to various places and marked several arrows onto the graph (*Fig. 4.3*). He first commented on the death rates and birthrates as 'interesting' [i], and then marked the two intersections as points where the population would, theoretically, remain stable [ii].

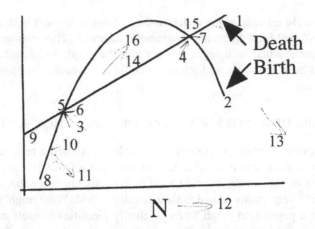

Figure 4.3. Rodney's marked-up population graph. The arrows make salient specific points, or show trends in some rate with changes in population. (Numbers are locations or directions referred to in the text.)

[i] So linearly increasing- (*points to [1]*) a very interesting (*points to [2]*) birth-rate curve. [ii] So these two points- (*Draws arrows [3], [4]*) theoretically, are where the population would remain stable, unchanging. [iii] The birthrate- wait a second- *N*? Rates? (*Points to [5], 15 seconds pause. Reads text silently.*) [iv] I don't know how much math I have to read into this. [v] On the surface, you'd assume that at these points- (*Marks lines [6], [7]*) the population all other things being equal is neither going to increase nor to decrease. *(4 seconds)* [vi] And if that it's true you would assume that here (*Points to [5]*), or I would assume, that here the birthrate- it's (*Points to [8]*) climbing dramatically but it (*Points to [9]*) is still- *(3-second pause)*. Oh Christ!

Rodney stopped short, apparently having doubts about 'how much math ... to read *into* the graph' [iv] and reiterated his interpretant of the two intersections as stable points of the population size [v]. But then the steep incline in birthrates to the left of the first intersection [5] became salient 'it's climbing dramatically' at which point he came to a halt [vi]. The situation became so complex that he uttered an expression of exasperation ('Oh Christ!').

So far, we see that Rodney perceptually isolated and made salient the two intersections where the population remained stable. That is, at a minimum he read the intersections as points where the two curves display equal values of birthrate and death rate, which leads to a stable population. At the same time, the equality of birthrate and death rate at the point of intersection [5] was not the only thing that was apparent. As Rodney began another interpretant, he appeared to be thrown off by something else: the slope of birthrate was much steeper than that of the death rate. As he placed his finger first on the birthrate and then on the death-rate curve, he began but never completed the contrast between a 'dramatically climbing' birthrate and the death rate, which is presumably higher ('it is still-'). That is, at this point, as the expression 'Oh Christ!' indicates, Rodney was exasperated by having to integrate a birthrate that is lower than death rate but at the same time (in the same horizontal position) is increasing more dramatically.

4.3.2. Population Crash: Is it Iconic Confusion?

Stopping repeatedly, Rodney returned to the caption and made several meta-level comments about his process of reading [vii]. These meta-comments and the use of conditional tense suggest that he was unsure about his reading.

> [vii] (*Reads caption*) 'Change in birthrate follows a quadratic function', is that important? (*4-second pause*), low densities (*points to [8], 17-second pause*). Oh maybe I'm reading more into this than there is. [viii] But I would expect that here- (*Points to [10]*) With a relatively high birthrate- (*Points to [9]*) Relatively low, sorry, death rate, relatively low... (*Points to [8]*) Birthrate, population is going to be going down. (*Draws arrow [11]*) [ix] And here (*Points to [2]*) as well, the birthrate well below the death rate (*Points to [1]*), and population is, density is increasing dramatically, (*Draws arrow [12]*) so there's a density effect. Population is also gonna be crashing here (*Points to [1], draws [13]*). [x] And here (*Points to [14]*) is where the population would be undergoing an increase.

Rodney pointed to the area below the first intersection [8] where the birthrates are 'relatively low' and the death rates are 'relatively high', the 'population is going down' [viii]. He drew an arrow toward smaller values on the vertical but toward larger values on the horizontal axis. He then turned to the area above the second intersection point where the same relation holds between birthrates and death rates, which he translated into the interpretant /population is also be crashing/ [ix]. At the end of this episode he pointed to the central area in the graph and associated with it a population increase [x].

On the surface, we see a scientist who did what we expect him to do: make inferences about a population given birthrate and death rate. A number of points are remarkable, however. First, his utterances of decreasing or crashing populations were made while drawing arrows that point diagonally from top left to bottom right ([11], [12]). Within the frame of this graph, the arrows indicate decreasing rate with increasing *N*, which is in conflict with Rodney's description. But because there is no indication that he perceived a conflict, we need to understand this piece of communication from a (phenomenological) perspective within which such a conflict does not exist. Thus, we are to see these arrows as indicating declining (or crashing) populations, which may be tacitly treated as an event, that is, associated with the progression of time. Time, however, is ordinarily plotted from left to right, which would therefore accounts for the direction of the arrow. These arrows make sense not within the frame of the diagram onto which they were drawn, but as indicators of the population dynamic in a population-time graph.

Second, Rodney drew an arrow to the right while uttering 'the density is increasing dramatically'. This interpretant of the graph is at odds with 'population is gonna be crashing', his inference from 'the birthrate is well below the death rate' [ix]. From the analyst's perspective, there appears to be a conflict and yet Rodney did not signal a conflict. Thus, from his perspective, there was no apparent link between population and 'density', so that he could say at the same time that the population is crashing while indicating that the density is increasing in the same part of the graph. Another possible explanation draws on findings in gesture research. Thus, when children are in transitional stages of knowing, their gestures and their words express different knowledge.[4] Here, Rodney's gestures and talk may express different understanding without that his awareness of an incompatibility between the different expressions.

Third, Rodney provided /population is gonna be crashing/ as interpretant for the sign /birthrate well below the death rate/. This interpretant is inconsistent with a normative reading, which suggests a decrease in the population size until birthrate and death rate are equal at the upper intersection [15]. This is not a singular reading, but six of the 16 scientists (i.e., about 38%) translated this part of the graph into /population crash/ (the incorrect answers in Table 2.1, p. 27). How shall we understand that Rodney and five of his peers, all trained and experienced scientists provided a reading that runs counter to that which ecology professors expect their students to say or write about this graph in introductory courses and seminars on ecology?

We notice that Rodney (and his five colleagues) read the relation between birthrate and death rate as a state of affairs rather than an aspect of a population dynamic. As part of a population dynamic, any change in population size entails a change in the associated birth- and death rates. In essence, these scientists read the graph in the way they might have read the sentences 'birthrate is larger than death rate', 'birthrate equals death rate', and 'birthrate is smaller than death rate'. Readers will recall that Todd also read the situation in the same way but then, when his own graphs contradicted the population crash, revised his reading. Other aspects of the graph that might have provided a context to their reading of other signs were not salient at this moment. Thus, as is apparent from the transcript above, Rodney did not articulate a link between his reading of the equality of rates and its translation into the stability of population size. Nor is the functional relation between the two rates and the population size salient in Rodney's or the other scientists' reading.

In Chapter 2, I already pointed out that such readings are consistent with a situation whereby the graph depicted birthrate and death rate as a function of $N_0 = N(t = 0)$. In this case, the values of the population at $t > 0$ are described by an exponential function:

$$N(t) = N_0\, e^{(b-d)t} \qquad\qquad (4.1)$$

where a and b are both positive numbers. For values $b > d$, the population increases exponentially whereas for values $b < d$, the population collapses. Increases (which scientists often suggested could be culled or harvested) and vanishing populations were just those that the scientists predicted here. In the following, I show that these interpretations are structurally similar to those that many high school students provide, a phenomenon that has come to be described as 'iconic confusion'.[5]

In one task often used to identify iconic confusion, individuals are asked to make an inference about the shape of a racetrack given a speed-time graph (e.g., *Fig. 4.4.a*). The task requires an inference about the shape (sequence of positions x) as a function of speed, mathematically the time-derivative of position (dx/dt). Those individuals who are said to commit iconic errors suggest that there are as many curves on the track as there are peaks *and* valleys in the graph. Scientists in this project were asked to make an inference about changes in population size (N) and rate of change (dN/dt). The transformation asked of the scientists was equivalent but the reverse to the one asked of the students (*Fig. 4.4*). Students are provided with an dx/dt–t graph and are asked to make inferences about a x–dx/dt graph; scientists are provided with a dN/dt–N graph and asked to make an inference about a dN/dt–t graph. The transformation from rate-system to rate-time representation is the reverse but equivalent rate-time to system-rate transformation. The students take each turn in the speed-time graph as an indexical sign to a curve in the racetrack (rather than only the minima). Equivalently, the scientists interpreted 'birthrate greater than death rate' to mean 'crashing'.

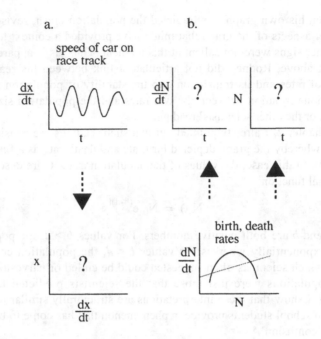

Figure 4.4. Structural equivalence between the transformations of (a) types of graph frequently presented to students, which leads to 'iconic confusion', and (b) the population graph in the present study.

4.3.3. 'I am trying to express it in a different manner'

Once Rodney had talked about what would happen to the population, he decided to draw a graph (*Fig. 4.5*). He set up the horizontal and vertical axes of a Cartesian graph, but then paused questioning how to draw 'this' [xi]. He added 'density' and '##' as labels to the abscissa and then stopped again [xiii].

[xi] So I (*Draws axes*) suppose you could- (8 seconds) How I would graph this? (8 seconds.) [xii] Population density? (*Writes 'density ##'*) or numbers. [xiii] What do I want to show here? [xiv] Population is crashing- (*Pen at ordinate, ready to write*) [xv] Oh forget it! (*Strikes out graph*) Now I'm confused. [xvi] I'm just trying to express this in a different manner and just showing the population growth curve but as I am having trouble with the upright axis.

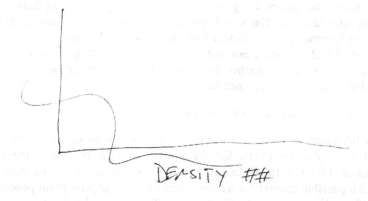

Figure 4.5. Rodney started drawing a graph that was to show how the population crashes, but never completed it and crossed out what he had drawn to that point.

He articulated /population density/ as a possibility (questioning tone) and labeled the abscissa (*Fig. 4.5*). He moved to a meta-level discourse asking himself what it was that he wanted to show [xiii] and answered the question with 'population is crashing' [xiv]. Then, all of a sudden he abandoned the attempt to draw a curve scratching it in a quick stroke. He described himself as being confused and then suggested that he had wanted to express the state of affairs in a different manner [in drawing] (i.e., an interpretant of the original graph). But, as he noted, he had trouble finding an appropriate dimension for the 'upright axis' [xvi]. With the meta-comment 'I am having trouble with the upright axis' Rodney demonstratively abandoned the attempt scratching the aborted attempt in producing a graphical interpretant of the population graph.

Rodney did not come through with his attempt of translating the given graph into a new graphical representation. In part, he may have constrained his progress by labeling the abscissa with 'density ##' that may have arisen in association with his interpretant /density effect/. This constraint operated in two ways. First, the variable was taken up and could therefore not appear as a label (contextual sign) along the ordinate axis. Second, it took the place on the abscissa that /time/ should take if he wanted to depict the temporal dynamic that was implied in the descriptions 'is crashing' and 'population growth curve'. In this situation, the earlier but latent conflict may have come to bear on Rodney's work of reading. Thus, although the earlier arrows standing for a population decrease indicated a sense for the population dynamic it came to conflict with another arrow and gesture along the x-axis standing for increasing density. As he further articulated and elaborated upon earlier statements in a new diagram, the conflicts became apparent. In this way, the drawing a diagram functioned in the way that it had in Todd's case without however permitting Rodney to come to terms with the conflict. Rather, he did not come to a resolution and, similar to Brandon (Chapter 2), abandoned his attempt.

4.3.4. Optimum Point and Unhealthy Level

After his attempt to construct a different representation failed by his own account, Rodney returned to reading the caption. He began by reading the sentence that defined the task [xvii]. In questioning the nature of 'level', he instantly invoked a possible context for /conservation of species/ [xviii] and proceeded to identify implications from the graph in regard to conservation.

> [xvii] (*Reads from caption*) 'Implication of the birth and death rates in the figure, as regards conservation of such a species'. [xviii] I guess it depends on what your, what sort of level you're trying to conserve them at. [xix] Obviously (*Points to intersection [5], Fig. 4.3*) there's some optimum point here (*Points to intersection [5]*) where births are exceeding deaths so and the population is increasing somewhat. [xx] It demands more knowledge of population ecology than I have. [xxi] At this point (*Points to intersection [15]*), you're reaching some sort of unhealthy level where effects presumably of density of the population is causing the birthrate to drop off dramatically while the death rate is still increasing. [xxii] (*Reads from caption*) 'What happens to population sizes in the zones of population size below, between, and above the intersection points?' I think I discussed that... (*Points to [8]*). [xxiii] Here, (*Draws arrow [16]*), population is going up, presumably. [xxiv] That's it for me, I'm not a population ecologist.

Rodney read the first and second equilibrium point as 'optimum point' [xix] and 'unhealthy level' [xxi], respectively. In both cases, Rodney did not just say that the birthrate and death rate were equal, but that the /population is increasing somewhat/ and /birthrate is dropping off dramatically while death rate is still increasing/. He elaborated /optimum point/ in terms of birthrate that exceeds death rate, which entailed an increase in the population. Similarly, he elaborated /unhealthy level/ in terms of a birthrate that drops off while the death rate continues to increase. Between the two interpretants, he uttered another meta-comment, which stated that the task required more knowledge of population ecology than he had [xx]. He abandoned the entire session concerning the population graph by stating that he was not a population ecologist.

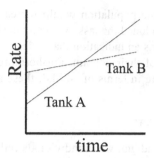

Figure 4.6. Type of graph used in studies with grade 7–11 students. Asked about which tank fills faster at the intersection, a large number of students respond with "Tank A" because its curve has a greater slope at this point.

What was salient to Rodney at each instant is not entirely clear from the videotape and transcript, for the statements are made about the state of the population at the intersection—where his elaborations were inconsistent with a normative reading of the state of affairs. That is, at the first intersection birthrate and death rate are equal rather than the former exceeding the latter. One possibility is that Rodney treated the graph as showing a development from left to right. In other words, it is salient that birthrate 'surpasses' death rate at the first intersection, but death rate surpasses birthrate at the second intersection. That is, at the point of intersection, the birthrate and death rate are not just equal, but to him it was also salient that the slope of birthrate is steeper than that of death rate. This then has an effect on what is happening to the population at that point: although the two rates are equal, the steeper slope of the birthrate will lead to an increasing population. Such interpretations were rather frequent among undergraduate students who wrestled with the problem of integrating readings that took account both of the rates at the point of intersection and their slopes.[6] Other researchers also showed that readings including both value and slope of two intersecting curves are quite common. These graphs were similar to those where students had been asked which of two tanks are filled faster at the intersection (*Fig. 4.6*). Large portions of grade 7, 9, and 11 students tended to suggest that 'Tank A' filled faster, based on the steeper slope.[7] That is, at this point in his interpretation, the greater slope at the intersection between the two graphs is consistent with the content «increase in population».

After reading the instruction for the task one last time and noting that he already answered the question for the section below the first intersection, Rodney tentatively stated that the population would increase in the central section. At this point, then, he abandoned the task. We can take his final comment 'I am not a population ecologist' as an indication that at least some scientists do not attend to graphs that are out of their domain. Furthermore, if we frame their competence related to graphing in terms of 'skills', these are certainly not domain independent.

4.3.5. Discipline as Context

In an expert-novice paradigm, when individuals fail to solve a problem according to some normative reference, investigators provide explanations that characterize performance (and results) as deficits. I do not subscribe to deficit explanations. Rather, I acknowledge that interpretants arise from the dialectic between the perceptual structures evident from the graph and the existential understanding of the individual. I attempt to understand how any differences arise from particular histories of scientists-in-context. This context is constituted, for each scientist in a different way, by the particulars of their workplaces and the concerns that reign in that niche of the scientific community that they most affiliate with.

Rodney, Daniel, and other scientists did not provide standard readings of one or the other graph in my research, although they successfully construct and work with graphs as part of their daily work. That is, if we studied their publications or observed them in their daily practice, we would conclude that they are very competent scientists, dealing with graphs in very appropriate ways. But in the present context, both Rodney and Daniel abandoned a graphical task that is part of undergraduate introductory courses and textbooks. It is difficult if not impossible to justify a deficit explanation; these individuals would not have been successful scientists if they had cognitive deficits as they have been used to characterize the performance of students who provided structurally similar readings as the scientists in my study. How then should we understand why some scientists were not successfully reading the (generic) graphs I had presented to them? In this and the following subsection, I begin to sketch possible answers.

Among the circumstances noted by the scientists as mediating their answers was the fact that they were not working in the particular disciplinary area to which they attributed a graph. For example, both Rodney and Daniel felt that they had not been too successful in the situation and abandoned the task. As Daniel in the first section of this chapter, Rodney invoked that he was not a population [forest] ecologist. That is, some scientists, represented here by Daniel and Rodney, implied that reading a graph requires familiarity with the particular discipline that has given rise to a graph. Rather than being a context-independent skill graph interpretation is, on these scientists own accounts, contingent upon understanding at least the phenomena that a graph is said to represent.

Another common complaint by scientists related to the construction of the graph and accompanying text. That is, the graphs that I had modeled after those used in undergraduate courses and textbooks appeared to violate the conventions that these scientists associated with the use of graphs. For example, Daniel and Rodney both felt that the graphs were poorly constructed. Thus, Rodney questioned 'Why do people choose to make graphs like this? Where [people use them] I turn them off. I find them quite confusing'. That is, Rodney (as Daniel and other scientists) found the graphs depicting models (population graph and isographs) to be bad practice and highly confusing.

> Well for me the philosophy of graphs is to present data in a simple fashion as possible. The more axes and the more stuff you're trying to cram into them, the more confusing they become. There are needs I guess for three-dimensional graphs and things that have more than x-y coordinates. But if they're not adequately explained in either in the figure legend or in the accompanying text, as far as I'm concerned they just confuse the matter.

In part, Rodney also reflected the position that Drew took (Chapter 2). These graphical models are simplistic and misrepresent the state of affairs that he is confronted with while doing field research. For example, Rodney suggested that 'many scientists will think statistically and in terms of graphs and data and all that sort of stuff when they're doing their research but then when they look at real-life situations that sort of analysis has no bearing'.

The scientists therefore articulated at least three dimensions along which the graphs did not allow competent reading. First, graphs require the familiarity and understanding of the natural phenomena that characterizes a particular discipline. Second, the conventions used to construct the graphs in the communities represented by the scientists differ from those that are employed in the construction of graphs for undergraduate courses. Given that the physicists and theoretical ecologists generally did not voice having problems with the graphs indicates at least a family resemblance between the conventions that go with graphs in textbooks and theoretical ecology rather than those in field and applied ecology. Third, a number of field ecologists voiced concerns with simplistic models expressed in the graphs. They talked about the complexity of ecosystems that is not expressed by graphical models of the type used in our research.

4.3.6. Unconventional Reading

Differences in disciplinary conventions and knowledge between the communities represented by the individual scientist and the source domain of particular graphs only provide us with a partial story for the non-standard readings. What we do not yet have is an explanation how a particular reading makes sense in the situation. That is, because the scientists did not purposefully engage in irrational activity, we need to understand what the world looked like that made each reading and each articulation of an interpretant a sensible one. Let us therefore return to particular aspects of Rodney's reading.

Rodney stated that at the intersections, birthrates and death rates were equal. Initially, he suggested that the population is presumably constant. But then he also made salient the slopes of the curves at the points of intersection. These should have an effect but Rodney was not sure whether the slopes should be interpreted (e.g., he was not sure whether he should read that much mathematics into it). We might think about Rodney's statement not in terms of conventional graph reading but as one in which each sign is taken typologically and contributing to some text. Rodney then would have read two statements pertaining to the same point of intersection. On the one hand, the two rates are equal. On the other hand, birthrate is climbing more dramatically than death rate at the intersection [5] and drops off dramatically at the second intersection [15] where death rate is still increasing. Thus, pertaining to the first intersection [5], the two rates are the same but there is a greater tendency for birthrate to increase than for death rate. This then becomes an optimum point where the actual births exceed the actual deaths and the population is increasing somewhat [xix]. The reverse is the case at the other intersection, which is an unhealthy level for the population to be at. Although the rates are equal, the tendencies of the birthrate and death rates are opposite.

We can interpret Rodney's statements and arrows referring to population changes in a similar way. If we abandon the assumption that there is a dynamic relation between the elements of a graph and see the paper simply as scratchpad littered with signs, then we can read the arrows independently of the other signs. In this case, then, the arrows [3] and [4] (*Fig. 4.3*) draw on conventional understandings of 'down' and temporal development. 'Down' will be in the direction from top to bottom of the page and change with time will be from left to right. Both arrows are then signs consistent with his interpretants that describe the population as decreasing, which always involves a decrease in time rather than an instantaneous one. Similarly, the verbal interpretant 'increase in population' is consistent with the arrow [16], where 'up' on the page conforms to vertical 'increase' and the left-to-right direction as temporal development. Finally, the arrow [12] pointing to the right and next to 'N' is consistent with 'increase', but this time in density. If each sign is taken as being of typological nature, as making a statement, then there is no inherent inconsistency. Rodney's statements are therefore incoherent and inconsistent talk only within a normative frame that is built on particular conventions. If we abandon the normative framework and think about the graph in terms of making individual (typological) statements in a graphical manner, there is no reason to characterize Rodney's reading activity as irrational or incompetent. I therefore suggest that to understand non-standard graph reading, we should consider graphs as forms of text constituted of arrays of signs, some or all of which may be of the same typological nature as language. That is, whereas scientific conventions treat graphs as representing dynamic aspects, arising from their topological nature, in the type of reading exemplified by Rodney, the graph at least partially constitutes an array of typological expressions.

4.4. GRAPHS AS OPEN TEXTS

In the previous two sections, we met scientists who abandoned their sessions relatively quickly with a certain amount of frustration over the absence of an immediately intelligible configuration and over the lack of sense they could make of the graphs. They attributed their difficulties to the lack of knowledge in the particular discipline to which they attributed the graphs, to the lack of fit between graphs and real-world settings, and to bad practices of using inscriptions. Although we could see that Daniel and Rodney, as others in such situations, at least referenced their understanding, partially articulated during the session, they expressed a sense that they had not achieved what they were supposed to achieve. That is, that they had not articulated what they thought that the unknown author was trying to communicate.

How does the reader of a graph (or any other text for that matter) know what the author was trying to say? How do they evaluate the relationship between their own articulated understandings and what is taken to be content of the graph? Earlier, I stated that graphs occasion a reader to articulate an existential understanding that always and already exists, that is, which comes with human nature. If this is so, scientists' assessments whether their readings map onto what they think the author wanted to say boils down to question of intersubjectivity. How do we know that we 'understand' another person? That is, how do we known that we are, as the popular adages go, 'on the same wavelength' or 'on the same page' with someone else? Even if, as in the case study that I am

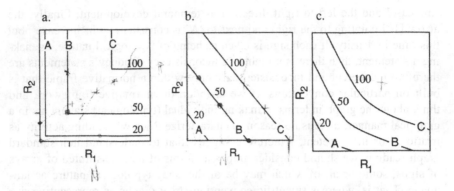

Figure 4.7. As Todd elaborated the original graphs, he marked them up in part to make certain points more salient, in part to assist him thinking through covariations, in part to show how a systems tracks if a parameter is changed.

presenting here, we extensively articulate our understanding, there is no guarantee that it maps onto the understanding of the author of a graph.

In this section, we return to Todd, the trained physicist who works for a federal research institute and is a very active researcher. Returning to this scientist who was generally more successful in dealing with the graphs than other non-university professionals presents the same person in a second context. This then also allows us to draw comparisons between the performances in two different contexts and to identify invariants. As Todd produced interpretant after interpretant, he articulated his understanding of mathematical models and of real systems to the session. But he ended the sessions indicating that the graphs did not make sense and that he would want to see references to the literature. As we follow his construction of interpretants, familiar themes encountered in previous analyses will resurface. However, Todd's interpretative effort (semiosis) was characterized in its extension, both in its broad scope and in length, which might lead us to think that he is 'more expert' than Daniel and Rodney. What, if any, are the differences between these three scientists?

4.4.1. Iteration 1: 'These are three mathematical models'

Todd began by covering the caption, attempting to read the isographs (*Fig. 4.7*) without drawing on the caption as a resource. On first sight of the graphs, he suggested that the task would be difficult. Without (detailed) inspection, he immediately characterized the three graphs as mathematical models.

a.
b.
Sampling sites
2 8 11 18 24 26 35
0
200
400
600
Depth (m)
3500
3000
2500
2500
2000
Iodine (10^9 atoms m^-3)
Fig. 1. Iodine concentrations in the upper 600 meters along a selection of the Atlantic. Dots show sampling depths.

Figure 4.8. a. This scanned image shows the graph Todd to show what he would expect in the real world instead of the idealistic isographs used in this task. b. The type of graphs that Todd used in a variety of research articles; there were six graphs of this type in the article he brought to the session.

> [i] That's gonna be real hard. [ii] OK, I would say that these are three mathematical models to represent the correlation between two rates, I can see this (R_2) as rates or populations. [iii] And the numbers are probably some scale, I mean there's obviously some scale but it would be a way, for example, of, it might be developing a model that relates birthrate to death rate. [iv] And, either you put it all in this kind (*Traces 100%-line in Fig. 4.7.a*) of a box category or you cluster your data (*Traces 100%-line in Fig. 4.7.b*) such that…

In this opening, Todd classified the graphs as isographs upon sight in the way we might classify an elementary school science book and then rationalize our classification by describing some of its features. Although this might surprise at first that Todd would classify these graphs that most of his peers found complex (50% of the scientists did not achieve the normative reading, see Table 2.1, p. 27). However, isographs constitute a standard representational format in Todd's work. The article that he had brought to the session contained six isographs (e.g., *Fig. 4.8.b*) similar to present task. After this opening, Todd began to draw a new graph (*Fig. 4.8.a*)—in the way he had already done in the context of the other two tasks (see, for example, Chapter 3). He did so by setting up a narrative about how he collected data that he plotted in his diagram. He hypothesized that the three diagrams were different ways actual data could be contoured.

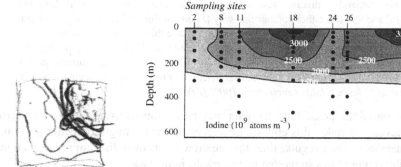

[v] See, in all likelihood you're dealing with some kind of a data (*Draws box in Fig. 4.8.a*) that, this is, I say, scattered (*Scatters points in Fig. 4.8.a*). Like the 'a' and you can either contour that way (*Points to Fig. 4.7.a*) or you group them up this way (*Points to Fig. 4.7.b*), or you assume this (*Traces 100%-line in Fig. 4.7.c*) quadratic power law. [vi] I think I would say I'd better be grouping my little x's in something that shape like this (*Draws lines through scattered points in Fig. 4.8.a*). [vii] And I am putting strictly in a box or lopping them off with a straight fence (*Holds pencil along 100% [ii]*).

Todd raised the possibility that data may be contoured using the three patterns displayed. He ended this episode by suggesting that it might be better grouping the data using the irregular lines that he drew in his own diagram rather than the regular contour lines suggested by the graphs in the task.

In this episode, we again see that Todd, occasioned by the identification of the graphs as mathematical models, articulated this understanding associated with collecting data, representing them in graphical form, and fitting them to existing mathematical models. In this same experience, simple curves such as those portrayed did not exist. We can think of this episode in terms of the relationship between signs and referents. Here, the graph occasioned the world of research to become present; Todd articulated and elaborated this world of experiments and the processes of data collection, data analysis, and data representation that go with it. Rather than taking something *out of* the graph, Todd articulated a familiar world *in front of* it. This world can be thought of as the referent that stands with the original sign in a mutually constitutive relationship. Rather than engaging in an inductive process by means of which the content of the graph is disclosed, Todd proposed a possible referent situation in which he collects data which he then attempts to contour by drawing on the three graphs. His suggestion that it might be better to contour (real) data in the way he showed (*Fig. 4.8.a*) rather than putting them 'strictly in a box or lopping them off with a straight fence' [vii] can already be heard as a criticism, which he elaborated further later in the session. Here, he articulated an understanding (arising from previous experience) of a world in which simple relationships, as portrayed by the isographs (*Fig. 4.7*), are not realistic.

Todd continued to detail his understanding of the relationship between mathematical models and data. He began the next episode by articulating what he thought that the author of the graphs wanted to show. From the perspective of the daily work at a research institute such as the one where Todd works, fitting observational data to existing models of phenomena is a common practice. It is therefore not surprising that Todd invoked the experience of fitting data using a contour program in the present situation [viii]. After reiterating that he was dealing with 'isolines' that he would approximate with contours, which in his own data are 'actually quite funny' [ix], he returned to articulating what in his view the author attempted to say with the three graphs. His statement that the contours for the field measurements represented in his own drawing (*Fig. 4.8.a*) were likely shaped in a 'funny way', in fact contrasted the regular shapes that are featured in the task.

> [viii] So, what I think the author is trying to do is show how observational data can be fitted to one of three models according to your perception of whether it's linear, power, logistic, or? [ix] These are isolines and if I were actually, take my field measurements and run a contour program it may actually contour in quite of a funny way, like that (*Draws more contours in Fig. 4.8.a*). [x] And what the author is trying to do is to take this, which is- [xi] These are isolines that actually enclose the data. All of these values up here would be 100, that's what- [xii] And he is trying to find out a math, a simple mathematical representation that approximates the contour. And he is saying that there're three ways I could simply approximate this wiggle (*Gestures the wiggles in Fig. 4.8.a*) line that I've drawn. [xiii] And the person would just make it piecewise boxing in, the other is to lop in and the other is to take into account that there is some kind (*Traces 100% line in Fig. 4.7.c*) of a power law that better does the grouping. [xiv] So, there're probably, what do they call that math? Norms of fitting- and you could either use least squares or least sums or least cubics. In the Western literature we usually use least squares but I don't think God ever came down and said least squares is better than least powers of ten, or that standard deviations are better than anything else. [xv] The contour identifies a relationship between two variables, and, this one is- I am giving up I have a hard time with it.

Todd further articulated his understanding with respect to data analysis. He attributed to the author the intent of modeling data using three mathematical models. Furthermore, he suggested that the fit of the mathematical models would be established by some norms, which he denoted by /least squares/, /least sums/, and /least cubics/. Todd also provided insights in his understanding of the arbitrary nature of the norms [xiv]. After reiterating that a contour identified the relationship between two variables, he seemingly announced the end of his attempt.

It has been suggested that 'novices' use contextual knowledge to make an interpretation task more concrete. The question is, of course, whether scientists (in contrast perhaps to mathematicians) ever discuss graphs in abstract terms without reference to experience. I am using 'abstract' here not as a property of specific objects. Integrating a function on the complex plane may be something abstract for most people, but it is something very palpable to physicists and mathematicians; doing a least squares fit may be abstract even for some scientists, but is a very concrete thing for most statisticians. So far we have seen that scientists do not interpret graphs 'in the abstract'. Rather, unfamiliar graphs occasion the articulation and elaboration of understanding, which, as an existential, is always and already a constitutive part of us and therefore always concrete. For mathematicians, physicists, and statisticians, this understanding includes entities that to most people are quite abstract, unfamiliar, but are very concrete and palpable entities for the former. The isographs and Todd's experience stood in a dialectical relationship occasioning the production of interpretants relating to his everyday work environment. He provided us with a window onto his daily work practices. Furthermore, rather than talking about 'just the graphs', he drew a set of coordinates, plotted a set of data, and showed how he might have to construct a set of contour lines in his own example, which in contrast to the task, also included individual data points. This in itself is an indication to the fact that most scientists deal with real data that are subsequently fitted with curves; these can be existing curves (mathematical models) with fixed parameters or curves with parameters that best fit the data collected. On the other hand, theoretical ecologists often deal with models independently of referents in the natural world. They are interested in the mathematical implications deriving from a model, and therefore in the transformation of representations into other representations. It is therefore not surprising that they talked about graphs without elaborating referents from the natural world.

Todd made an indication that he wanted to abandon the interpretation. *This* should surprise. Here we have a successful scientist, who regularly publishes research articles. Furthermore, he uses isocline graphs in his own research including, among others, six such graphs in the research article that he subsequently talked me through (*Fig. 4.8.b*). That is, through his research he had substantial experience with isoclines. It is therefore not surprising that he immediately categorized the isographs in the task. What then hampered his interpretation of these graphs? Why did he not 'transfer'? One answer, which I am advancing throughout the book, is based on the contention that graph reading requires an understanding of the natural domain to which the graph is said to refer. In part, context specificity includes the use of familiar measures, units, and variable names. These allow the scientist to understand the graph in terms of the processes by means of which it is generated. Thus, field scientists often cited lack of scales, units, and specific variable names as problematic aspects of the task graphs.

4.4.2. Iteration 2: 'This could be nitrogen and phosphorous'

At the end of the episode, Todd had not talked about a potential referent situation in the subject domain of biology proper but already wanted to give up. As I was interested in finding out about the referents in the natural world that scientists associated with the graphs, I encouraged Todd to persist.

The next section of the transcript shows how Todd moved to talking about particular resources rather than talking about the graphs in terms of generic resources R_1 and R_2. He developed throughout the remainder of this session scenarios from the natural world involving specific substances as resources. Again, the graphs gave rise to an articulation and elaboration of existing understanding rather than constituting inferences to unknown referents.

In this first episode after my request to continue, Todd placed nitrogen and phosphorous in the context of plankton in a lake. That is, he developed a scenario in which the productivity levels (high concentrations of) plankton is high at high levels of nitrogen and phosphorous [xviii]. He then looked at the first isograph graph (*Fig. 4.7.a*) in terms of this scenario.

> [xvi] (*Reads caption silently*) [xvii] This could be the (*Traces R_1 axis in Fig. 4.7.a*) nitrogen and (*Traces R_2 axis*] phosphorous that are taken out by plankton in a lake and those that grow the- form a part in the food chain. [xviii] If you have high levels (*Points to high R_1 levels*) of phosphorous and high levels of (*Points to high R_2 levels*) nitrogen, you expect high (*Points to 100%*) productivity rates. [xix] This (*Points to '100%'*) would have to be single species response, because if you're dealing with different types of plankton you would find that some actually don't do all that well at the high nutrient levels, and some others kick in. [xx] That's why, for example, adding phosphorous to the Great Lakes during the sixties and seventies led to nuisance blooms and kicked out the- [xxi] So, this (*Points to the three isoclines*) must, this is some, this is a total growth, a total bio mass response. (7 second pause)

He modified his claim about the productivity by suggesting that the graph depicts a single (plankton) species response [xix] because different types of plankton respond differently to nutrient levels. (At this point, we do not know about the extent of Todd's experience in this domain. However, given that he works in a large research institute focusing on marine systems, he is very familiar with the research conducted there even if it is not his own specialty.) He then provided a specific example of high productivity levels of plankton: the nuisance bloom in the Great Lakes during the 1960s and 1970s. Again, Todd elaborated the graph in terms of a narrative related to the research at his work site rather than talking about the graphs in context independent terms, resources R_1

and R_2 and their different relationships depicted in the three graphs. Existing understanding is the starting point for interpretation; it is elaborated by the production of interpretants. The unfamiliar (i.e., abstract) graph is appropriated into this existing understanding by associating specific graphical structures (signs) to articulated structures of understanding. This process, by means of which decontextualized graphs are developed in terms of, and tied into, existing understanding, therefore characterizes scientists as well as 'novices'.

4.4.3. Iteration 3: 'This is a little bit unrealistic- It's just too crude'.

During his initial reading, Todd appeared to be less concerned about the status of the graphs as realistic models of natural phenomena. His graph (*Fig. 4.8.a*) provided only a glimpse of his own understanding of natural system and how they are modeled. Now that he was encouraged to articulate natural referents, Todd elaborated his understanding of isoclines that would result from research and therefore denote 'actual' (natural) systems.

Following the first reading of the graph from the perspective of a specific scenario—plankton productivity in response to nutrient availability—Todd produced a number of statements and explanations for his claim that the graphs are not as realistic as the caption claims them to be. That is, although these graphs are presented in undergraduate ecology course and in textbooks as depicting realistic scenarios, Todd and many other field scientists questioned their realism. It will appear that the conflict between his and the text's claim about the realistic nature of the scenarios contributed to Todd's final intuition that the graphs did not make sense. That is, they could not be accrued to the network of familiar significations, which he disclosed and elaborated in the course of the session.

[xxii] These (*Reads caption*) 'graphs depict three different biologically-realistic scenarios of how two nutrients'. [xxiii] Well this one (*Points to Fig. 4.7.a*) is a little bit unrealistic, usually (*5-second pause*), if I actually measure some- (*5-second pause*) [xxiv] Again we're talking about a family of curves in here (*Points to origin of graph*), no nutrients, no growth, I mean dead. [xxv] Hum, for, if it's a corn field or whatever it's gonna follow some curve (*Traces curve he has drawn in Fig. 4.7.a*) to get where it's going. [xxvi] I don't like it when it jumps to fifty because, only because of crossing a particular level of say phosphorous and these things have to work together, I mean. It's unrealistic that you, yeah, you had, you kick in step functions. [xxvii] Another thing about it is that natural systems are never in such equilibrium that they have a vertex (*Points to vertex of 20% line*) in the corner. Usually, aquatic systems are such that I would expect these lines to be more like inclined (*Draws a line below 20%, Fig. 4.7.b*) this way or...

He began by elaborating the origin of the graph as a point where nothing happens [xxiv]. But then he stopped and began to talk about corn (or some other crop) and suggested that it would follow some curve, which he traced as it crossed the isoclines [xxv]. When he crossed the 50% isocline he again critiqued the graph for using step functions. This critique is surprising if one thinks about isoclines as indicating the contour of a function $z = z(x, y)$ at different levels of the dependent variable; it is in this way that Todd had earlier talked about the graphs and that Todd used them in his own work. It seems that he perceived the graphs as showing the contours of three slabs lying on top of each other ('when it jumps to fifty'). Step functions would therefore constitute the transition from one level to the next. At this point, the referent of a natural system moving along the line is not clear—he elaborated this point considerably a little later.

In the last chapter, I discussed the importance to pay close attention to what a graph reader *actually* rather than *possibly* attends. What the individual perceives as salient constitutes the objects and events that he or she acts upon and therefore the true domain ontology (rather than the one a graphing researcher might design a priori). In a way analogous to seeing the earlier portrayed cube (*Fig. 3.9*) as a wire frame, box, or glass aquarium, the isographs can be viewed as the contours of imaginary sections through a three-dimensional surface or, as Todd does here, three superposed slabs. How a drawing perceptually appears has consequences to subsequent elaborations and interpretant production. With unfamiliar graphs, even scientists may perceive structures that lead them away from standard interpretations. That is, although the graphs are structurally identical from my social science researcher perspective, they clearly were not from Todd's perspective.

He then suggested that aquatic systems (with which he is very familiar from the research conducted by his colleagues at work) do not exhibit vertices that lie on vertices lined up in the way shown in the first of the three graphs (*Fig. 4.7.a*). This comment is quite evident in the context of the type of isographs that are the result of Todd's own work (*Fig. 4.8.b*). At this point, it is important to reiterate that for understanding Todd's (or any other scientist's) reading, we need to know less whether the graphs are reasonable from a normative perspective and need to know more about what makes Todd say what he says. That is, from a normative perspective, Todd is incorrect in stating that the isoclines are unrealistic. (For more on the use of such resource interactions see the Appendix.) Isoclines such as the ones featured in the task (*Fig. 4.7*) are central to successful models in behavioral ecology (e.g., nutrient choice or foraging patterns). However, if we take the interpretive process as an elaboration of existing understanding 'in front of' the graph, which is more or less the site where this elaboration occurs, then Todd's answer tells us about the kind of systems that he actually is familiar with.

Todd subsequently elaborated two issues: first, there is the question whether the isocline is realistic and second, he provided a content that his own curve might denote. As an example, he proposed to elaborate the growing of corn without adding nutrients to the soil. He marked points where the levels of each nutrient available to corn might be in subsequent years. He ended the episode by saying that he would not expect his curve to end in the origin.

> [xxviii] For example let's say we're growing corn and in one year we've got this amount (*Draws top right cross on his curve in Fig. 4.7.a*) stuff in the soil. And the next year we grow corn without putting in fertilizer we've got less nutrients in the soil, and it drops it down to here (*Draws middle cross on his curve in Fig. 4.7.a*). And the next year it drops it (*Draws bottom left cross on his curve in Fig. 4.7.a*) down more- [xxix] Well, I'm gonna expect it's just gonna run out of one nutrient before it runs out of the other. I'm not gonna expect it tracks (*Draws upper branch to the origin in Fig. 4.7.a*) its way right down to the corner, I don't think. Hum, probably...

He provided an interesting articulation of his understanding of a realistic system, such as growing corn, if no nutrients were made available other than those that exist in the soil at the outset. Consistent with his earlier claim that natural systems are not likely to track along the vertices he suggested that his own system would end when one of the resources is depleted—with still some left of the other. That is, he identified a situation of some agricultural system and hypothesized how it might change over time if the resources were taken out until one of the resources is being completely exhausted [xxix]. Although Todd did not articulate a specific example, it corresponds to the slash-and-burn–type agriculture characteristic of traditional farming.[8] Trees in primary or old secondary forests are cut and burned, and crops are planted in the ashes. After only a few years, the resources are depleted in the way Todd suggested and the fields are no longer productive; they are left to regenerate or left permanently.

Todd then continued to talk about how he expected natural systems to track and thereby elaborated why the second and third models (*Fig. 4.7.b, c*) were more realistic than the first model. Again, rather than making inferences that would lead him from the graph to some context that it can be said to stand for, Todd articulated a specific (concrete) situation and mapped it onto the existing space spanned by Cartesian axes.

> [xxv] This one (*Fig. 4.7.c*) is this is telling me that as one or the other nutrients gets small, the population is compensating a bit by recycling, if that's- (*Pause*)

[xxvi] I don't think these (*Fig. 4.7.b*) aren't really any different. [xxvii] This (*Fig. 4.7.a*) is just too crude. They're, what they're doing I think here is taking something that's somewhat realistic (*Fig. 4.7.b*) and simplifying as a box (*Fig. 4.7.a*). [xxviii] This (*Holds pencil parallel to lines in Fig. 4.7.b, moves from 100% to 20%*) is a little bit more realistic 'cause it points you back down toward the corner. [xxix] This (*Fig. 4.7.c*) one has the bowing in because as you start to draw down one or the other it allows- Wait a minute, is that right? [xxx] It allows other processes to come in, more recycling, perhaps another species starts to dominate. So, this implies a little bit more efficiency.

In the first part of this episode, contrasted the second and third models, which he characterized as more realistic, with the first model (*Fig. 4.7.a*). He suggested, without elaborating this point, that the first model constituted a crude simplification of the second (*Fig. 4.7.b*) [xxvii]. In this context, he perceived the other two graphs as little different [xxvii], because both describe the relationship of two nutrients when the system is left on its own [xxv]. The additional point made in the third graph (*Fig. 4.7.c*) concerned the response within the system, which occurs either through compensation [xxv], recycling [xxix], or by allowing another process to come in [xxx].

Todd expected a system without additions to deplete its resources along a trajectory that is perpendicular to the lines drawn in the graphs. The understanding underlying such a conception can be expressed in different ways. First, assuming that the system initially is at some point $(R_1, R_2) = (x, y)$. In subsequent years, the system would be found at $(x - a, y - b)$, $(x - 2a, y - 2b)$, and so on. This means, the system would track in a straight line with a slope $m = b/a$ and an intercept of $y_0 = y - ax$. For a particular system, a and b are related such that the trajectory is perpendicular to the isoclines. This is equivalent to saying that the system, beginning at the point (x, y), tracks along the steepest slope of the $z = z(x, y)$ surface.

In this view, the isoclines represent the composition of resources required by a plant. The system would track toward the origin, that is, depletion of both resources, in the two graphs that Todd found more realistic (*Fig. 4.7.b, c*). In the first graph (*Fig. 4.7.a*) this would only be the case if the system started on the vertex of the curves, and end up in the origin. The likelihood for this to occur is small, making this an unrealistic model. Now Todd did not fully articulate such an understanding; it is not certain that in his structuring effort he would have articulated a network of signifiers that corresponds to this description. It is certain, however, that many models of natural phenomena and many statistical processes are modeled in exactly this way, making it possible that Todd was attempting to articulate this understanding, which he presumably shared.

4.4.4. Iteration 4: 'I can still grow fifty tons of alfalfa with no phosphorous in the soil- It's just ridiculous- It doesn't make sense'

Once again, Todd came to a halt. As I had considered the formal part of this graph completed, I suggested to him to consider the plant growth as the dependent variable. Each line then represented a certain amount such as 50 millimeters, which leads to different patterns of growth as a function of the independent variables R_1 and R_2. That is, I provided a specific interpretant for the dependent variable, a particular measure, and measurement units. These hints allowed Todd to make one more attempt.

> [xxxi] Well, this (*Fig. 4.7.a, 50%*) one is saying that, for example if I want to grow fifty tons per acre of corn, all I have to do is, is get to this point (*corner*) in terms of putting this much fertilizer on the field. It wouldn't do me any good to put twice as much phosphorous on because it's gonna be wasted and not used. [xxxii] This one (*Fig. 4.7.b, 50%*) tells me that if I want to stay on fifty tons or above fifty tons- Hum. (*Pause*) I can do that by putting on this much nitrate (*Writes 'N' at abscissa*) and (*Writes 'P' at ordinate*) this (*Draws lower 'box' with corner at 50% in Fig. 4.7.b*) much phosphorous. I can also get that much growth by using more, that doesn't make, that doesn't make sense to me, and, you know, I'm really not sure, I'm not happy with these curves. I can still grow fifty tons (*Draws upper 'box' with corner at 50% in Fig. 4.7.b*) with a different nitrogen-phosphorous mix and I don't see why that would be. [xxxiii] When I go to the hardware store and buy bags of fertilizer they are very careful to put twenty to one or whatever the phosphorous-nitrogen mix is. This (*Fig. 4.7.b*) tells me that I don't need to worry about it, I can still grow grass at the same rate, I can still even grow fifty tons of alfalfa with no phosphorous in the soil, and it's just ridiculous. [xxxiv] There are certain principles and a plant growing has to have just the right assembly, a collection of all the nutrients, the major, the micro, it has to take those in to grow. And to say that it can grow a hundred (*Circles 100%, Fig. 4.7.b*), a hundred tons per acre simply by just dumping nitrogen on and not putting any phosphorous, to me that doesn't make sense. I don't see why, if that's even biologically realistic. [xxxv] Like I say and- But I'm also coming at this much more from an aquatic system's point of view. And I know that when, in the ocean, once you run out or if in a lake once you run out of phosphorous nothing would grow because and then you can have a soup of nitrogen and you know, unless you kick in another compensating organism.

Although Todd had previously described the first graph (*Fig. 4.7.a*) as not making sense, he did not indicate any problems with his current interpretant [xxxi], in which each corner point represented the minimum amount of each resource required for achieving the particular production rate. However, as he

moved to the second graph (*Fig. 4.7.b*), adding two points and its associated resource values, he began to doubt [xxxii]. It did not make sense that the growth
rate should stay constant although the mix of the two resources was changed. He
then further elaborated his understanding in terms of a visit to the hardware
store, where he would find specific mixes of different fertilizing compounds
rather than varying mixes. There are principles that suggest fixed ratios rather
than varying ratios of compounds. His second example made the same point by
moving the situation to an extreme point on the curves, where the fertilizer
would consist of one compound only. This did not appear to be realistic, a point
that he underscored by drawing on yet another familiar example. He knows that
aquatic systems would not be viable if one resource was completely depleted.

Todd interpreted the first graph in a way consistent with the normative
reading. He used the same structure of starting at a particular point on an isocline and inferred that a different mix of resources would leave the production
constant. This inference, however, was inconsistent with his understanding. He
iterated three examples to show why the graph did not make sense, that the
graph seemed to indicate something that 'is just ridiculous' given his understanding of how the world works. Even though he made up the scenario, Todd is
not a farmer, going to the hardware store and buying fertilizer, selecting one that
has a particular composition, is something that could happen in the world that he
is familiar with. That is, even if a scientist makes up hypothetical scenarios,
these are reflections of understanding of how the world could possibly be in his
experience rather than some world out there independent of his experience.

4.4.5. Iteration 5: One More Graph

Todd appeared to be at the end of what he could do. Using different approaches,
he had repeatedly attempted to perceptually structure the graphs and interpret
the resulting signs. But in each case, he had come to a point where the graph appeared to be saying something that was inconsistent with his existential understanding of how a section of the world worked. That is, his interpretants as a
collection provided a description of a system incompatible with the world in the
way he understood it. As this session drew to its end, Todd constructed yet another graph (*Fig. 4.9*) that would be more consistent with his understanding of
the resource requirements of natural systems. He began by drawing the axes of a
new graph, labeled the axes with /N/ and /P/, and then took another look at the
graphs that I had provided [xxxvii]. As he gazed back, he suggested that his
lines should not intersect with the axes as second and third graphs, but should be
in this respect more like the first graph. He constructed his own isoclines and
then suggested that the system would track along the vertices. He completed the

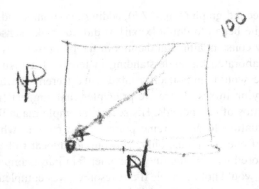

Figure 4.9. Todd produces a second graph of his own to show that he would not expect the isoclines to intersect with the axes.

session by stating that the graphs were tough, did not make sense, and that he would have to check on the references, presumable to see whether they were accurate.

> [xxxvi] I don't think I really even draw these graphs. (*Draws axes of graph in Fig. 4.9.*) So you got N and P, I like, I don't mind having 0 down in this corner, a hundred up here, and some kind of transition. [xxxvii] But I don't know that (10-second pause), I don't know that they can intersect these axis like they do here (*Fig. 4.7.b, c*). So, what will this one (*Fig. 4.7.a*) tell me? If I want to, this one is, if I have, if I want to go twenty tons per acre. Yeah, it doesn't intersect the axis like that, like these two (*Fig. 4.7.b, c*). So it recognizes the fact that this needs little phosphorous in the soil to do it. [xxxviii] So, probably I would be happier with curves (*Draws curves in Fig. 4.9*) that look more like that. And then, what probably you'd find in natural systems, in that equilibrium, is that the actual data would follow the vertices of, then you get correlation plots (*Draws line through vertices*) and, something like that. Ugh, that's a tough one! It doesn't make sense, I mean, and I'd like to check the reference to see.

Todd classified the graphs as mathematical models and, after having repeatedly tested what they seemed to say against his own understanding, ended in frustration. In the course of his session, he provided two alternative graphs that were structurally consistent with his understanding of natural systems. The first of these graphs (*Fig. 4.8.a*) also expressed the messiness of real data and the non-linear relationships among variables. The second graph (*Fig. 4.9*) further elaborated this understanding but expressed simpler relationships than the first graph. As in the previous cases, we saw that the graph occasioned Todd to articulate and elaborate his understanding of research and a variety of natural systems. His interpretants of the graphs were inconsistent with the understanding of both domains.

4.4.6. Limited versus Extended Semiosis

When I initially looked at the interviews, I thought that Todd most closely corresponded to the stereotypical image of the expert scientist. He took the task seriously, and though he was being prodded, willingly spent twice the amount of time that other scientists spent elaborating on the graphs. When Todd is compared to Daniel and Rodney, we might be tempted to say that he was more knowledgeable, more successful, or perhaps, more resourceful. Although he ended saying that the graphs did not make sense, he produced more graphs and real-life scenarios than the other scientists did. He engaged in a much more extended semiotic effort than his peers did, without however exhausting possible readings. Does this make Todd a greater expert than his peers?

Based on the research presented here, the expert-novice classification scheme turns out to be rather limited. My study seems to suggest that generic graphing expert does not exist even with relatively simple graphs. Rather, graphs occasioned the scientists to elaborate their understandings; the extent to which this process unfolded differed being contingent on graph and individual. One difference lies in the extent to which particular aspects of graph are associated with a particular domain of understanding, and therefore occasions the production of a series of interpretants that articulate this understanding. In this case, an interpretative session gets started and becomes framed by a dialectic tension between existing understanding and the ability of a sign to trigger the person in articulating this understanding. The overall production of interpretants is also a function of the 'stick-to-iveness' of a particular individual.

4.5. ARE SCIENTISTS EXPERTS AND OTHERS NOVICES?

Given that scientists and students are often contrasted and take opposite poles in an expert-novice paradigm, we should reflect on the question of attributing expertise status to this group of people. Let us take a look at a research project conducted in Britain on 14–15-year old students' interpretation of story problems in which graphs constituted the predominant (multimodal) text.[9] The students were presented with a graph (similar to that in *Fig. 4.10.a*) that shows the amount of oxygen and shrimps in a river. At a certain village, there is a sewage treatment plant. The students are asked what the graph tells them about the oxygen and shrimps. This task is interesting because it shares similarities with the

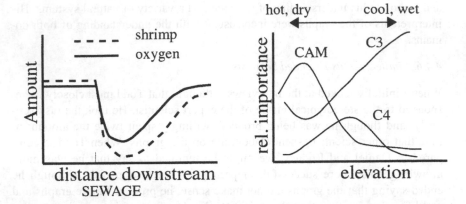

Figure 4.10. The graphs used in a study with 14/15-year-old students (a) and professional scientists (b) display distributions along a location dimension.

plant distribution graph (*Fig. 4.10.b*). The abscissa and ordinate produce a space characterized by location and frequency. Both therefore display distributions of two (three) different entities and require a comparison to arrive at the standard answer—the positive correlation of oxygen and shrimp amounts and the relative shift between the distributions of different plant types. Sewage and adaptation are the causes of the particular distributions, respectively.

Many students in this study focussed on the middle part of the graph (*Fig. 4.10.a*) but did not relate different parts of the graph to each other; this was particularly the case for 'poor interpreters'. About 43% related the shrimp to oxygen distribution. About 38% suggested that shrimp were moving or diving to escape the sewage. The sewage graph was more difficult for students than another graph in which the amount of foxglove under a tree increased after the tree had been pruned. (The distributions were similar but inverted parallel to abscissa.) Unfamiliar with the context, students tended to provide literal graph readings in which they described features in detail with little reference to the context. Some students provided pictorial readings of the graph. It should be clear by now that these results bear striking similarities with those from the scientists. Granted, scientists have had more experiences and a greater repertoire of situations that they can elaborate in front of the graph. Yet when the question is whether individuals arrive at the normative interpretation, unfamiliarity with graphs, modes of data collection and transformation, and familiarity with the domain that is said to be represented in the graph are more important characteristics of normatively unsuccessful interpretations.

In the British study, students were expected to arrive at a causal relation between oxygen levels and the presence of shrimp. Only a small number of students, however, related the changes in oxygen levels and shrimp frequency. Preece and Janvier noted that their students did not relate information from one part of the graph to the information in another. But this is just what I observed among my scientists. For example, the scientists who suggested that a population crash above the second intersection of birthrate and death rate do not relate this proposition to population size and to earlier propositions about the population equilibrium at the intersection. Furthermore, most scientists in the database did not make statements that two graphs (CAM, C_3) are inversely related because the different plant types compete for space. It is not surprising that these scientists did not invoke differences in photosynthetic mechanism as an explanation for the different distributions. It is therefore even less surprising that many children do not invoke oxygen as an explanatory resource to account for changes in the frequency of shrimps. Rather, the children focused on the sewage. Finally, students are reported to have more difficulties with interpreting the graph in the context that was more unfamiliar to them. Furthermore, students provided pictorial readings, for example, drawing on 'shrimps [that] are diving then later are coming back up again' as a resource in the context of the graph that sharply dips down before asymptotically returning to its earlier levels.

In the present chapter, I showed that when scientists are unfamiliar with a graph or a content domain, they also engage in a variety of interpretative and graphing practices that do not fit within the normative paradigm. For example, Rodney used arrows pointing toward smaller rates and larger N to indicate that the population was decreasing. This is not unlike the children's answer of diving shrimp—it is only when these answers are read against the backdrop of the graph and its normative interpretation that these answers are wrong. Before making inferences about particular interpretations, we need to better understand just what the signs are that individuals attend to from moment to moment and we need to assess the degree to which it is structurally linked to the graph rather than just generally about the phenomenon. In the case of Rodney, I showed how an alternative reading does not make his actions irrational or inconsistent. Some propose that graph readings involve a process of depersonalization and stripping graphical representations of their context. The present studies suggest otherwise. The less a person is familiar with the context to which the graph refers the more problematic graph reading becomes. Whatever a person understands as and about the referent situation of a graph therefore does not interfere with graph reading but is a central resource. The nature of the context (as viewed by the individual) may predispose particular content and structure of the interpretants coming forth.

When students do not arrive at the interpretations that researchers establish as the norm, 'mental capacities', 'misconceptions', or 'deficiencies in understanding' are frequently invoked as explanatory resources. For example, based on data with seventh- to eleventh-grade students, whose graph interpretations shared some of the features highlighted in this chapter, a previous study on graphing concluded that 'Mental structures provide a framework for understanding the difficulties that students have with science content. [...] Most of the students in elementary and many in junior high and secondary schools do not have the mental structures to understand line graphs'.[10] If we did not know the source of the interpretations in my database, we might conclude that, using the reverse argument, the research participants have insufficient (or altogether lack) the mental structures required for the task. Given that that the participants were successful scientists, such an interpretation immediately strikes me as outlandish. There are therefore other factors that we need to invoke to explain why my participants engage in non-standard practices when they read graphs.

In the same vein, I argue that we do not progress in our understanding of science if we jump to the quick conclusion that students have mental or cognitive deficits because they make use of particular (perceptual) resources in their interpretations. Rather, graph-related competencies should be evaluated in terms of the familiarity with graphs, phenomena, and the translation between the two. Thus, eighth-grade students were shown to exhibit a tremendous competence in producing and reading graphs and other mathematical representations. When these students were compared to those of teachers in training, all of whom had previously obtained science degrees (B.Sc. or M.Sc.), eighth-grade students turned out to be more competent (a statistically detectable difference). What distinguishes these eighth-grade students from those students participating in other studies is that they planned, conducted, and reported their own research with the purpose of convincing others. That is, these students worked in a context that was not unlike that in which scientists find themselves in their everyday work. In this context, then, even young students are highly competent rather than producing answers that lead to the attribution of cognitive deficits.

What is important to me in the present context is the notion of expertise and how we use it to distinguish individuals. One the one hand, we might argue that these scientists are not experts, certainly they were not experts (in the way the term is traditionally used) in the present context of the current graphs. On the other hand, the same scientists have proven repeatedly that they have considerable expertise at work, including graphical representation of their research for academic (publishing) and other communicative purposes (commercial clients, advising policy makers in the public sector). This then would lead us to an interpretation of expertise as highly contextual even within a domain of inquiry rather than as a general ability. Graph interpretation is not a context independent skill but rather a complex competence that arises at the intersection of a number of different aspects. These aspects include familiarity with the referent domain, as is evident not only from my model but also from repeated comments of the type 'I am not a population [forest] ecologist'. Other aspects include familiarity with the particular graphs used (e.g., many scientists were not [very] familiar with isographs—despite their 'structural equivalence' with geographic maps). It helps when the reader is familiar with the (research) contexts in which graphs are the outcome of protracted sets of actions that begin with designing a study, collecting data, transforming the data, and ultimately presenting them in graphical form. Finally, readers have to be familiar with the conventions ruling sign use of the sub-discipline and with association of familiar phenomena and their representation in the form of graphs. A recent study conducted in middle-school mathematics classroom confirms these contentions, as the competence with which students drew conclusions from statistical data was a function of their understanding of the data creation process.[11]

We can also look at the problematic readings produced by scientists in terms of meaning, here understood as a structure of existential understanding as it is evidenced in the lived experiences of the participants. When scientists aborted their readings before arriving at what feels like a satisfactory conclusion, they in fact felt (and often explicitly stated) a lack of meaning. I am emphasizing here that it is not graphs that lack meaning but that the graphs could not accrue to existential understanding that is inherently meaningful. Meaning is an existential, a background that comes with being human and against which intelligibility becomes possible. It is therefore *in terms of* meaning that something can be conceived as possibly existing. It is in the projection of a graph against this background that the very possibility of the graph is disclosed. In other words, interpretations arise from the dialectical relation of graph and background; the resulting network of interpretants articulates that part of the background that makes the graph possible.

FIVE

Articulating Background

Scientists Explain Graphs of their Own Making

This chapter is concerned with scientists' explanations of graphs that they had produced as part of their work. Invariably, my scientists began this part of the session by talking about the context within which their research was situated, how they collected the data, what the underlying science concepts were, and so on. It was only then that they pointed to their graphs and said what these were about. That is, scientists articulated and elaborated the background, a web of signification, required for any graph, as sign, to become intelligible. As previous chapters showed, the existence of such background cannot be taken for granted by the researcher on graphs and graphing but its status has to be an object of empirical study in its own right. A case study is presented of one scientist talking about bar graphs, the comparison of which required considerable knowledge of organic chemistry and ecology. At the end of this chapter, I articulate the work of reading graphs in terms of activity theory. This analysis highlights the differences between reading or interpreting unfamiliar graphs as well as familiar graphs that come from the scientists' own work.

5.1. FROM INTERPRETING TO READING GRAPHS

In the past, research on graphs and graphing has assumed that experts isolate certain features and then interpret them or make inferences about states of affairs. In this approach, the world is taken as a source of information taken up by an individual and then processed by the available mental machinery—often modeled in production systems that are implemented in a sequentially processing computer (see Chapter 2). In the humanities literature, competent reading is treated very differently. Researchers have come to understand that competent reading does not take letters or words as signs that need to be interpreted, or from which inferences have to be drawn. Rather, transparent reading leaps beyond the material basis of the sign (e.g., the book or journal page) to the thing

Figure 5.1. A channelized stream as described in the local newspaper no longer suitable as salmon habitat.

that the sign is said to stand for. In transparent reading, letters, words, and sentences are fused with the things and situations that they stand for in the same way that hammers and saws are transparent in the everyday use of a competent carpenter. Because these tools are extensions of their selves, the carpenters have to cogitate their use as much as they have to cogitate placing their feet in walking. Their tools and their feet are simply part of an extended way of being in the world. Similarly, words in everyday use are extensions of our ways of being in the world. Thus, when we read the following ecology-related excerpt from a regional newspaper, we engage with the things talked about rather than with the text.

> The damage [of the salmon habitats] was caused mainly by channeling the creeks and removing gravel from the area. Straightening the creeks (ditching) not only makes the water move through the remaining culvert too quickly to support rearing beds, but removes the surrounding vegetation. That, in turn, erodes the environment on which birds and other species depend for survival.[1]

looks like a ditch or culvert. Here, the text deals with the damage done to salmon habitat by practices related to the straightening of creeks. Straightening, which, the paragraph tells us, is accompanied by the removal of gravel, leads to an increased flow rate of the water. We can picture that such increased flow is desired by farmers who want to get rid of excess water on their land or by engineers who want to prevent flooding. Such flooding might itself be caused by an increased number of impervious surfaces such as paved roads, driveways, roofs, parking lots, etc.

The text then provides us with an occasion for integrating it to our background understanding about possible effects of ditching and channeling practices. First, the increased flow rates create conditions that adversely affect the rearing of salmon. Increased flow rates might carry away smaller gravel, might lead to a leveling of the substrate, might wash away salmon eggs and roe, and might not provide a stable habitat for salmon fry. Second, the text also allows us to construct other effects. Straightening removes 'surrounding vegetation'. In part, of course, this vegetation consists of grasses. However, grasses often are the first plants to return after an environment has been damaged. Other plants are more deeply affected. Anyone who has walked outside and along a creek has noticed that there are frequently bushes and trees that grow along traditional beds along meandering streams. When these are channelized and straightened, the bushes and trees no longer exist to provide shadow-throwing overhangs. The summer sun heats up the water, which entails a decrease in the oxygen levels (the solution rates of gases are inversely related to temperature) and thereby further effect the habitats. We can easily imagine that the ditches, carrying fast flowing streams during the rainy season or after the snow melt, become slow-moving, putrid algae- and weed-supporting places in which fish no longer survive.

When we read the newspaper about familiar and even less familiar topics, our usually tacit background understanding allows us to make sense; that is, the article allows us to articulate structures that are consistent with our understanding of the world. The text talks about the familiar, and in my reading I can produce other texts, elaborations, which the original paragraph appears to imply. In fact, understanding the paragraph implies that I am familiar with the world. The particular shapes and sounds of words no longer matter, they become transparent, as the text transports me to real or imaginary places with creeks in their natural and straightened form.

When the scientists read and interpreted the graphical tasks, they began by inspecting them and then, in a dialectic movement oscillating between articulating their own understanding (referent) and isolating possible signs in the graph, they elaborated both understanding and graph. On the other hand, when they talked about their own graphs, they often talked at length about a particular phenomenon, experimentation, data collection, instrumentation, and so forth. Only much later did they point to the graph and suggested what it said. That is, even with a simple bar graph or plot of means in an analysis-of-variance paradigm, scientists provided first a lot of contextual background before articulating exactly what the graph expressed. This practice implies that a substantial and specific background understanding has to be articulated before the graph can be understood; throughout this book, I suggest that this background understanding is *required* for a competent reading of graphs—it derives from and consists of the massive, mostly unarticulated experience that we have of being in the world. When graphs are used in research on graphing, very little background is usually provided.

In this chapter, I provide a representative example of scientists' reading of their own graphs. Readers will notice that even the background provided by the scientist is insufficient to make sense of the phenomenon. I elaborate the scientist's reading to facilitate my readers in appreciating the complexity of the background understanding that makes possible competent use of *these* graphs. The competency scientists' exhibited when they talked about graphs from their own work often stood in stark contrast to the problems that they had with the graphs that I had brought to the session. This therefore raises the question how these scientists could both exhibit great competency and, when it came to graphs that are standard fare in undergraduate courses, fail and abandon a task. After presenting a case of reading familiar graphs in the first part of this chapter, I begin to articulate the differences between the two situations, interpreting graphs from the workplace and interpreting unfamiliar graphs, unfamiliar because they do not insert themselves in the everyday practices of their work. (For university-based scientists who articulated successful explanations, the graphs from the undergraduate courses that I used also were part of the familiar [teaching] environment.) I introduce and use activity theory as framework for articulating doing interpretation tasks and talking about graphs that are integral part of the work environment.

5.2. TRANSPARENT GRAPHS IN SCIENTIFIC RESEARCH

When the graphs were from scientists' familiar worlds, they were no longer interpreted or read but constituted signs that stood for this world in a metonymic way. The graph occasioned an articulation of an understanding of the world that had given rise to the graph—through series of translations—and which was therefore said to be the content (referent) of the graph. In this section, we follow one scientist into his world, which is constituted not only of natural phenomena but also the conceptual domains of physiology, organic chemistry, and ecology.

5.2.1. Background

In this particular study of a scientist's elaboration of his graph, I chose the interview with a scientist (Glenn) who does research on the migration of toxic substances such as polychlorinated biphenyls (PCBs) and dioxins through ecosystems. These toxic substances have been in the news every now and then for the past decades so that readers are likely to be familiar with some aspects of the problems surrounding their presence in natural environments. In the particular study, the scientist talked about polychlorinated biphenyls. PCBs were produced in relatively large quantities for use in such commercial products as dielectrics, hydraulic fluids, plastics, and paints. Since the 1990s, PCBs are no longer produced (at least in western nations) but continue to be released into the environment through the use and disposal of products that contained them. They also make their way into the environment as byproducts of various industrial processes.

Some PCB compounds are highly toxic. Studies in monkeys and rodents indicate that *in-utero* exposure to PCBs (as dioxins and other estrogenic chemicals) can alter behavior and learning ability. There also exist human epidemiological studies that show similar findings. Some longitudinal studies of children born to mothers exposed to PCBs prior to and during pregnancy demonstrate developmental delays, including reduced IQs. At the moment, there still exists very little specific information on the environmental transport and fate of PCBs, particularly those that exhibit dioxin-like toxicity. Available information on the physical and chemical properties of dioxin-like PCBs indicates that these are associated primarily with soils and sediments and are thermally and chemically stable. PCBs are likely destroyed by photo-degradation, which is followed by slow aerobic or anaerobic biodegradation by microbial organisms.

Glenn is a senior research scientist and professor in an applied-research university program. Among his specialties he counts environmental physiology and toxicology and the analysis of quantitative patterns. He conducts his research for small businesses, big industries, and governmental agencies both locally and throughout the world. His publication record averages about two refereed articles per year. He is active in the national funding agency and collaborates on national and international reports on environmental health, the uptake of dangerous substances (e.g., PCBs and dioxins) by organisms, and the spread of these substances through ecosystems.

Glenn's intent was to explain a graph that showed differential bioaccumulation of various PCB compounds in marine organisms. The graph was part of
an article that he had published three years earlier in an international journal.
However, before he could come to the particular (bar) graph, he felt compelled—as all his peers in this study—to articulate the context of the study and
many aspects of what he had done to get to the graph. There was a sense among
him and his peers that the graphs could not be understood (comprehended) unless the context was explicit to a certain degree. Given that both my graduate
student Michael Bowen (with whom I shared the interviewing task) and I had
Master of Science degrees, the scientists generally talked about their work from
scientist to scientist rather than at a level that characterizes the popular science
literature for more general audiences. This, then, required additional work of
explicating just what the scientists assumed the audience (interviewer) already
understood.

5.2.2. Getting Started

Glenn, as all other scientists, had been asked to bring at least one of his own
articles or reports to talk me through one graph with particular attention to what
the reader should be able to learn from it. Glenn began by stating the general
goal of his research, the movement of PCBs through the environment, and then
provided a first explanation as to the nature of PCBs [ii].

> [i] The overall purpose of this project was to actually look at how PCBs move
> through an arctic environment, an arctic marine environment and how- [ii] The
> first thing one has to understand is that PCBs aren't a single chemical but in
> fact a family of chemicals formed of up to 209 possible individual chemicals,
> which are called congeners. [iii] So, the neat part about that is that if we ana-
> lyze them as [sets of] congeners instead of a single PCB number, the relative
> concentration of each individual congener provide us a fingerprint. [iv] And we
> can watch how that fingerprint changes over space and time in the environ-
> ment. And it tells a lot about how PCBs move through the environment, peculi-
> arities of the change in chemical composition and chemical toxicity as they are
> taken up by various different biological organisms and so on and so forth.

Each family of PCBs has a 'single PCB number' and is, as becomes clear only
in a later remark, designated as an 'arochlor'. The significance of 'fingerprint'
and 'relative concentration' became clear somewhat later, when Glenn pointed
to the first set of graphs (*Fig. 5.2*). The relative concentrations of the different
constituent congeners in an original PCB mixture (e.g., Arochlor 1256) form a
'fingerprint', whose movement and change can be monitored as PCBs are taken
up by some organisms and subsequently move, by means of predation, through

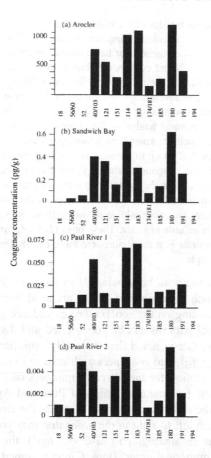

Figure 5.2. Graphical plate of the type interpreted by Glenn. (Proper names and compounds were changed to conceal the identity of the participant.)

the food web. At this point Glenn referred to the graphs but, rather than continuing to talk about the fingerprints, he talked about the conventions of graphing and how these conventions are related to the details of data collection.

[v] So, what happens here (*Points to graphs in Fig. 5.2*) in fact is that you are seeing a histogram of the concentrations with respect to the y-scale and there are different scales here. The scale actually isn't important but the relative concentrations for each of the individual samples, [which] have a whole bunch of different congeners. [vi] And so, the numbers that you see here on the x-axis

corresponding to each point of the histogram is in fact the IUPAC nomenclature number for each individual congener. It's basically- it's just the name they are called. [vii] So a congener that has a couple of chlorines attached to it in various positions is called congener eighteen. [viii] And the ordering of the congeners, along the axis here, is here, in fact isn't linear, about eighteen, from eighteen to fifteen, to seventeen, to fifty-four, to thirty-one. [xix] That's simply the order that congeners come off in gas chromatograph column—that's the way you do the chemical analysis—the order of elution on the G-C column. And the important point to know here is that lighter congeners with fewer chlorines come off first, right up to the most heavy congeners which come off last. [xx] And there is a corresponding change in the chemical properties of the PCB congeners as you go from the lighter congeners with fewer chlorines to heavier congeners with more chlorines. And the change in the properties says- These ones (*Left of plate*) are more soluble in water, these ones (*Right of plate*) are very, very poorly soluble in water. These ones (*Left of plate*) are more volatile they tend to evaporate into the atmosphere more readily than these ones (*Right of plate*) for example.

Glenn pointed to a figure in his article that in fact constituted a plate including four histograms. Such collections of graphs are used frequently in the scientific literature and, depending on the context, are considered good graphing practice because more global patterns can be constructed and displayed.[2] Although the scales are different, Glenn noted that the really important issue is the relative concentration of the different congeners with respect to one another [v]. He then articulated an explanation for the different numbers on the abscissa, which are the numbers that the International Union of Pure and Applied Chemistry (IUPAC) assigned to the different congeners. Because the frequencies of the congeners are plotted from left to right in the order that they come out of the analytic instrument, the gas chromatograph ('G-C column'), the numbers of the compounds are not in numerical order. Here, Glenn assumed that the listener is familiar with this analytic technique, which separates substances based on their different solubilities (a process called elution), which in turn are associated with different rates of travel along a column. These numbers are not in numerical order because the order of the IUPAC numbers does not correspond to the order of solubilities.

Glenn articulated more understandings of the chemical make up and physical properties of the congeners. Some of his elaborations are not consistent with standard accepted scientific knowledge. For example, he suggested that the lighter congeners with fewer chlorine atoms come off first and the heavier congeners with more chlorine atoms come off last. Whereas this is approximately the case (congener 1 has one chlorine atom, congener 209 has 10 chlorine atoms), it is not true for nearby congeners. For example, Glenn read off his graph the order from congener 18 to congener 15; however, congener 18 has three chlorine atoms and congener 15 has two chlorine atoms. Thus, whereas there is an overall trend from fewer to more chlorine atoms, Glenn's statement about the ordering of congeners does not hold for the correlation of ordering and number of chlorines in general. Associated with the ordering of the congeners in the graph are physical properties of solubility, which, in water, tends to be high for congeners with low IUPAC numbers and low for congeners with high IUPAC numbers. Finally, the trend in solubilities parallels that for volatility.[3]

So far, then, Glenn had done a lot of setting up. He elaborated on the nature of PCB and its compounds, hinted at the chemical composition of the congeners, provided information about the labels and the scaling of the axes, associated the order on the abscissa with specifics of the analytic method, and articulated the relationship between the ordering of chemicals in the graph and some physical properties.

5.2.3. Movement through the Environment: First Iteration

Glenn then provided a first indication of the movement of PCBs through the environment, based on inferences from the fingerprints of PCBs in different geographical locations, more or less distant from the original spilling site. This spill occurred at some place (which I call Sandwich Bay here) and is associated with the fingerprint of the original PCB mixture, Arochlor 1254 produced by Monsanto in the first graph (*Fig. 5.2.a*).

> [xxi] And now what we can do is start to look at the relative fingerprint if you'd like for the congeners in each of individual sample as say PCB mixture moves through the environment. So in Sandwich Bay, for example, there was a radar site that introduced PCB transformer fluid into the environment. And the technical mixture used was something called Arochlor Twelve-fifty-four. [xxii] This was a Monsanto mixture that actually had a fingerprint that kind of looked like that (*Fig. 5.2.a*) if you examine a subset of two hundred and nine congeners. So, it had a bunch of mid-range congeners which primarily had five and six chlorines attached to them in some range and some kind, the, the ones with the mid-level chlorination dominate. [xxiii] Now, if we go from what we actually know is the source input into the environment, what was put into the water column through the release into the Sandwich Bay sediments as kind of primary point of loading, then we can see, just by kind of squinting at the relative size of the bars in these two graphs, that Sandwich Bay sediments kind of look like Arochlor Twelve-fifty-four. And therefore the primary source of PCBs to the Sandwich Bay sediments was the original spilling of this Arochlor Twelve-fifty-four transformer fluid into Sandwich Bay. [xxiv] Now, this in fact is a

sample collected in Paul River, which is outside of Sandwich Bay, just about three to four kilometers removed. This (*Points to a map in the article*) is a figure of the bay here itself which is so small that no readers can see. Squint if you can, right? So, again, it (*Fig. 5.2.b*) shows a relative composition within Sandwich Bay right near the point of discharge to further away (*Fig. 5.2.c*) where in fact you seem to have a slightly lower predominance of these heavier congeners here (*Right part of Fig. 5.2.c*) and perhaps a few more (*Left part of Fig. 5.2.c*). But for all intensive purposes they kind of look like the same. In arguments about, in terms of pattern recognition, about similarities versus differences, at that stage, it kind of looks like hand-waving exercises. [xxv] There (*Fig. 5.2.d*) is a second sample from Paul River where you seem to have lots of lighter congeners and heavier congeners and less in the middle. And again to try at that level interpret any similarities or differences in pattern of Arochlor Twelve-fifty-four in these is a bit of a hand-waving exercise one has to take their interpretations with a grain of salt.

Glenn not only provided a description of the original source of the spill, but also details about the specific PCB mixture used, its producer, and the analysis—not all congeners but only a subset of them were analyzed in this investigation, particularly those with a 'mid-level chlorination' [xxii]. In three steps, he then asked the interviewer to compare the fingerprints of samples collected at three different sites (Sandwich Bay, Paul River 1, and Paul River 2) with the original fingerprint. From this comparison, he had inferred that the sites at Sandwich Bay and Paul River 1 were contaminated by the spill, but Paul River 2, which was more emphasized in his article, had a very different fingerprint, suggesting some other or additional contamination. With his elaboration of the interpretation that he had reported in the article, Glenn also made salient that the evaluation of similarities and differences between the fingerprints 'is a bit of a hand-waving exercise'. As in my representation (*Fig. 5.2*), the original graphs exhibited similarities and differences between the fingerprints in the three different locations in their relation to the original fingerprint that were not easily apparent. Even Glenn, despite the strong claims in his research article, suggested '[taking] the interpretations with a grain of salt'.

In this episode, Glenn provided a reading of a set of graphs in which perceptual similarities were used to make inferences about the source of PCBs in different geographical locations. While this informs us about one set of graphs and the inferences that had been drawn from them, it only constituted a preparatory aspect for the real story. This story concerned graphs that displayed the differential bioaccumulation of PCBs in organisms, which themselves are part of a complex food web and an even more complex ecosystem. But before getting to this part of this story, Glenn had to prepare his audience by elaborating on physiology and organic chemistry.

5.2.4. Physiology and Differential Bioaccumulation

Before Glenn came to the interpretation of a graph that had as its content the differential bioaccumulation of PCB congeners, he articulated issues related to the toxicity of certain PCBs and to the relationship between solubility and bio-accumulation.

> [xxvi] Now there is a difference in toxicity, which relates first of all to the bioavailability. So in fact these (*Right end of Figs. 5.2*) are the ones that to tend to be more hydrophobic and more lipophilic, so they're more readily taken up into biological lipid-rich tissue so they're more readily taken up across membranes and accumulate in livers and lipid rich tissues. So, those (*Left end of graphs*) are more hydrophilic. These (*Right end of graphs*) ones are the ones that are more readily taken up through diet through biomagnification at this end of the scale. [xxvii] And these (*Left end of graphs*) are the ones that are more readily taken up directly, say for example through gills and through external epidermis because they are the ones that are water-soluble. They tend to accumulate less in the tissues. [xxviii] But then on top of that there's an issue of the molecular, the specific molecular actions that influence toxicity. Some of them in addition have their chlorines in places that they look like dioxin and furan molecules, like two-three-seven-eight T-C-D-D, and have the same mechanism of action. They fit to the 'A-h receptor' almost irreversibly time turn on the N-F-O induction system and cause great damage internally in the cells and cause cancer in invertebrates and things like that. [xxix] And the other thing is that there's a whole range of different congeners in terms of availability for animals through their metabolic machinery to metabolize them and actually excrete them.

This episode consisted of four main points surrounding the issues of toxicity and bioavailability of PCBs to organisms, and how the latter is correlated with the physical property of congener solubility. First, Glenn reiterated that the congeners appearing on the right end of the abscissa, the higher-chlorinated congeners, dissolve less well in water (hydrophobic) and better in oils (lipophilic). These congeners are therefore readily absorbed through cell membranes and accumulate in lipid-rich tissue such as the liver. Because these congeners are stored in the animal, predators, which are higher up in the food web, they will preferentially accumulate these by lower-order organisms stored PCBs [xxvi]. On the other hand, the water-soluble congeners not only can be taken up directly through the skin (external epidermis) or gills but also are more easily excreted [xxvii].

In his third point, Glenn suggested that the effect of the PCBs is not merely related to solubility and bioavailability but also to specific molecular actions that contribute to the toxicity of a material. Here, some of the PCB congeners resemble other families of toxic substances, including polychlorinated dibenzodioxins (PCDDs), one of which is 2,3,7,8 tetrachlorodibenzo-p-dioxin (TCDD), and polychlorinated dibenzofurans (PCDFs). Endocrine-disrupting chemicals can interfere with the way in which steroid hormones and thyroid hormones normally work. As a consequence they may upset the delicate balance of hormones in the body, which, in turn, may lead to adverse effects on health. The 'Ah' receptor is not a hormone receptor but another kind of receptor in cells. Some chemicals, notably the dioxins and certain PCBs, can bind to these receptors and in so doing trigger many different biological effects, among which can be the disruption of hormones. Some scientists think that these chemicals can cause anti-estrogenic effects and alter levels of thyroid hormones. The final process involved in differential bioaccumulation pertains to the ability of animals to metabolize and change particular PCBs, and thereby change their solubilities so that they can be excreted rather than accumulating in the lipid-rich tissues.

At this point, Glenn was ready to tell the main point of his story, the differential bioaccumulation of PCB congeners. He turned to a page with another set of graphs (*Fig. 5.3*) and began to refer to groupings of congeners [xxx]—which already constituted an interpretant of the groupings readily available in the figure.

> [xxx] So in fact, if you take all the congeners again and I show them here (*Fig. 5.3*) but I've done different groupings with the congeners. What I'm doing here is looking at- [xxxi] Again one of the nice parts about this particular experiment real-world experiment is that we knew what the source looked like for PCB congeners. We could track it through the food chain in particular [organisms]. And these are the congeners- this is the congener composition in some bottom-dwelling fish, called four-horn sculpin. [xxxii] What is intended here is to actually get at which of the congeners were most selectively accumulated in the tissue of the fish and which ones were amenable to excretion, OK? Based on whether the chlorines were around the two biphenyl rings right.

This part of the reading was again characterized by false starts and backing ups. Before Glenn further elaborated the issue of the grouping of congeners, he articulated the source of the sample (sculpins) and the intent of the experiment to trace the (differential) congener composition through the environment. He framed the intent of *this* graph as one of showing which of the congeners accu-

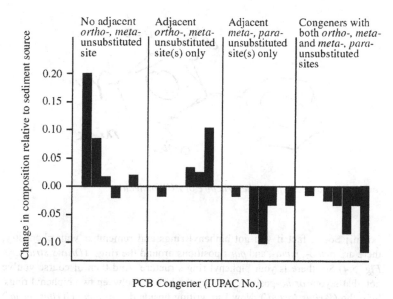

Figure 5.3. A second graph Glenn drew on while explaining the graph in Fig. 5.2.

mulated in an organism, here the bottom-feeding sculpins. But before he could make this point, there was still some ground to be covered, some understanding to be explicated—which, as it turned out, pertained to the different positions that chlorine atoms can take in the main PCB (biphenyl) structure. That is, Glenn moved to make explicit what the signs /ortho/, /meta/, and /para/ in the headings of the four groupings (*Fig. 5.3*) actually referred to.

5.2.5. A Lesson in Organic Chemistry

In order to understand the content of the graph (*Fig. 5.3*), it is necessary to know something about the chemical structure of PCB congeners. Glenn linked the three terms *ortho, meta,* and *para* to biphenyl rings and quickly sketched two linked hexagonal structures, each containing a circle. Each of the three terms signified a different position on the right hexagon, whereby the horizontal constituted an axis of symmetry. He then suggested that the second hexagon contained the equivalent positions, so that the two hexagons constituted mirror images.

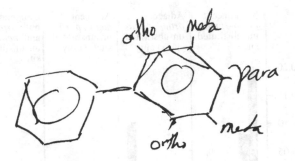

Figure 5.4. Diagram to explain the structure of PCB molecules.

[xxxiii] So, in fact if you got biphenyl rings and remember your chemistry, there are *ortho-*, *meta-*, and *para*-positions around the rings. (*Draws structure, Fig. 5.4*) So, there is your biphenyl ring structure. And then of course you've got, that's your *ortho*-position, right by the linkage between two biphenyl rings. So ortho (*Writes 'ortho'*). Now I'm writing upside down- meta (*Writes 'meta'*) and para (*Writes 'para'*), right? And similarly you've got *ortho, meta, para* (*Points to bottom of right ring*) so that (*Points to meta-position*) one there is *meta* (*Writes 'meta'*) again and *ortho* (*Writes 'ortho'*) and again the same in the second ring (*Points to left ring*). [xxxiv] So we can envision actually having like hydrogen at each of those corners or we can put a big fat chlorine, chlorine is a lot bigger and has got greater volume as an atom if you like in terms of the electron shells than hydrogen does, right?

In the final part of this episode, Glenn suggested that there might be hydrogen or chlorine atoms in different positions. Readers who have not had chemistry for a while may find it difficult to follow Glenn's presentation, which really has to be understood against the background of a more elaborate model (*Fig. 5.5.a*). The basic structure of a phenyl ring contains six carbon atoms linked into a hexagonal structure, containing both single bonds (one line) or double bonds (two lines). These double bonds are not fixed but can move around the hexagon. Two phenyl rings are joined to form the double-hexagon structure known as a biphenyl structure, which is the constituting feature of PCBs (polychlorinated biphenyls). Because each carbon atom can form four single bonds (a double bond counts as two single bonds), there are a total of ten possibilities for atoms of groups of atoms to be attached to the rings. In PCB congeners, one to ten chlorine atoms (therefore, '*poly*chlorinated') take up these positions. Each time a chlorine atom is bonded to one of these positions, chemists speak of a substitu-

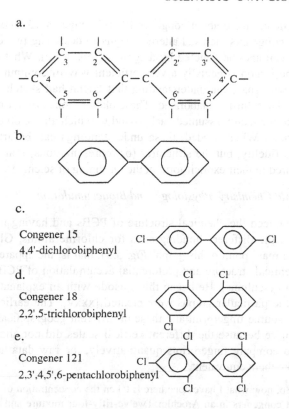

Figure 5.5. Structure of biphenyls. a. Expanded form, indicating the two carbon hexa-gons, all electron pairs (lines) and double bonds, and bonding sites (open lines). Up to ten chlorine atoms can be accommodated. b. Reduced representation, in which the free-moving sets of electron pairs are indicated by the ring. c., d., and e.

tion, that is, a substitution of the normally present hydrogen atom by a chlorine atom. Because the chemical and physical properties differ depending on the po-sition(s) of the bonded chlorine atom(s), the carbon atoms are numbered or, equivalently, named, whereby positions '2' and '6' as well as '3' and '5' are equivalent. For brevity sake, this more complete structure is often abbreviated in a diagram showing the two hexagons and indicating the free electron pairs by circles inside (*Fig. 5.5.b*). This second, abbreviated structure constituted the starting point for Glenn's explanation and therefore was assumed by him as the necessary, shared background for the explanations that followed.

When Glenn talked about congener 18 and congener 15, about heavier and lighter PCB congeners, he used alternate signs to designate two chemical compounds that can also be denoted by diagrams (*Fig. 5.5.c*). What on the surface appears to be jargon is actually a very efficient way of communicating among researchers who share the understanding that Glenn had sketched and that my explanations have further elaborated. These elaborations are not normally made but constitute the tacitly assumed background to which all the other explanations can be accrued. When needed, these understandings can be articulated, with more or less fidelity, but the structures (or representations) that are articulated are not required in their explicit form in the reasoning of scientists.

5.2.6. Organic Chemistry, Physiology, and Bioaccumulation

Having introduced the chemical structure of PCBs and having made a distinction between the different bonding sites for chlorine atoms, Glenn moved to articulate the main point of his graph (*Fig. 5.3*), that is, the apparent relationship between chemical structure and preferential accumulation of PCB congeners by one organism (sculpins). He began this episode with an explanation of method about how the particular graphs were created [xxxv]. The earlier comparisons between the source fingerprint and those in different geographical locations had been qualitative because the different vertical scales did not allow direct comparisons. To compare fingerprints quantitatively, the raw data had to be processed to make them equivalent.

> [xxxv] So, now what I have done here is taken the concentration of all of these different congeners in an Arochlor Twelve-fifty-four mixture and normalized them to an overall concentration of one. OK. And then I did the same thing with the congener concentrations in liver tissue of sculpins. And then what I said is I'm gonna just subtract what I found in sculpins from the original Arochlor Twelve-fifty-four mixture, then I'm gonna divide these congeners into various subsets depending on how I find chlorines wrapped around these two biphenyl rings. [xxxvi] So, in this first one here (*right-most panel in Fig. 5.3*) are all those PCB congeners that don't have any sites here (*Circles para-meta site, Fig. 5.6.a*), they are not filled by chlorines. OK, so there's two parts to this. If in fact there are chlorines in a way that have two hydrogens here right and that position or either that position there here (*Circles meta-ortho site, Fig. 5.6.b*), they're found in this group. And in terms of relative composition then for that group overall there's on average a higher percentage relatively of congeners that don't have a free available site where you put at least two hydrogen site by site, the *meta-* the *para-*sites or the *ortho-*, OK. [xxxvii] So, in other words, what that means is that sculpins are preferentially bioaccumulating

a.   b.   c.

Figure 5.6. Sequence of additions to the original diagram. Each circled pairs of position corresponds to one of the conditions in Fig. 5.3.

those ones and, and don't seem to be able to preferentially excrete them relative to the congener mix as a whole.

After describing how the fingerprints from the source and sculpin liver were normalized and subtracted and the differences grouped (according to the position of chlorine atoms) and plotted, he moved to articulate the relationship between chemical structure and presence off a congener in the first, left-most panel [xxxvi]. All of these congeners, so the explanation goes, at least one of the hydrogen atoms in the *para-meta-* and *ortho-para*-combinations have been substituted by chlorine atoms. One such example is congener 121 (*Fig. 5.5.e*) because it does not have a pair of hydrogen atoms in either *para-meta-* or *ortho-para-* combination. Glenn then made one central point—this group of congeners, which he had just circumscribed in terms of the lack of adjacent, unsubstituted sites, is preferentially bioaccumulated in sculpin liver tissue. The fact that these congeners were accumulated rather than excreted can be inferred from the positive values of most bars in the histogram. Glenn then moved to articulate corresponding relationships for the other three groups.

> [xxxviii] So now, this group over here is (*Second panel, Fig. 5.3*), those congeners where, rather than having neither of the free *meta-para* sites or *ortho-meta* sites, there is only an adjacent *ortho-meta* site. So, what you have here then is that the congener in this particular group might have an *ortho-meta* site free (*Circles ortho-meta site, Fig. 5.6.c*) where in fact you don't have a chlorine atom in those positions. But there is no free *meta-para* site in either of the two rings, right? [xxxix] So again, having only an *ortho-meta* site as an area of enzymatic attack, in other words, not having a chlorine [atom] to block it, doesn't work for sculpins. Which means that they actually also preferentially bioaccumulate those congeners that only have an *ortho-meta* site, right.

In the two parts of his explanation, Glenn articulated and elaborated both the structure of the PCB congeners that fit into the second group [xxxviii] and the importance of having pairs of neighboring sites that do not contain chlorine atoms [xxxix]. It is at these 'free' *ortho-meta* or *meta-para* sites that enzymes can attack and therefore change the PCB molecule. Furthermore, the diagram shows that sculpins bioaccumulate congeners that only have an *ortho-meta* site but not a *meta-para* site for enzymatic attack; therefore, sculpins cannot secrete this group of congeners, including congener 15 (*Fig. 5.5.c*), which is identified by its particular chemical structure. The importance of the *meta-para* site for enzymatic attack and therefore for metabolism and ultimate excretion appeared in the descriptions of the remaining two groups of congeners. Congeners in both groups have at least one *meta-para* site that is not occupied by a chlorine atom, such as congener 18 (*Fig. 5.5.d*).

> [xxxx] But now, any of the congeners that have an adjacent *meta-para* site, in other words, any of the congeners where at least somewhere around these rings, there's an hydrogen there and an hydrogen there (*Writes two letters 'H', Fig. 5.6.a*). In any of the permutations or combinations thereof, rather than having a chlorine there, in fact, occur at relatively lower concentration in the liver tissue or the whole tissue of the fish than the original mixture that's taken up. [xxxxi] And the same thing- (*Right-most panel, Fig. 5.3*) if it's got both a free *ortho-meta* and *meta-para* site. But again it's the *meta-para* site that's driving it.

Glenn first described the free *meta-para* site in terms of his drawing, where he added the letter 'H' twice, in a neighboring configuration, and then elaborated that this group included any other congener with another *meta-para* site alone or in combination. He added that the right-most group of congeners (*Fig. 5.3*) in addition had a free *ortho-meta* site. He emphasized that the important aspect for these congeners was the free *meta-para* site. The description 'lower concentration' was accompanied by a pointing gesture (indexical sign) to the histograms, in which all components extended into the negative range of the graph (thereby constituting a consistent interpretant of 'lower concentration').

At this point, Glenn had articulated the background against which the graph makes sense, that is, the background understanding required for allowing the graph to be integrated into a pre-existing, inherently meaningful world. From a scientific perspective, however, he had not yet finished, for the data and inferences had not yet been compared to the existing theory. This then was the next step in Glenn's elaborations. According to the theory, the enzyme added a new compound from a functional group [xxxxii]; in his diagram, Glenn exemplified the addition of a functional group in the form of the hydroxy group ('OH'). When he added the second hydroxy group, a slip up occurred that was not caught in the situation. He added the second 'OH' to the same carbon atom ('5' position), rather than replacing the hydrogen in the '4' position, which is not

a.   b.

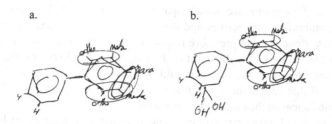

Figure 5.7. Two further stages of the diagram adding hydrogen to meta and para positions (a) and adding, one after the other, two hydroxyl groups to a meta position.

consistent with standard organic chemistry because there is only one possible bond available at each of the carbons (*Fig. 5.5.a*).

[xxxxii] That actually fits in with theory that in fact in terms of enzymatic attack. The enzyme can get in and can add basically (*Adds 'OH', Fig. 5.7.b*) either a hydroxy group or (*Adds second 'OH', Fig. 5.7.b*) a dihydroxy group, or various other functional groups to the PCB congener if in fact it's got two adjacent hydrogens in that position around the biphenyl ring. [xxxxiii] And then the implication is that if- if there is an area that's amenable to enzymatic attack, then there are various more hydrophilic constituents added to the molecule and that decreases the overall polarity of the molecule. And therefore the molecule has greater aqueous solubility, it can be carried in body fluids more readily including things like urine and is more readily excreted. So in other words, what that means overall is the fish can excrete certain congeners that have an adjacent *meta-para* site but not any of the other combinations.

The enzymatic addition of the functional groups increases the solubility of a congener in water ('aqueous') which leads to higher levels of excretion via the urine. Glenn elaborated the increased solubility by stating that the addition of the hydrophilic functional group made the congener less polar—which is the direct opposite what he had stated in the article. In fact greater polarity is required for higher solubility.

Researchers on graphing usually do not investigate the background with which their subjects come to the interpretation session. Researchers appear to assume that an individual can read graphs pertaining to the speed of an object or the density of shrimp without ascertaining that their participants (e.g., students) are actually familiar with speed or shrimp. I have gone through the transcript at some lengths to articulate the chemistry associated with the graph. All of these elaborations are required because the characterizations (articulated contexts) of each of the four groups are not transparent from Glenn's talk, in part because of the accumulation of adjectives and double negatives. The classification of congeners was done on, and therefore required an understanding of, the basis of the chemical structure of the different PCB congeners. Furthermore, the inferences and comparison with the theory would not have been possible had it not been understood that enzymatic substitution changes solubility and therefore the fate of the substances in the body of an organism.

Glenn could have stopped here. But then, the very purpose of his research would not have been sufficiently appreciated, and the implications for the presence and role of PCBs in the natural environment left underdeveloped. He felt obliged to address the question about how differential bioaccumulation affects organisms in higher trophic level, that is, larger predators that ingest with their food only those PCBs that have been accumulated earlier in the food chain.

5.2.7. Movement through the Environment: Second Iteration

Glenn began this episode by stating the problem as a question of congener concentration in any type of organism as the food chain is traversed from lower to higher trophic levels, taking as starting point the sculpins via ringed seals to polar bears [xliv]. He then provided a specific numerical example how the relative concentrations in PCB mixtures change, from sixty congeners in the initial spill to only five or six in the polar bear species, which constitutes the uppermost trophic level, as the bears have no predators [xlv].

[xliv] Well the interesting thing is if you now go through the food chain. Let's go to the next step. The sculpin is kind of along the base of the food chain, sucking up worms and clams and things living in the sediments. They are not quite at the base of the food chain but pretty close. The next question is what happens when a ringed seal eats four hundred sculpins and what happens when the polar bear eats ringed seal and each step of the process there are these congeners that can be lost. [xlv] So, for example if we go back to start with something, if we go back to the original diagram we were talking about before. Then you start looking at an Arochlor Twelve-fifty-four mixture that- Actually out of the two-o-nine possible congeners, not all are analyzed here, there are only about seventy-nine analyzed out of the two-o-nine possible ones. But I mean, let's say you start there and you've got maybe sixty of them represented in the overall PCB mixture, by the time you get to the polar bears there are really only five congeners left, five or six congeners. [xlvi] So measuring overall concentration of PCBs in the environment, which I have heard of being done, isn't a kosher way of actually figuring out PCBs. There are different relative toxicities in each one of these things and the relative proportions change completely every time from one compartment [trophic level] to the other, but apparently so

as you go to the food chain, where there's some potential metabolic modification and excretion. And that's the whole point. [xxxxvii] Now, if we look at
environmental regulation of PCBs, they're all based on toxicities of overall
Arochlor mixtures, right, you can't even get human toxicity data except as
based on total PCBs as Arochlors. I don't think, by and large, that's good
enough for regulatory purposes. So, we got some really neat environmental science but when you actually look at the level as to where the science integrates
with environmental management, it's a bit depressing.

Included in the examples was a further elaboration that Glenn and his co-authors
had in fact not analyzed samples for all of the congeners but only for a subset of
the 209 existing congeners. Glenn then launched a critique—which may have
well been a factor motivating all of his research—questioning the measurement
of overall levels of PCBs when in fact only some of them are bioaccumulated
and end up in higher trophic levels. Thus, whereas his science is 'neat', he characterized the applications made of this and similar science by regulators and
environmental management as 'depressing'. We can understand the present
elaboration as a statement of the significance of this research. It is important to
note that there are changes in the relative composition of PCB mixtures and the
increasing elimination of PCB compounds as mixtures move to higher trophic
levels. This research therefore implies that the toxicity of PCBs has to be evaluated differently for different trophic levels and accounting for different proportions of each congener characteristic for the trophic level.

5.2.8. Wrapping Up and Summary

Glenn ended his session by recontextualizing the study as one that had led to a
simplification. Because this simplification expressed more general concepts of
toxic substances in the environment, it could lead to inclusion of the graph in
textbooks—a reference to my own research agenda concerned with the interpretation of graphs in textbooks.

> [xlviii] Anyway, that's not an intuitive figure diagram, that one actually takes a
> fair bit of figuring. But again, the nice part about it is that it analyzes the data in
> a way that leads to that simplification (*Points to each of four headings in Fig.
> 5.3*). This is appropriate in terms of what larger concept we can derive from the
> work and what, it's the same kind of thing one might envision leading you to
> put it into a textbook. [il] If you're talking about how PCBs are metabolized or
> physiological-based environment kinetic models for PCBs, then this (*Points to
> histogram, Fig. 5.3*) is what drives which of the PCBs are taken up into the
> food chain and which ones do not get passed on through the food chain. [l] The
> other way to do it is completely ignore this level of things (*Covers histograms,*

Fig. 5.3) and just actually show (*Points to structure diagram, Fig. 5.6.b*) the different kinds of congeners based on the theory that are amenable to enzymatic action. It's still almost easier to explain this from cause and effect mechanistic principles (*Points to structure diagram, Fig. 5.6.b*) rather than the underlying data (*Points to histogram, Fig. 5.3*) that support the theory.

This episode included an interesting comment by a scientist about what should be represented in science textbooks. Here, the simplification achieved in understanding differential bioaccumulation by means of the histograms (*Fig. 5.3*) and its articulation of general principles is judged worthy an inclusion in textbooks. This comment is noteworthy because Glenn himself judged the diagram as being not intuitive. Furthermore, my extended elaboration was required for making the graph and Glenn's talk intelligible, which shows just how complex the world is within which this graph exists as an unproblematic referent and tool. In this contradiction, the apparent time and effort that go into figuring out the referent (content) of the graph and its recommended inclusion in a textbook (which is used by students with much less background understanding), lies a phenomenon of historical *amnesis*, a term marking forgetting in contrast to the process of remembering termed *anamnesis*. This amnesis allows the effacement of traces of the processes that led to a graph. This amnesis then leads to a dichotomous situation where we have a natural world on the one side serving as a referent to graph before us on the other. When all the various contexts (first articulated by Glenn and subsequently in my reading of the interview) have been abstracted to leave only the diagram, we may have in fact brought about the particular condition that made graph reading so difficult.

5.2.9. Emblematic and Transparent Nature of Graphs

By the time he completed talking about his graph (*Fig. 5.3*), which required the articulation of a lot of background understanding including another graph (*Fig. 5.2*), Glenn had spent twenty-seven minutes. In this, he was in fact at the lower limit of the range of time (25–45 minutes) it took scientists to talk about their own graphs. They talked without breaks or stops, continuously articulating the background against which the graph acquired the referential relation to its content. That is, the scientists elaborated not only the content of the graph but also the history of the referential relation (how the graph came about), context of the content, and details of the methods by means of which the data were created. There were only occasional and minor slip-ups, mumbles, and stumbles, as these often occur in real-time conversation, which is not edited in the way scientific articles are. All this talk was occasioned by the graphs, which always consisted of relatively few lines or elements—though even the scientists sometimes suggested that their graphs are not necessarily 'intuitive'. That is, the small number and relative simplicity of the sign elements available in the graphs stood in contrast to the amount of background understanding they occasioned. A relatively simple object—in terms of the constituent elements (signs)—stands for, and requires understanding of, rather complex worlds. I therefore propose that scientists' own graphs have the status of *emblems*.

An emblem can be defined as a visible object representing something else by a natural suggestion of characteristic qualities, or a habitual and recognized association. An emblem is always something simple, and in the past, often consisted of a picture accompanied by a short text (e.g., motto) that was intended as a moral lesson or meditation. In the scientists' case, their graphs had a similar function: The graph indexed, pointed to, and stood for a whole range of objects, events, and processes. When asked to 'talk me through the graph', scientists articulated some of these objects, events, and processes. That is, the graphs also had a metonymic relation to the processes that led to their creation, that is, a relation in which some entity stands for a more complex thing of which it is a part—'red-head' may come to signify a person with red or reddish hair. In the present context, the graph is the outcome of an often longish process of research, involving many entities (objects, tools, and instruments), scientific concepts, and skilled scientific practice. In a transparent way, the graphs stand for the entire work history, from which they have become detached as the scientists usually omit their own agency when they publish their work. Familiar worlds, however, are not usually represented in explicit ways but, in everyday cognition, constitute the taken-for-granted background that allows problem identification and problem solving to occur. This background does not have to be represented as explicit content of (information in) the mind. When requested or in situations of breakdown, this background understanding can be articulated. However, the articulation should not be mistaken as a faithful representation of the cognitive processes underlying graph use. Throughout this century, researchers in very different disciplines have shown that there are considerable differences between tools and processes *in ongoing practice* and the *descriptions of* tools and processes, even by the practitioners themselves. Theories of graphing that model the articulation of graphs rather than graph-use therefore fail to capture expertise evident in everyday practice.

The extent to which scientists articulated relevant objects and events was a function of the audience present: they knew that Michael Bowen and I had training in the natural sciences and in the statistical analysis of data. Thus, they never used the kind of discourse used for the popularization of science. The scientists used many words that are not in everyday use outside of scientific labo-

ratories and journals, so that, in some contexts, they would have been charged
for use of jargon. The connotation that is implied with this term is one of unin-
telligibility.

The scientists in this research did not intend to be unintelligible. Rather,
they used words as they would use them in the context of their work. These
words, such as 'congener' or 'biphenyl ring' are useful signs because they allow
scientists to communicate in efficient ways. When required, these signs can be
'unpacked', that is, elaborated by other, interpretant signs such as when Glenn
provided a diagram of the biphenyl ring structure. Much like the graph itself, the
words used in the talk about the graph have complex referents, which were
elaborated through interpretant signs as part of the talk about graphs. That is,
when asked, scientists articulated a complex structure that appears to be packed
in or 'behind' the signs (words, graphs) but invisible in the scientists' normal
use. These signs, graphs and words, functioned like transparent tools for doing
science, or transparent windows onto the scientists' worlds.

Some aspects of the graphs did not easily lend themselves to articulation.
Thus, Glenn found it difficult to articulate the qualitative difference between the
fingerprints characterizing the Paul River 1 and 2 samples. Yet (and without
providing more substantive argumentation) his research article suggested that
the two fingerprints were 'markedly different'. During the interview, however,
he said that distinguishing these fingerprints as coming from the same or from
different sources required 'a lot of hand waving'. That is, he can perceive a definite difference that appeared to resist verbal articulation; these similarities and
differences were not apparent to me despite my scientific and statistical training.
How is this possible? Of course, Glenn is very familiar with fingerprints and
their analysis not only from the study that he presented but also from his research more broadly. This familiarity, which had arisen from having done many
comparisons, allows the perceptual identification of differences even if similarities and differences resist verbal articulation.

When a scientist is unfamiliar with a graph and the background understanding necessary to make sense of it is absent, then tools, resources, time, and
energy are required in establishing the necessary background. When tools, resources, time, and energy are available, scientists can construct and articulate
sufficient background for all parts of the graph to make sense, that is, make connections between the interpretants they generated and their existential understanding. Thus, my own reading of the graphs presented by the scientists and my
reading of the associated protocol required time and effort to conduct sufficient
research for familiarizing myself with the domain of research. My own prior
understanding associated with my educational background, professional experi-

ence, and prior research provided a sufficient starting point so that I did not have to begin at ground zero. Furthermore, because I normally work in my familiar environment, I have access to the Internet to find resources related to scientific concepts and to mathematical and statistical tools for modeling purposes. I also have time and sufficient interest to pursue an issue so that I can achieve a sense of having understood. On the other hand, if the tools and resources are not available, I do not expect scientists to do much about the lack of familiarity with the graphs or their content and context. I do not expect that they capture and remedy any slip-ups, errors, or misunderstandings. Although there would be differences between individual scientists' interpretations, access to their everyday tools and resources (including time) necessarily change the interpretants that they can produce. Only the physicist Nelson was willing to engage in an extended inquiry (see Chapter 3). To understand graph- and graphing-related competencies, it is clearly not sufficient to present scientists with unfamiliar graphs in situations that isolate them from their work world.

5.3. GRAPHING AND ACTIVITY SYSTEMS

This book is concerned with scientists' graph-related competencies. In Chapters 2–4, I present analyses of the talk produced over and about three graphs that represent materials typical for the undergraduate-level introductory training in ecology. In this chapter, I analyze an example of the kind of talk produced by scientists over and about graphs from their own work, graphs that they had produced themselves. Before conducting the study I assumed that scientists would exhibit considerable 'expertise', whatever this would turn out to be, which could be used as a standard for assessing the competencies of science student along the trajectory that takes them from middle school to university and job. Contrary to my expectations, the scientists did not exhibit general expertise when it came to reading graphs that were not directly related to their work. Here, I understand 'work' as pertaining both to research *and* to lecturing activity conducted by university-based scientists. Once we disregard those situations where scientists were very familiar with a particular (type of) graph, there remains little general expertise with respect to making inferences that bear a semblance with the norm, here understood as the interpretation given to the graphs in standard textbooks. Chapter 4 exemplified the considerable struggles many scientists experienced, and even Chapters 2 and 3 provided glimpses of performances that were not as smooth as the expert-novice literature usually portrays it. After I had conducted three ethnographic studies of scientifically trained individuals in their everyday work, I began to realize the need to frame graphing in relation to their daily activities—which led me to the notion of activity systems and activity theory.[4] Activity theory has become an important tool in understanding complex work situations.

Traditionally, the sign-reference relationship was the primary object of research in mathematical cognition. More recently, it has been recognized that this relationship is not independent of the community of interpreters, which leads us to a semiotic approach (and, equivalently, to the theory of community of practice[5]). But even considering the community of interpreters in itself is insufficient to account for mathematical understanding; that is, a semiotic analysis is therefore only part of the story. To capture other parts of this story, I draw on activity theory, because it captures the context of activity in non-reductionist fashion. Activity theory does not simply focus on *subject-object* (e.g., scientist-graph) relations but makes thematic how other entities—including *tools*, *community*, *rules*, and *division of labor*—mediate the primary relationship. Besides making thematic the mediated nature of activity, activity theory also forces researchers to attend to the historical changes within a system and the contradictions that both interfere with the activity and serve as drivers of change.

As a way of introducing activity theory, consider the following situation that is discussed in some depth in the next chapter. Karen is a water technician employed on a grant that assists local farmers and community activist groups in understanding the fluctuations of a local creek, from which the farms draw water for irrigation. On the farm, it is Karen's job to operate a water level monitoring station consisting of a pen chart recorded that keeps a complete record of the fluctuation of water levels in the creek. In activity theoretic terms, Karen is the subject, whose relation to Henderson Creek, the object, is of interest to the study. The tools that she has at hand for monitoring water levels and working in the creek mediate her relationship. However, the tools do not just exist in and of themselves; the tools already exist in the culture, which has bequeathed them to Karen. That is, the community, which already has established certain rules about the use of these tools, mediates the relationship between Karen and the tools (*Fig. 5.8*). The relationship between tools and community is itself mediated by historically evolved rules that describe the ways in which the tools are used.

Activities are oriented toward some entity and driven by something. This entity, the object or rather the motivation related to the object, is constantly in transition and under construction, and manifests itself in different forms for different participants in the activity. Objects appear in two fundamentally different roles, as objects and as mediating artifacts or tools. In both activity-theoretic and semiotic terms, there is nothing in the material constitution of an object that would determine which of the two roles it has in an abstract and decontextual-

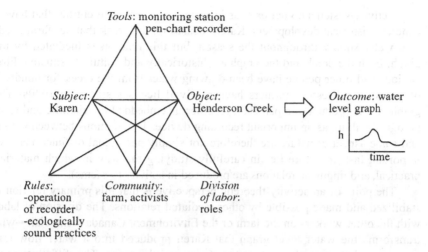

Figure 5.8. Activity theory articulates how the primary relation between subject and object is mediated by other entities. Moreover, each mediated relationship exists in a context that includes all other mediated relationships that can be constructed on the basis of the entities.

ized sense. The activity as a whole determines the place of the object, and, because of the relevant background understanding that comes with settings and situations, the activity also determines 'meaning'.[6]

Water monitoring station and the pen-chart recorder attached to it mediate the relation between Karen (subject) and Henderson Creek (object). There are other tools, too, that mediate the relation between Karen and Henderson Creek, including her knowledge about creeks and their ecology or her embodied skill of riffle construction—in cultural sociology, such factors are considered as schema that mediate the relation between agency and structure.[7] This relation is equivalent to the semiotic relations that have been considered so far. Here the subject, object, and mediating artifact find their correspondences in interpretant, referent, and sign, respectively. In classical semiotic theory, however, there is no exact correspondence, because the individual subject has been left out from the triadic sign-referent-interpretant relation. On the other hand, in an activity-oriented approach, the intentions and agency of a human subject (i.e., the scientist) is are the necessary ingredients that bring about the relation between tools (language, measurement instruments) and nature (of inquiry object).

Activity systems do not exist as ideas but are contingent entities that have a concrete historical development. Karen focuses on the creek that her farm needs as a water supply throughout the season, but this relation is mediated by the graph. Both the creek and the graph are historically and culturally situated. First Nations and other people have been drawing water from the creek for hundreds of years, and Western farmers have received licenses since the 1940s. The graph, as the product of a pen chart recorder, is embedded in scientific and technological culture as an important recording device. The relations between graphs and some aspect of nature are therefore not simply perceptual or functional. An important task therefore lies in carefully studying the way in which material, practical, and linguistic relations are produced in activity (systems).

The point of an activity-theoretic perspective is that this primary relation is stabilized and made possible by other mediated relations. The division of labor with the other workers on the farm or the Environment Canada technician—who transforms the water level graph that Karen produced into a water flow rate graph—mediate the relation between Karen and the creek. The different mediated relations (*Fig. 5.8*), as an ensemble that includes material and social dimensions, constitute the basic unit of analysis of human behavior, that is, an activity system. The fusion of the material and the social (discursive) produce relations of signification and the individuals that are positioned, qua subjects, within practices.

In Chapter 6, I show that Karen's reading of her graph in terms of the water levels of Henderson Creek is inscribed in many other relations, some of which are made thematic here. For example, the primary relation of subject (Karen), object (Henderson Creek), and mediating artifact (graph) is mediated by another which exists between herself, the community (here represented by the Mayor Walter) and the graph. Yet another relation made thematic in the next chapter is that between the Mayor Walter, the Henderson Creek watershed, and Karen. Still another mediated relation is that between Karen, the water levels, and the Environment Canada technician who comes to establish the calibration curve that allows Karen to read her graph in terms of liters per second although it really displays water depths. In another triangle, the graph replaces Henderson Creek as the object and is, in turn, mediated by other graphs. Activity theory allows us to understand graphing as a situated activity that is mediated by many other entities, relations, and systems. In Chapters 6–8, I use activity theory as a frame to describe graph production and graph reading as historically constituted processes that arise from the everyday work of the individuals concerned. In this context, graphs provide the basis whereby particular physical relations are inscribed as relations within the organization of practice. In such cases, we cannot simply speak of 'representation' because signs represent more than physical relations. As beings, we always come to a world where graphs, as signs in general, are always and already social. In Karen's work, the graph taps many other practices (signifying and material) within the valley where she works, irrigation, damming the creek, fishing, habitat maintenance, building impervious surfaces, building riffles, planting trees in riparian zones, oxygenating the creek. It is in relation to these other practices that Karen's graphing exists. Looking at her graph reading in the absence of everything else, we could come to the conclusion that she competently reads it. But this is of little help for understanding the relations that make this graphing competence possible in the first place. But let me return now to explaining the scientists' different graph readings using activity theory as a framework.

5.4. DOING PUZZLES VERSUS ARTICULATING WORK

To prevent misunderstandings, I want to clarify that my videotapes recorded situated activities both when scientists read unfamiliar graphs and when they talked about their own. In many instances, the participants were in the familiar surroundings of their offices. My video and transcripts are records of these activities. However, each session is further situated in a larger context. Here, the differences between the two types of reading become apparent. When scientists talked about their own graphs, these were continuous with all the other things they normally do as part of their everyday work. That is, because the graphs are emblems that bear an indexical relation to the processes and contexts of their creation, the sets $\{R, S, c, r\}$ pertain to those unique solutions that allowed them to get their days' work done. On the other hand, when scientists read unfamiliar graphs, they faced a different kind of task, one that was discontinuous from their everyday work and private life. They had to bootstrap readings by structuring the graphs and seeking suitable instances from their past experiences, such as to find plausible solutions to the relation $R = f_r(S,c)$. I want to consider at least two accounts for explaining the problematic reading and the nature of the tasks to be 'decontextualized'. These explanations are (a) the discontinuity between the sets $\{R, S, c, r\}$ that characterize everyday scientific practices and those that scientists had to generate during the task and (b) the indeterminacy of the results of the bootstrapping process.

5.4.1 Doing Puzzles

Graphing research frequently uses tasks that involve descriptions of situations. These descriptions are said to make the tasks contextual. When participants fail to provide standard interpretations, deficit arguments are rallied as explanations. The research presented here shows that not only the general public but also highly trained scientists struggle with such task even though these were culled from undergraduate sources in their own discipline. Providing textual information does not itself guarantee successful standard interpretations, and many errors occur that are similar to those committed by students. Given that my participants were successful scientists, explanations based on cognitive deficits are certainly inappropriate. To better explain the crass differences I observed when scientists read my graphs versus their own, I unpack the situational differences in the two tasks as they relate to Ted and Glenn.

Ted was interviewed in the familiar surroundings of his work, his office. Ted agreed to participate in my research and understood his interpretation as one that would help me better understand graphing and how to teach it (better) in school and university contexts. More than anyone else, Ted took the task as a challenge with the fervor of a puzzle aficionado. He began each task by reading the Cartesian graph without attending to the accompanying text. Only when he felt that he had said as much as he could did he read the text. But it was clear that I, the investigator, was (somehow) in control of the standard interpretation. Ted and the other participants were complicit, playing the game of being interviewed for the benefit of future generations of students, who might receive improved instruction in graphing. In activity theoretic terms, this aspect is a rule (of conduct) that mediates the relationship between researcher and scientist (bottom left in *Fig. 5.9*). Furthermore, the researcher (me) mediated the relationship between scientist and graph rather than by the specific scientific community in which the scientists are members (on their own accounts).

In this situation, Ted (like the other scientists) engaged in a task in which the central elements are not from his everyday (work) life. Although his work with graphs on a regular basis provides him a familiarity with general graphical conventions (rules) in the scientific community, Ted was unfamiliar with the particular graphical relationships and with the standard referents of this graph within the ecology community. But because Ted wanted to help me, he did what he could to produce a reasonable interpretation. In the process, he had to find possible values for R, S, and c to satisfy the relation $R = f_r(S,c)$. Because different sets of values can satisfy the relation, Ted had to bootstrap his interpretation, using what is available as resources and constraints.

Ted drew on everyday and mathematical resources and tools to get the process started. He generated 'limits in the environment', 'competition', 'trapping', 'fishing', and so forth as situations that have a relation to the particular shape of the graphical sign. He also generated 'nonlinear systems', 'chaotic be-

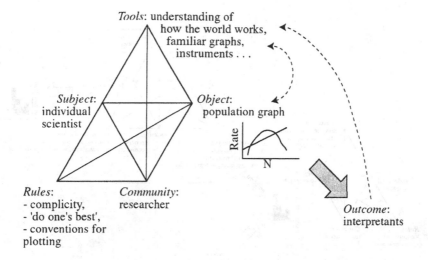

Figure 5.9. The activity of reading unfamiliar graph framed in terms of activity theory.

havior', and 'family of curves' as interpretive resources from his work of modeling natural systems. These resources were therefore among the *tools* (*Fig. 5.9*) that mediated Ted's relationship to the graph. Among these tools are also the language taken-as-shared by Ted and I. That is, for Ted to draw on particular tools, he had to work under the assumption that these tools were also familiar to me. At the same time, scientists may actually interrogate their understanding during the activity itself by using a graph; double arrows are used to indicate that tools and object may actually be exchanged in the overall activity of interpretation. Furthermore, each interpretant produced can itself become part of the tools that scientists subsequently referred to and used to mediate subsequent (and changing) relations between scientist and the aspects of the graph that were salient to them.

We therefore read the transcript of Ted's reading as the protocol of an individual engaged in this bootstrapping process. This process involves entities that he assumes to be signs without his knowing what they signify. This bootstrapping process is a tentative and sometimes hazardous (because of the arbitrary nature of signs) tracing of a system of signification rules by means of which sign-reference relations are established. The protocols are therefore full of the mumbles and stumbles of real-time activity in an unfamiliar environment.

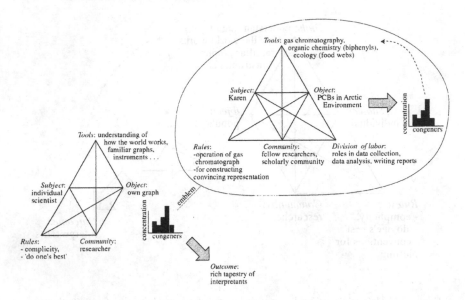

Figure 5.10. Activity system when scientists talked about their own graphs. Here, the graph functioned as an emblem, indexing the familiar lifeworld of the scientist, which was articulated in the setting. This situation is therefore characterized by very different entities (rules, tools, community) and mediated relations than in the other situation.

5.4.2. Explaining Work

When the scientists presented, explained, and elaborated the graphs, much of what they said consisted of thick descriptions of the contexts to which their graphs referred and the phenomena for which the graphs stood. I exemplified this situation in the case of Glenn. His work on PCBs and his knowledge of electrical generators and industrial practices that lead to PCB use in their production, as well as the knowledge of physical, organic-chemical, geographic, and ecological characteristics of the research problem provided Glenn with a rich tapestry of an experiential and embodied understanding. Even if unarticulated, this experiential and embodied knowing constituted a background that is the source of meaning against which signs, objects, events, and activities become intelligible. The graph(s) that Glenn talked me through constituted emblems that have an existential, indexical relation with the familiar context in which the graph was generated (*Fig. 5.10*).

Glenn had intimate understandings of a variety of domains. Through his work, he is familiar with how particular events and conditions will or may modify the traces a PCB mixture leaves as it moves through the environment. It borders on self-evidence to say that Glenn is very familiar with all aspects of his everyday work-related world. His understanding of this work world had arisen from the dialectic of contradictions, when things did not seem to make sense or when graphs derived from evidently different sources are nearly indistinguishable. Each aspect of the graph (emblem) has its place in a network of significant events and practices, characteristics of the recording device (order of chemicals from gas chromatography column), ways trophic levels affect the PCB fingerprints, and the relationship of a fingerprint detected in an organism to the fingerprint of the source spill. Therefore, when Glenn talked about his graph, I witnessed an individual operating in his everyday context. Signs functioned against a background that the signs presupposed. In such situations, signs cannot be understood as mere relations; rather, signs and their signification are an integral part of the context, and they can be signs only for those who dwell in this context. In fact, signs are no longer apparent individually but holistically and in an unarticulated fashion and even in transparent ways. In the context of tools, 'transparency' means that a tool is not being consciously attended to but simply taken for granted. It does no longer take attentional resources much as we do not have to attend to the stability of the floor when we walk into a building. That is, the meaningfulness of graphs arises out of multifarious connections within a network of familiar and transparent sign relations.

Glenn's relation to the chart had become so transparent that he was no longer concerned with the material details of the graphical signs themselves, but his reading leapt beyond to the world he knows so well. The reverse movement from the world he knows to the expressive domain also happened continuously. Provided any situation description, Glenn could project without hesitation what the graph would look like. Through their work, scientists are familiar not only with the graphs as entities that signify but also (and especially) with the particulars of the setting that led to the graphs. Through their work, many if not most of the practically possible sets of values satisfying the relation $R = f_r(S,c)$ have become so familiar that they no longer distinguished between signs and what they stand for. Thus, scientists pointed to the graph and made statements such as 'this is a clogged pipe' or 'this is a blue cell' although what could actually be seen was a line graph. That is, scientists' activities were situated in the sense that they dealt with familiar entities including signs, on the one hand, and objects and events to which these signs referred, on the other, and the conventions that regulate sign use.

Although such questions and elaborations did not elucidate the graphs qua representations, they did serve to elaborate the referent domain. They had become pieces of equipment that were 'ready-to-hand' and scientists used them transparently. As equipment, the signs pointed out the context of practical activity that the scientists shared with others. That is, through questions and elaborations there was considerable intersection between the scientific expert on the graph and the persons in the audience. They then discussed these questions rather than the questions of *S-R* mapping or the nature of *S*. I learned about the understandings my participants had developed in the respective referent domains from these interactions.

5.4.3. Pervasiveness of Ongoing Concerns

The graph-related interpretation practices documented here are far more context dependent than it is theorized in existing research. I was able to discern that some of the dispositions characteristic of different disciplines and specialties are transported from the work to the task worlds. For example, the ongoing concerns of scientists in a domain are an important aspect that researchers need to take into account if they want to understand human activity. Concerns include common missions, interests, and fears, which together constitute the disposition that scientists take to their work. Some of the scientists' concerns leaked into their readings of the unfamiliar graphs. Differences in concerns led scientists to draw on different resources and engage in different practices. For example, the physicist Nelson initially approached the problem by engaging in the process of critiquing the 'poor' graphical representation of stable and unstable equilibria. He then used mathematical/graphical modeling via software/computer (a tool ready-to-hand for him both conceptually and physically) to re-represent the original graph in a form that to him better communicated the idea of stable and unstable equilibria. His concerns were not about the ecological 'realism' of the graph (such as animal populations that paralleled the graph) nor immediately about implications of the graph/model to conservation issues. Rather, he drew on existing mathematical resources and images of potential energy curves for a variety of systems to better represent the model. The practices he engaged included multiple transformations of inscriptions (e.g., graph and equations) by drawing on differential calculus and integral calculus as linguistic resources and a mathematical modeling program as material resource.

One of the theoretical ecologists had an academic ecology background (in contrast with many others in the field of theoretical ecology who have physics backgrounds) having started to specialize in it at the beginning of his second year of undergraduate studies, but has not conducted any field research. His ongoing concerns centered on modeling rather than on the relationship between the population graph as a model and actual populations. My analyses revealed considerable similarities between his approach to the population graph and that taken by the physicist. The theoretical ecologist viewed the graph as a model that represented a 'piece of conceptual scaffolding around which to sort of hang my thinking about [theoretical] population ecology'. Early in his discussion of the problem, he stated some generalities about the theoretical reasons for birthrates and death rates to follow the indicated patterns and how such an idealized model could be used to 'actually collect data'. He then began a critique of how the quadratic birthrate depicted in the model was actually rejected by a number of theoretical ecologists, 'a lot of people challenge it as being completely unsubstantiated empirically. There's no evidence for it at all'. Despite his statement about implicit error ('[There is] error around these relationships that I don't need to seem to be aware of. At least, I assume it's there'.), he concluded that for conservation of such a species its population would need to be kept above the unstable equilibrium point so that the species is kept extant. From the beginning, he treated the model as deterministic and not probabilistic; as such, if the population dropped below the threshold indicated by the first intersection, the population would go extinct 'without some form of intervention'. Intervention, in his usage, meant some form of human intrusion or effect on the system.

The different concerns were also related to different ways of interpreting and using the signs provided in the graphs. Thus, different scientists used 'N' that appeared in the population graph task in different ways (Table 3.2, p. 97). The theoretical ecologist interpreted 'N' as population density but, when questioned, rejected any possible behavioral response by the animals to low density such as through their congregating in a smaller geographic area (thus raising their density above the critical threshold but not altering their population size). He rejected the argument that animals responded to lower density with changes in behavior ('well, that's not in the model'). He was therefore arguably viewing the model as a static rather than dynamic representation of a population. His concerns were about the validity of the given model (as represented by the mathematical functions) and its utility as a foundation for other models, particularly multi-species models, and any consideration of influencing factors drawn from field ecology appeared to be of secondary importance.

An experimental ecologist, who conducts field work with non-commercial species, was concerned with the 'realism' of the model—which he discussed by drawing comparisons to animal populations which possibly fit the scenario shown in the graph—and the management of animal species. He talked about the mathematical function in the problem—as were those of the physicist and the theoretical ecologist—but instead involved graphical representation of the model and comparative species. In distinct contrast to the physicist and the theoretical ecologist, his discussion of the question scenario was often tied to 'real' species (fishes, birds, and so on). However, despite contextualizing the graph in this manner, his discussion frequently referred to conservation *through* human management of the species. In that sense, his interpretation of the word conservation could be read to mean 'maximally, sustainably exploit'. His treatment of the graph was a deterministic one, in that he viewed it as a tool used to predict what would happen to a population at a certain size. He defined 'N' as 'population size', underscoring his interpretation of the graph as deterministic—unless some intervention occurred, either accidental or human, the population of animals could not grow once the critical lower threshold of N was passed. He viewed the lines in the graph as being misrepresentative because they did not incorporate natural variations as these normally find their expression in real data. The inclusion of 'noise' ('stuff that we don't understand') in the system would change the interpretations that could be drawn. To him, the idea of stable and unstable equilibrium points were misleadingly represented and that for any population, if such theoretical points did exist, that they would more likely be found in a wide range rather than at a single point. He therefore suggested that the theoretical unstable point 'would be very close to zero'.

One ecologist had conducted fieldwork with previously harvested, now non-commercial and/or endangered species. He initially viewed the population graph problem as a theoretical construct that could be useful tool for thinking about natural populations. His ongoing concern was one of conservation with little or no human interference not as a term meant as management for harvesting. As other field ecologists, he viewed the model as the graph itself, not as the juxtaposition of algorithms; as such he considered the problem graphically and not mathematically. He interpreted 'N' as 'population density' rather than population size. He argued that animals 'behaved' and as such could re-adjust their distribution if population size approached the critical threshold. He viewed the threshold represented by the unstable equilibrium differently than the other scientists in that it was less deterministic of population extinction and instead represented a warning about when to start worrying about a species. During his discussions about the ecology problem he drew on examples of animal populations with which he was familiar. He appeared to articulate a dynamic view of the model in that he suspected that the N at which the equilibrium points were found would shift somewhat with an interaction of population size and density. As such, he treated the graph as a 'snapshot' of what happens at some moment, not as a static image that will hold into the future. His understanding and use of the phrase 'conservation of such a species' (in the question) was not founded in 'conservation as management of an exploited species' but represented a desire for understanding the natural population cycling and behavioral responses of a specific organism.

These examples show how ongoing concerns that arise in their daily work frame the relationship between scientists and graphs. The ongoing concerns, although they are not really part of 'ability' or 'competence' nevertheless mediate scientists' perception of graphs, tune them in different ways to salient features, and color the interpretants that they produce. But because many scientists were unfamiliar with the graphs, the full impact of these ongoing concerns did not come to bear on their interpretations. In the context of their daily work, however, these concerns constituted an important aspect that shaped graph use. In the following, second part of the book, I present three case studies of graphs and graphing competence at work.

PART II

GRAPHING IN THE WILD

SIX

Reading Graphs

Transparent Use of Graphs in Everyday Activity

This chapter focuses on graphs as they are used and talked about in everyday settings. Here, graphs become apparently fused to the things or contexts that they describe. Graphs are no longer interpreted but are read in the way one may read a newspaper article about a familiar feature in the community. Graphs become transparent; they are present to hand in the same way as familiar words and physical tools. The present case study focuses on the graphs used by a water technician on the farm where she works. As part of her and her employer's engagement with an environmental group, Karen gets to interact over and about the graphs with a variety of people in the community. The chapter highlights contextual features that allow graphs to be transparent in use. In such situations, Karen does not 'interpret' a graph but 'reads' it. Her reading leaps beyond the material sign to a world that lies beyond.

6.1. FROM CAPTIVITY TO THE WILD

In the first part of this book, I present the results of my study of 'graphing in captivity'. The result show that even when presented with graphs from introductory courses in their own field, only some scientists compare favorably to the image of the general expert that they are often portrayed to be. The activity-theoretic analysis in the previous chapter articulates the differences between engaging with unfamiliar graphs and talking about graphs that issue from scientists' own work. When they talked about their graphs, scientists were not only highly competent, but also revealed the vast extent of the background understanding against which their explanations became intelligible. To better understand the competencies observed when scientists talked about their own graphs, I conducted three ethnographic studies of environmental activism, ecology, and experimental biology in order to observe and subsequently theorize 'graphing in the wild'. In this chapter, I present data derived from my work among environmentalists, and in particular of Karen, a water technician.

The data derive from a three-year study on the representation practices among environmental activists. This environmental activist group focuses its activities on the Henderson Creek watershed and the community of Oceanside.[1] Its primary intent was to promote a change in the community with respect to those practices that affected the environmental health of the watershed. For example, in Oceanside, there exists a small industrial park. Some of the companies in this park release their wastewater into 'stinky ditch' (as it is affectionately called), which itself drains into Henderson Creek. Furthermore, because of extensive farming in the valley, many agricultural practices directly affect the creek. This includes diminishing its levels by drawing water for irrigation purposes, contaminating the creek water chemically through the use of fertilizers, contaminating the water biologically through cattle grazing immediately next to the creek, straightening its course, removing shade-providing trees, and many other ways.

One of these activists was Karen, a water technician employed by a local farmer with funds from a government grant. The farmer had environmental concerns related to the water resources in the Henderson Creek watershed where his farm is located, and especially in regards to the creek from which he obtains the water for irrigating his fields. Monitoring the water levels, building riffles to increase oxygenation levels, replanting riparian areas with shade-providing plants and so on was part of Karen's daily work. She operated several stations (one of them on the farm) that continuously monitor the water level in the creek by means of a floating device whose up-and-down movements were transmitted to a pen chart recorder (*Fig. 6.1*).

In the course of my ethnographic study, I videotaped Karen in four instances where she explained the water level graphs and talked about the graphs. In addition, I spent considerable time with Karen walking in and alongside the creek, studying the habitats from the mouth of the creek to its beginnings. We spent time talking about the watershed while standing far above the creek overlooking the farm or while walking the fields surrounding the farm. I also spent considerable time working with seventh-grade students in the creek, teaching them how to collect data, make observations, and how to understand the watershed as an ecosystem. Karen occasionally helped me in this activity, further allowing me to record her talk, and therefore to learn what she knew about the creek and watershed. Our conversations were recorded in the form of videotapes, audiotapes, and fieldnotes. Some materials on which I draw derive from grant proposals written by the activists to garner funds for their activities. I fur-

Figure 6.1. Henderson Creek has been turned into a ditch so that it drains the surrounding fields, which are situated in an area that used to be a bog. To the front and left, a water monitoring station operated as part of her work for the environmentalists.

ther draw on understandings deriving from my ethnographic work among the activists.

6.2. PRACTICAL COMPETENCE IN EVERYDAY SITUATIONS

Over the past decade, it has almost become a truism to say that the mathematics in school and the mathematics in everyday settings are considerably different, and often incommensurable. However, it is much less evident what competencies that people bring to (implicit or explicit) mathematical tasks in everyday situations. Contrary to the research in high schools that appears to indicate a lack of competence in the use and interpretation of representations and particularly of graphs, I found the activists and other community members associated with them quite competent in this respect. Even without scientific training, many people come to use graphs in the same transparent ways in which they use physical tools to work in the creek or words for talking about the weather. Having worked on the farm, in the watershed (*Fig. 6.2*), and with the environmental group for four years had allowed Karen to develop an intimate understanding of the geographical area, climate, topology, geology, history, agriculture, politics, and so on. She also had developed close working relationships not only on the farm but also with the technician from Environment Canada that processes the water data she produced and also came to the farm several times each year to take measurements of the creek profile. When Karen talked about the water level graphs just as they came off her monitoring station or when the graphs for an entire year were displayed during the environmental group's annual open-house event, all this understanding was latently present, providing the background that enabled her competencies. Whenever required in some situation, Karen articulated and elaborated the existential understanding that she had developed as part of her daily work in the valley. Because the graphs were always associated with social and material relations, it makes little sense attributing graphing competence to mental operations. Rather, graph use arose from the dialectical relations of individual agency and social and material structures salient in the situation.

There are moments when Karen's transparent use of the water level graphs was quite apparent. In one situation, she pointed to a vertical spike in the graph and said, 'This is a clogged pipe'; in another instance she pointed to a peak and said, 'This [water] comes from the north arm'. In both instances, she pointed to a structural feature of the water-level graph but talked about things in the world. In the first instance, the 'pipe' is part of the monitoring station, taking the water away from a container below the station and to which it is connected by means of a vertical pipe. In the second instance, Karen referred to the fact that Henderson Creek is the result of the confluence of two creeks, the smaller one of which she calls 'north arm'. For geographical reasons—steeper slopes, less water-retaining areas—the north arm drains rainwater away faster than the second tributary, which is associated with a small peak that precedes a larger peak. In both instances, therefore, Karen's reading leapt beyond the material basis of the sign, the ink trace on the paper, into a world that lies beyond.

The referential function of the graph as sign is not one of its properties. Rather, the nature of the graph arises from the questions of why and how it is used. At best, a graph (sign) is useful or not useful and its properties are tied to the ongoing concerns. These ongoing concerns, as we have seen in Chapter 5, are of central importance in understanding how a scientist is tuned during graph reading and interpretation. In order to understand the practice of graphing, we have to highlight (a) *what for a* graph is employed (which pertains to its service function) and (b) *wherefore* it is used, pertaining to its usability. These charac-

Figure 6.2. View of the Henderson Creek watershed from east to west toward the inlet, which lies between the peninsula (foreground) and the mountains in the background.

teristics have to prefigure the question of the referential function of graphs in everyday use. The graph is relevant together with the world—Henderson Creek watershed, farm, Oceanside community, and so on—which Karen inhabits *together with* other people and objects that surround them. The referential function of a graph arises out of the relation *together with* people and things. In this, the graphs are not just handy as other useful things but as signs, they make the world (here Henderson Creek water levels and causes for its changes) explicitly accessible for circumspect inquiry. Graphs are therefore not meaningful in themselves or acquire meaning for this or that reason. Rather, graphs are part of and accrue to an existing meaningful totality. Individual's competent reading always presupposes knowledge of the things and places that the graphs are about, how such places work, what kinds of things can happen ('clogged pipe', closing of dam), and what actually happened in the context of the graphical feature at hand. The opposite then constitutes a case where the graphs function merely as other things that surround us of the kind that we perceive as mere things that we do not know or understand because they are of no relevance to our lives. Thus the significance of graphs is always closely related to their relevance, which, in turn, is always discovered on the basis of a previously disclosed relevant existential understanding of the world.

At the moment when the following episode was recorded, Karen and a group of interested environmentalists and teachers stood next to the small water monitoring station situated on the bank of Henderson Creek about three meters above the current water level. From the pen chart recorder that is visible behind the open door, Karen had pulled about one meter of chart with a graph representing the water level of the creek over the past few days. While pointing to different places on the chart, Karen read the chart. The following excerpt begins after she had provided an indication of the scale and had explained that to get the volume, the chart has to be read in terms of a calibration curve (see below) that she has not ready to hand—which nevertheless constitutes part of the background against which the current reading was made. ('[*Gesture*]' marks where text and gesture depicted in the video-based drawing coincide.)

Karen: In the summer, we're down here (*Gesture*).

We're about, the first square (*Gesture*) doesn't mean anything, that's just a bit of leeway, we're at about two squares up. We have about twelve liters per second.

In a really, really incredible rain event, we get right to the top here (*Gesture*) and it's five thousand liters per second. [iv] So, we have a creek that really increases its volume, as the smaller creeks really fluctuate. [v] And then we have all of our storm drains in the watershed funnel into the system. [vi] So, on top of the rain fall event, from the areas along side the creek and in the creek itself, we get all this water augmenting this flow. [vii] So, this peak [points], which maybe hypothetically, twenty years ago would have come up to three thousand liters per second (*Gesture*).

Now after a rainfall event, it's coming up (*Gesture*) to five thousand liters per second because we're getting the run-off from all the pavements, rooftops,

Karen: roadways, and we're getting all the contaminants with that as
            well. If that low (*Gesture*) in the

            summer (*Gesture*) gets below

            that first square (*Gesture*), we dry up, farming cannot occur and
            a lot of farmers depend on the creek.

This episode begins with Karen's projection of a state in the world, 'water level in the summer', onto the chart. That is here, beginning with her understanding of the world, and from past experience with the chart, she enacted a reverse movement from the referent to the sign, the graph. She then elaborated the previous statements by translating 'down here' into a measure of flow (volume/time). In the next utterance, we observe again a movement from the referent domain (the experientially real 'incredible rain event') to the sign domain and into another sign domain. By indicating where she expected the curve to be on the paper, the amount of flow in the creek thereby came to be articulated. In the next three utterances, Karen provided situation descriptions from the referent domain. She talked about the creek and its smaller, fluctuating tributaries, storm drains that funnel into the watershed, and the water from the areas alongside the creek all of which increase the water flow to the earlier described levels. That is, she did not merely read or project a graph, but had an intimate knowledge of the situation and the conditions that bring the water to the high levels of 5,000 liters/second. Furthermore, Karen is also familiar with the historical situation. The same rain fall event which now produces 5,000 liters/second would have produced only 3,000 liters/second some twenty years ago which she attributed largely to specific characteristics of the landscape (pavement, roof tops, roadways) with which she is very familiar. She then returned to the summer situation and provided both situation description ('dry up' 'farming cannot occur') and a corresponding state in the sign domain ('below the first square').

Karen moved effortlessly back and forth from the chart, the sign domain, to the local setting, the referent domain. She read the graph transparently, much like we read the newspaper: a simple glance at the material basis of the text suffices to instigate an interpretant that elaborates a situation description from which the visitors can learn something about the watershed. However, the process is even more complex than it appears at first. Because the cross-section of the creek is not square, the water level curve does not directly indicate the amount of water in the creek. The relationship between the amount of water running past the station and the height of the curve is strongly non-linear. Thus, when Karen reads the graph in terms of a specific quantity, she does this based in terms of a calibration curve established by a government agency. Thus, in her reading of the graph, not only the pen chart record but also the calibration curve were transparent.

In the past, much research on mathematical knowing has focused on models of mind irrespective of context. This approach is being questioned, for it neglects the contexts that enable cognition to exist in the first place. Here, context is viewed as consisting of the historically constituted concrete relations within and between situations. Knowledge is not an entity that can be acquired but rather knowing is equivalent to acting in the world; knowing is a process rather than a state. Knowing arises from historically constituted (concrete) relations within and between sociomaterial situations and involves the individual body as much as the individual mind. I choose to link the social and material rather than the social and historical, or the social and cultural because I view all social practices as historically contingent and embedded in some culture. However, the material aspects of cognition are seldom enough emphasized or completely left out of the modeling of knowing and learning. The body-mind ensemble cannot be dissociated: the sociomaterial subject is part of and formed by the sociomaterial world. Graph reading therefore has to be understood as involving a relational dynamics between sign and practice that is 'not created inter-subjectively in any simple sense, but are produced in relation to aspects of social practice which are culturally and historically specific'.[2]

6.3. CREEK, COMMUNITY, AND HISTORY

Karen and the graph she read to the visitors on the farm and at the open-house event organized by the environmental group in the community hall of Oceanside do not exist in a vacuum. Rather, Henderson Creek and the community located within its watershed boundaries have their own political, social, and economic histories. Commuting into the community and working with the people that inhabit the watershed provides Karen with many opportunities to find out about past events and the historical evolution of water-related contexts. It becomes part of her lived world, in which she gets around without attending to her every move. Sometimes, such moves are virtual travels backward in time to a world that once existed, and which still exists in the stories of the seniors and First Nation elders. Thus, although she had been around for only four years, and although the water-level monitoring station has existed for about the same time, she can make then-and-now comparisons of the amount of water coming through the creek after a specific rainfall event.

Karen: So, this peak (*Gesture*) maybe hypothetically, twenty years ago would have come up to three thousand

liters per second (*Gesture*), now after a rain fall event, it's coming up to five thousand liters per second

(*Gesture*) because we're getting the run-off from all the pavements, rooftops, roadways, and we're getting all the contaminants with that as well.

For Karen, the graph as it comes off the pen chart recorder does not exist in and of itself, disconnected from anything else she might experience and in the way unfamiliar graphs appear to participants 'in captivity'. Rather, for Karen, the graph is part of an activity system that, though characterized by the cyclical time structures of annual changes in nature and farming practices, irreversibly evolves in an expanding cycle of development. Four years before this reading, neither Karen nor the water level recording device existed in the community. Now, having worked in the community for some time, Karen not only read the graph as a historical record of what has happened during any of the four years of its operation but also projected what she and her audience might have seen if the device had existed some time back. Karen has become so familiar with the community both through personal (see below) and vicarious experiences that her projections of the differences between then and now are quite reasonable. They are particularly reasonable given her own observations of water levels that went beyond the calibrated area on her chart and which she estimated to correspond to 7,000 and 8,000 liters per second in a creek that in some summers only carries 10 liters/second. In part, she accounts for these increases in terms of 'the run-off from all the pavements, rooftops, roadways'

Yeah, it actually drains Johnny Park Estate we get lots of storm water from there that is actually in another watershed but it's being funneled this way, not just the way it used to. So, I don't know when that estate went in but- This (*Points to almost horizontal part of graph*) is our baseline that's built into it so. If, say, the Estate expanded and we got more storm water, that (*Points to baseline*) would be higher.

Karen attributed the increases at least in part to the estates that sit on the mountainside bridging two watersheds but funneling all its water in to Henderson Creek. Furthermore, if there were future increases in the size of these estates, the necessary bylaw changes for which are currently at debate in the community, even more water would arrive quickly and in large amounts at her monitoring station. That is, Karen's reading of the graph was mediated by the historical evolution of the watershed and the communities that exist in it. It is further situated with respect to other, larger watersheds and with respect to the season (winter) when such rainfall events occur contrasting (extremely) low water levels in summer when the farms need it most for irrigating their fields. Furthermore, elsewhere in the transcript Karen situated the increase from 3,000 to the projected 5,000 liters per second flow rate, on the one hand, relative to small and large watersheds, on the other hand. Smaller watersheds suffer from flash floods, especially with the large number of impervious surfaces and straightening of the stream that Henderson Creek and the community as a whole have experienced in the past.

Finally, the graph does not just show 3,000 and 5,000 liters per second or the estimated 7,000 and 8,000 liters per second but these values exist in relation to the physical characteristics of the watershed, the grass cover and pavements. With respect to the possible expansion of Johnny Park Estates in particular and run off in the entire community more generally, Karen also knows that if more than 12 percent of a watershed are covered by impervious surfaces (e.g., pavement), its health will be seriously affected. In this watershed, the impervious surfaces have increased tremendously over the past two decades.

The mediating relation of the community and its history is further highlighted in the following excerpt. As a member of the environmental activist group, Karen also accesses the historical records that speak of plenty in terms of water resources (e.g., people used to canoe in the creek) and trout sizes and quantities no longer heard of.

> And there's some talk, like the First Nations people are very interested in becoming involved in the creek. They would like to replant some coho in the creek. And that was done in the early eighties, actually. And talking to other fisheries biologists the coho and cutthroat can very well cohabitate in the creek provided that there's enough food supply. That's what we have to know about,

Figure 6.3. Only small stretches of Henderson Creek resemble the creek known to the local First Nations people and the settlers in the mid-1800s.

> if there is enough or if there isn't how you can increase that. And then if everyone is okay with introducing some coho, we can do so in the next step.

In this community, First Nations people were thinking about introducing coho salmon, and residents (and activists) were interested in a revitalization of the creek and its resident stock of brook trout. In the stories told by the elders from the local band—as well in the preserved notes of a priest during the latter part of the nineteenth century—there are trout as large as twenty inches in size and so plentiful that a morning's catch provided sufficient for a family. At that time, the creek looked very different; only minute stretches along the creek resemble what it once looked like (*Fig. 6.3*). Now summer levels are so low that it is difficult to imagine even small fish to survive (more on this see next section). These historical data are further enhanced by the stories of individuals who recall canoeing in the creek during the summer months—further evidence for the existence of sufficient water in the hot and dry summer months for trout to survive.

Here, Karen's work with graph inscribes itself in a situation that has historical roots in the 1940s. At that time, her farm as well as eleven others received their water licenses, although, as she emphatically pointed out, there was little known about the ecological complexity of watersheds, their watercourses, and the underground aquifers that feed them. Furthermore, what happens at her monitoring station is also linked to and interacts with what happens at other water monitoring stations. Thus, Karen's reading of the graph accrues to an already meaningful ensemble of mediated relation to other currently existing practices and to the watering and communal water distribution practices from which they have evolved. Karen's activity inscribes itself in a historical context. The years she has spent as a water technician in the watershed have given her many opportunities to talk with farmers (other than her employer), First Nations people, and other local residents who have been living next to the creek for more than half a century.

From an activity-theoretic perspective, Karen is the subject in a system characterized by the creek as its primary and activity-constituting object; the available tools including such entities as the water level graph, ecology discourse, and historical narratives mediate this relation. The previous semiotic analyses already brought to the fore that there is not merely a one-way inferential relation from one type of segmentations of matter (here signs, graphs) to another segmentation of matter (here natural objects). Rather, in the everyday work of graph users, signs and referents, tools and objects are continuously exchanged (*Fig. 6.4.a*). The present and subsequent analyses shed further light on the interchangeability of objects and the tools that mediate subject-object relations and that there are always multiple and not always compatible and interchangeable tools that mediate the relation to the other entity (*Fig. 6.4.b*). It is therefore more useful to think about networks of entities, any one of which can take the role of the object, all others being available as potential resources for mediating the primary relation. Meaning exists for the individual subject in so far as there is a dense network that, in a holistic manner, each entity can be accrued to a web of signification with a sufficiently dense neighborhood of entities.

Some readers may think that I flattened the landscape of cognition by turning everything into a semiotic entity, much as postmodern scholars (and ethnographers) have turned everything into text. On the one hand, these readers are certainly right—but I see no way around considering cognition in relation to entities that are 'signi-ficant'. By their very nature, significant entities imply the making (Lat. *facere*, to make) of segmentations in the continuum that can refer to something else and therefore are signs (Lat. *signum*, sign). On the other hand, these entities do not all have the same quality, that is, they are not 'just text (words)'. Rather, as I will show in the next section, in Karen's experience, Henderson Creek is a quite palpable entity associated with physical, material, and emotive qualities that other entities (e.g., graph) by themselves to not have. The graphs, as material things, exist in the context of a totality of useful things

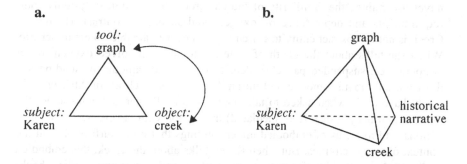

Figure 6.4. a. Karen's knowledge is not simply characterized by a tool-mediated subject-object relation. b. Rather, graph, creek, and historical narratives are exchangeable objects of activity, the other entities serving as resources and tools mediating the primary relation.

with which they stand in a variety of relations. As such, they have the same qualities of being as other familiar and unfamiliar things that people notice in their surrounding world.

As will become clear throughout the chapter, other entities also mediate Karen's relation to the creek and graph. Thus, we can imagine an expanded version of the mediating relations (*Fig. 6.4.b*) in which there is a network rather than a triangle of entities. This approach provides us with a hint to the nature of 'meaning' of a graph. It consists in the enacted relations of any aspect of or the entire network of entities—where I stress enactment, which can be in the form of talking and gesturing, imagining, or remembering.

6.4. GRAPHS AND THE CONCRETE LIVED-IN WORLD

In much of the research on graphing, the things that are said to correspond to the graphs at hand are more or less abstract entities, only barely described and, especially among students, hardly ever familiar. (This was the case for a number of scientist participants, especially in the context of the isographs.) Thus, in Chapter 4, I discussed one research project the children are shown a graph containing two lines one pertaining to the amount of shrimp in a river the other to the amount of oxygen, both amounts being plotted as a function of location; there was also a factory. The children are to infer that the factory decreases the amount of oxygen available to shrimp, thereby decreasing the amount of shrimp. This effect is larger near the factory than further down river. The researcher never ascertained the familiarity of the children with biological systems, their requirements and dependence on oxygen, and so on. In contrast, Henderson Creek is not an abstract entity to Karen but is a every palpable object in her life. When she talks about the depth of water, amount of dissolved oxygen, water temperatures, suspended particles, density of trout, or amount of shading she does not talk about removed and unfamiliar entities, such as an African child might have to do when asked to talk about 'snow'. Rather, she talks about the outcomes of her physical (tool-mediated) interaction with the creek. These inter-actions lead to forms of embodied understandings that often surface only in the context of these activities. But when Karen talks about the creek, this embodied understanding provides the sometimes-implicit and sometimes-explicit back-ground against which the graph becomes intelligible.

6.4.1. Embodying the Recording Device

Knowing the source of the data and the instruments by means of which data were collected was an important aspect of scientists' determining the level of competency. Visual aspects of the instruments and hand movements involved during data collection were subsequently expressed in the form of iconic ges-tures that were employed together with language. Karen made available her un-derstanding not only in verbal form but also in the form of iconic representations available in her gestures (*Fig. 6.5*). For example, in the transcripts above, Karen can be seen to move her finger across the paper. This already constitutes an iconic representation of the pen chart inscribing the line that is the focus of the present interaction.

> So, the way this works is there's pipes going across the creek and the water comes into the (stilling?) well area and this is, the fluctuations in that water level drive this wheel here and then this pen works.

Karen knows the instrument that records the water level graphs and in particular the mechanism by means of which the curve is being taken. She knows the in-strument so well that her body participates in communicating the functioning of the device. As Karen talked, her hand-finger ensemble (sequence in *Fig. 6.5*) moved along a trajectory similar to that of the pen. The trajectory of her finger, therefore, stood in an iconic relation to the trajectory of the pen. Interrupted by the gesture that indicated the distance on the paper that amounts to one day, her finger moved, similar to the pen, across the paper. In the same way, the to-and-fro movement of the pen was embodied in her hand movement up and down along the temporal dimension of the chart. This became further evident when Karen talked about this dimension of her graph during one open-house event

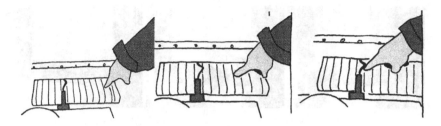

Figure 6.5. Karen's finger moved across the paper from right, corresponding to higher water levels, to left, corresponding to lower water levels. Her finger made the same movement as the pen that recorded the trace.

(*Fig 6.6*). Here, Karen's hand moved from right to left, along the horizontal direction of the paper, while uttering 'hours go this way'. Her hand follows the direction of the paper in the pen chart recorder, taking the same trajectory. The gesture therefore also stands in an iconic relationship to the paper as it moves in time. Time is not just an abstract label on the axis, but is something that continuously unfolds and is indexed by the turning wheels and moving paper. Karen has a very embodied understanding of time as it pertains to her graphs. Interestingly enough, indicating the time through a bodily movement may depend on the particulars of the circumstance, for in another situation, Karen's gesture described a trajectory in the opposite direction. Thus, in one situation her gesture described the visual perception associated with the formation of the graph as it unfolds in real time on paper under the pen. In the other situation, Karen's gesture in the opposite direction was similar to way in which the movement of the paper is perceived. In both situations, time was something embodied in the (apparent) motion of the paper or pen, and embodied in her gesture. Karen does not merely know the relationship between some sign (point on graph) and the amount of water. Her gestures embody the working of the pen and the movement of the paper, which allows the trace to unfold horizontally and thereby representing time.

There are other embodied movements that stand in an iconic relation with features of the world from which the graph derives. For example, as Karen talked about the well in which the water fluctuations drive a floater, her arms formed a circle (*Fig. 6.7*), she bent her knees, so that, as a whole, she described a drum-like object. Right after, Karen enacts the graph recording with her entire body, hand and body standing for (i.e., representing) the pen and floating device, respectively. She began to bend her knees so that her body moved downward

Figure 6.6. As Karen explained how to read the horizontal axis, her hand moved in the direction that the paper is pulled through the pen chart recorder.

like a piston 'inside the equipment' in 'the upright column,' while her right hand follows in and amplifies the downward and upward movement of her body. But in this, Karen's sensori-motor actions are not just indexical to the floater, but to the fluctuations of the water level (the *object* in *Fig. 6.4.*) throughout the year. Karen performs the recording: her right hand embodies the pen movement, her entire body enacts the up and down of the floater in the 'stilling well', and her right hand stands in an iconic relation to the recording device.

In these examples, Karen did not just talk about the graph but moves (parts of) her body through trajectories that stand in iconic relations to the graph, paper, recording device, and monitoring station. The movements of these entities, which she frequently observed over the past four years, exist as sensori-motor representations in Karen's experience, and were available in public to her listeners. The movements of pen and paper, which exist in the material trace of the graph, were literally embodied. The meaning of 'time goes this way' and Karen's up and down movement with her body, followed by her hand simultaneously exists in two ways. First, it built on the sensori-motor action involving Karen's finger (hand) over (in front of) the plotting paper and the graphical space it defines. Second, Karen drew on the symbolic associations with the marks and lines on the paper of the conventional graphical signs. These indexical gestures are sensori-motor and symbolic re-enactments of material events and therefore an integral part of the discourse through which Karen articulates her understanding. The sensori-motor processes therefore constitute an important aspect of collective processes of meaning making, and the witnessing of the other's subjective understandings. The graphs constitute subjects and objects in referential ways as simultaneous, co-existing participants in the described events.

In contrast to most classical research on graphing, where the 'subject' does not know the data source or manner of data acquisition, Karen as the scientists

Figure 6.7. 'This is inside the equipment the upright column'

in the subsequent chapters, is intimately familiar with the recording device. Water level and time are not just abstract coordinates but deeply experienced phenomenon-related images that have become integrated into Karen's physical comportment as she talks about the graphs.

6.4.2 Henderson Creek as Lived-in World

In ways similar to the recording instrument, Henderson Creek is not some abstract, non-descriptive creek but an entity and place with which Karen has a very concrete relation. Karen is very familiar with the creek. We have walked along its bed numerous times, sometimes including other individuals interested in the restoration of Henderson Creek or consultants. Karen has constructed riffles from local rocks to improve the oxygenation rates, and planted trees that eventually would shade the creek to provide habitat for trout. She has used a dissolved-oxygen meter to determine the oxygenation of the creek water both above and below the riffles she single-handedly built on the farm property, or in collaboration with the other activists in other parts of the creek. She has also taught together with me seventh-grade children, especially how to collect data, the relevance of the creek to farmers such as her employers, and the current environmental and human pressures on creek health. When she talks to the students about fish in the creek, these are not fish in general but rather these are fish that she catches in a trap, holds in her hands, temporarily brings to the surface by stunning them with electroshocks, makes rise using paper spitballs. She knows how the fish feel in her hands, how they react to her spitballs, and they change their reactions when she throws too many. This embodied understanding underlies the following explanations that she gave a group of teachers visiting her on the farm.

> This summer was pretty good, about twenty liters a second, but it can go down
> to about eleven, and that's, for fish to survive in this creek, we need about

twenty. So, when we get low, the fish will find a pool somewhere to hide and if it weren't for these deep areas, in little pockets, throughout watershed, they are sort of hiding and laying low until volumes have come up.

Karen spends a lot of time in and along the creek. So when she says 'when we get low' it is not just a general statement about water levels corresponding to the curve on her graph; rather, it is a situation that she has experienced physically. In the summer, there are many places where one can cross the creek dry-footed whereas the same places will be a raging creek three or four feet deep. 'When we get low' is the moment that there is no water for placing the trap, where there is insufficient water for the fish to hide to come out after her spitballs, and where the electroshocks do not bring a single fish to the surface. 'When we get low' is a concretely lived experience first that becomes associated with specific discharge volumes only subsequently. The discharge volume graphs become associated with the understanding that she has previously developed in and of the creek. Karen also continues to measure the temperature of the creek water and knows that it gets too warm in some places, particularly those without riparian bushes and trees and therefore exposed to the sun, for trout to survive.

> Certainly, as the flow gets lower the temperature would get higher. Because they only got this much water to heat up, it's gonna all get warm throughout it. But if it's this deep and there's a pocket that is covered into the bank you can- You gonna have a nice little hiding spot. I think they need anywhere from eight to eleven parts per million of oxygen. Most fish do. We have little sticklebacks that, actually they eat them. They can survive at two parts per million oxygen completely exposed sunlight areas. So where there is a food supply, there's still no cutthroat trout because they are not gonna follow them into those areas. In terms of temperature, anything around thirteen is a really nice temperature, they will, the bigger grand-daddy-kind-a cutthroat that hang out here. In the summer, it's like twenty-seven, in this fully exposed area. But down at the bottom, they've got a seven-foot depth to hang out when the dam is in. And they got cool temperatures down at the bottom.

These explanations allow us to catch glimpses of Karen's familiarity with the creek as an ecological and a physical system in which the flow rates affect water temperature, rising temperatures cause lower dissolved-oxygen levels, and there are temperature gradients in deep water. All of these factors affect the places where cutthroat trout and sticklebacks (trout food) can live during different parts of the year and especially during the summer months. Thus, Karen's reading of the graph was not independent of her understanding of the creek, the physical characteristics of heat capacity, and ecological relationships between species that have different requirements on their physical environment. Karen's reading of the graph during the summer months existed in and as of the mediating relations in respect to the creek.

Karen knows the creek in many other respects as well. As part of her work, she actively constructs 'riffles'—rock structures designed to create turbulence in the water and thereby to increase the oxygenation levels. These are being built in Henderson Creek and in its tributaries. Karen monitors the oxygen levels throughout the creek an in particular in the vicinity of her structures both before and after she has built them. The quote shows that Karen is familiar with the covariation of water levels and water temperature. As the increase in water levels are to a great extent out of her hand, and mediated by economic dependencies in the community, one way of improving creek health is by strengthening the riparian zone and by shading it. She therefore replants the riparian zones along Henderson Creek and its tributaries, though the growth times involve a considerable delay before the impact of shade plants on the water temperature can be established. All these activities are an integral part of Karen's work on the farm and, as far as other areas of the creek are concerned, of her work in the environmental activist group. They are part of the understanding that arises from the involvement in, and familiarity with, the circumstances under which the graphs were created, an understanding that is prerequisite for a knowledgeable reading or for an interpretation. An individual can knowledgeably read or interpret a graph only when he or she has recourse to such circumstances or can appeal to knowledge about what can conceivably happen in such circumstances. For some of the scientists featured in the first part of this book, such knowledge and background understanding apparently was not available. It therefore should come as little surprise that they were not as successful as I had expected them to be prior to the investigation.

6.4.3 Correlating Events and Signs

Things that people do are closely related to their perceptions. In the present situation, such correlations between actions (adding or removing boards in the dam, starting irrigation systems) or events (rain, clogged pipe) and perceptual features of the graph are more incidental, but nevertheless lead to a deeper familiarity with and understanding of the graph that comes of the water monitoring station. It appears almost trivial to say that Karen comes to construct covariations between the actions and events and particular features of the graph, which in turn become signs for the actions and events. Just a few hours after it rains, Karen sees the pen move to a position that she knows from experience to be associated with higher water levels. Rain and a perceptible change from 'base flow' in the form of a rounded peak *always* go together. This covariation is

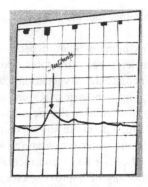

Figure 6.8. A part of the graph Karen collected two years earlier and currently displayed at the Open House of the environmental activists. To the original graph, Karen has added all the rainfall events in the form of bars. Visitors can see that each bar is associated with a peak of the 'natural-event' type.

therefore not just something that Karen detects from correlating numbers such as the amount of rainfall as measured by the and the height of water as indicated by the pen position. Nor is this a covariation that Karen has to calculate from juxtaposing her graphs and the amount of rain plotted in the form of bars on top of the graph (*Fig. 6.8*). Rather, the covariation is similar to that experienced by a child that hears a sound pattern transcribed as [ʃeizlõg] according to the International Phonetics Association, which it comes to associate with an object that we name *chaise longue*, a reclining chair with a seat that supports the sitter's outstretched legs. 'Non-natural events', as Karen called them, such as adding or removing boards in the dam at the farm, also arose from covariations. These events have the structure of an experiment in which an action, adding or removing boards, has an effect on the water level. Again, it was by means of noticing that the act of putting a dam into the creek were correlated with specific changes in the graph that occurred immediately after that Karen came to know this particular aspect of her world. She described this aspect of her work in the following way to visitors at the open-house event.

Karen: We have a license to install a board across the creek.
          To get our storage amounts we put in the first board
          (*gesture*) and the

Karen: second board (*gesture*) water and that's all we did for
           now.

Karen pointed out that the farm has a license to take water from the creek—an
environmentally precarious practice given the little water that there is during the
summer months. Drawing water from the ground water stock is even more tax-
ing on the health of the watershed. So, when the farm workers add or pull
boards, she observes a correlated response by the pen and in the graph it produces. There are other times as well when Karen and her fellow farm workers
install boards, such as during the dog trials.

> Around this time we have the dog trials that happen on the farm. And what they
> like is deep water for the dogs to swim across the creek. So, we have a dam that
> we operate and we put in two of those boards in the dam and they got their
> water level increased and it took about a day and a half before it started over-
> flowing the boards.

That is, depending on the time of the year, planks are installed or taken out. Correlated with such activities are sudden changes in the water level, pen chart
movement, and ink trace on paper. Karen reads these (abrupt) changes in the
trace not just as adding or removing planks, but in terms of particular events that
are salient on the farm. Thus, she sometimes does not even talk about the actions
of adding or taking away planks but about events such as dog trials or the beginning and ending of the permit for a dam. In May, planks are added to the dam to
make a reservoir that provides water for irrigating the surrounding fields later in
the summer. These, too, are concretely lived events in Karen's work on the
farm. Each time her coworkers do something to the dam there is an associated
event inscribed recorder as trace on the paper by the pen chart. Both natural and
'non-natural' events come to find expression in the graph so that such graphical
aspects come to be equivalent to the event itself. In the lived experience of
Karen, these two events always come together.

6.5. DIVISION OF LABOR

To understand Karen's transparent reading of the graphs at work, we need to
take into account the various social relations of which her graphs and other related ones are a constitutive aspect. I already pointed out how the actions of her
coworkers on the farm, placing planks to dam the creek, were associated with

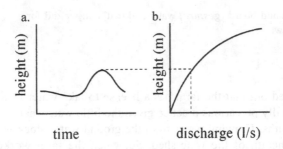

Figure 6.9. The Environment Canada technician produces a calibration curve that allows Karen to convert the water height recorded by her strip chart recorder (a) into a corresponding discharge (measured in liters of water per second).

changes in the water level and corresponding changes in the graphs that Karen recorded. The workers' actions of unclogging the outflow pipe similarly had a correlated feature on the graph. In addition to this division of labor on the farm, there were other work relations that constituted the nature of the graphs. For example, Karen collaborates with a technician from Environment Canada to produce representations that subsequently become part of the set of artifacts mediating the relation between Karen (subject) and her graphs (object). These activities have outcomes that change the relations in the mediational triangle with Karen as the subject. The graphs are themselves objects mediated by other graphs (mediating artifacts). The graph Karen explained to farm or open-house visitors exists in relation to other graphs. Furthermore, because the environmentalists also produce graphs based on the work of the Environment Canada technician, additional graphs exist in and through their relation with Karen's graph and with respect to each other. One of the curves that the Environment technician produced is a calibration graph for translating the water level into water volume:

> What the Environment Canada technician does is- he or she, comes down three or four times a year, gets into the creek, and measures the area across the creek. So, we get an area, and then we multiply the area times the velocity and that's the discharge. And the discharge is in liters per second. And based on the corresponding water levels- we eventually get a calibration curve which means that someone like me can come down and say that means 'X' volume. And for example, at one point two, the line was here, we got seventy-one liters a second. So, he's done this, he's got a calibration curve (*Gestures a curve as in Fig. 6.9.b*) so I can look at some point compare to the calibration curve, and say OK that's about, it's about eighteen hundred liters per second.

Karen did not literally convert height into discharge. She read her graph in terms of the discharge, volume of water flowing by the water house every second, when it really displayed the level of the water as if she was internally converting the height into discharge amounts. However, I prefer to think of the situation in the following way. As she works with the charts on a daily basis, she has come to conflate the two measures so that the calibration curve has become transparent to her reading. I use transparent here in the way Martin Heidegger and Gregory Bateson used the term, when they discussed the carpenter's use of a hammer or the blind person's use of a cane. In transparent use, we do not think 'hammer' and do not represent it in mind. The blind person does not touch the surrounding world with the cane but is in direct contact in the way another person touches the world with his or her hands. In a similar way, Karen reads the graph in terms of water level or discharge amount. It is as if, correspondingly, an individual has come to the point of using the Celsius and Fahrenheit scales without having to translate. Here, as in children who learn two languages from early on, no conversion (translation) is actually ever involved. Rather, two different signs systems have become interchangeable. That is, at this point in the development of the activity system in which she is the cognizing subject, water levels and water volumes are not longer independent representations but different aspects of the same situation.

The Environment Canada technician also gets into the creek, establishing cross section data, maps these against water levels, and constructs a calibration curve. These curves are themselves outcomes of an activity system and, after being transferred into her activity system, become mediating tools in Karen's work. Karen is familiar with major markers established by the calibration curve. When she reads the graph, she talks about water volumes rather than the water levels displayed. But she does not actually produce the water volume graphs (*Fig. 6.10*) herself. This, too, is done by the Environment Canada technician, and takes about a year to get done, as she repeatedly emphasized. The resultant graphs come back to the community to be used by Karen, the farmer, and the environmentalists for a variety of purposes. For example, graphs have become part of a proposal that seeks further funding so that Karen can continue her work.

Karen's relation to the Environment Canada technician can be characterized as a classical example of division of labor, in which one subject (individual, group) produces the tools that another subject requires. Here it makes sense to view this division of labor as occurring at a societal level and to characterize the interactions between the two individuals as those that occur in neighboring and interacting activity systems. What we then observe is that the different artifacts

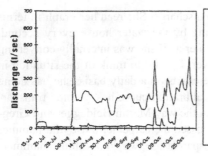
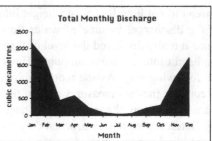

Figure 6.10. Using Karen's charts the Environment Canada technician produces new graphs displaying discharge versus time or total monthly discharge. These graphs, too, are resources that Karen may use as (transparent) tools when she reads her own graph.

(graphs) move within and between activity systems (*Fig. 6.11*). The graphs qua social objects embody the material properties of inscriptions. Inscriptions can be layered, transformed, juxtaposed to other graphs, and inserted into documents. For example, there are relations to the third graph in that the discharge rate curve (*Fig. 6.10*) have maxima and minima that correspond to the water level graph from a local well (*Fig. 6.12*), both being related to the amount of rain fall onto the watershed. The graph that plots the depth of water in local wells (*Fig. 6.12*) therefore inscribes itself in the practices of drawing water for irrigation purposes. It arises from integrating the transformed graphs over time for one-month periods. These graphs derive, in part, from operations that characterize inscriptions, including translations (non-linear), layering, scaling, and integrating. Some graphs (e.g., *Fig. 6.10*) can be constructed from Karen's original graphs by transforming the latter using the calibration curve and translating the water levels into volume.

The fact that to Karen these new charts are themselves tools often surfaces unexpectedly during her conversations with the visitors to her farm or the annual open-house event organized by the environmental activists. In the following conversation with two visitors to the open-house, Karen points out that the base flow (which she has earlier explained as the value indicated by the line without a rain-induced peak) has increased to 1.2 meters since the summer at which time it was only 0.1 meters.

Karen: And if- You'll know from the summer, base flow is looking around point one, so one point one meter above. Base flow is back up to almost one point two

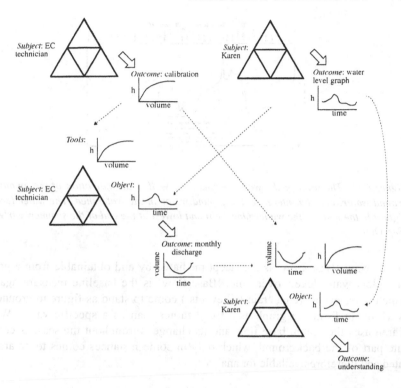

Figure 6.11. Exchange of artifacts in neighboring activity systems constituted by Environment Canada, on the one hand, and the farm operated in an environmentally conscious way on the other hand.

meters. So that base flow, the ground water component is coming back, so from-

Visitor 1: It's recharging a bit

Karen: Yeah, the ground has recharged.

Visitor 2: It's November.

Karen: Yeah, it started to recharge. Basically, the recharge is from October to May, then drops, and then starts back up in October. So, we've seen this start back up with this base flow component.

Although Karen may have been talking about the actual water levels during the summer and winter when she specified the levels 0.1 and 1.2 meters and the difference between the two measures, 1.1 meters, 'base flow' cannot be obtained

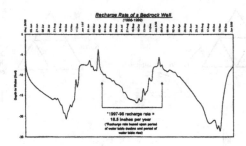

Figure 6.12. The level of the water surface in a well is an indication of the amount of ground water, and, by inference, the ground water in the watershed. The water level is highest in the dead of the winter (Jan-Feb) and lowest at the end of the summer/ early fall (Sep-Oct).

in this way. 'Base flow' is a concept mediated by and obtainable from a graph that plots water levels over time. 'Base flow' is the baseline measure against which other events (e.g., 'rain fall events') come to stand as figure to ground. It is available from the graph as a trend rather than as a specific value. When Karen uses the graph, base flow and its changes throughout the seasons constitute part of the background, which only in some instances comes to be articulated and therefore available for analysis.

6.6. GRAPHS AS SITES OF STRUGGLE

My description of graph reading so far largely focused on different relations involving Karen and other aspects of the setting in which she works. These descriptions and analyses revealed the multiple ways in which Karen's reading is linked to other material things and embedded in other practices. Because these things and practices involve other people as well, some even more familiar with the context, there is the potential that Karen's understanding comes under scrutiny. This directly leads us to the role of graphs in face-to-face interactions.

In face-to-ace interactions, graphs can become sites where social interactions occur over issues that are relevant to the lives of the people living in the area. One reason for the low frequency of face-to-face interactions is the type of situations in which Karen was recorded. With the teachers on the farm and with the visitors to the annual open-house events, Karen explained the graphs and her work. She was in what we might term a 'knowledge display' mode. However, there were instances in my transcripts with substantial interactions in which what she articulated came to be contested by others in the community. In such moments, Karen was no longer the sole master of her graphs but their relevance and significance to the community was a contingent and emergent result of the interaction.

The following excerpts derive from videotape recorded at one open-house event of the environmental activist group that operates the community. She intended to show and explain visitors what her work consists of, and how it inscribes itself into community life and into the plans that the activists have for improving water quality and quantity. The following episode begins when Karen talked about irrigation and vertical jumps in the graph that stand in a reflexive relationship with irrigation practices. As the opening line of this episode show, Karen currently was in display mode. As the owner of the graph and co-organizer of the open-house event, she led the visitors along the posted graph and read aspects to the visitors—in the way a parent might read a book to a child.

Karen: These (*Gesture.*) very, you know, ninety degree angles in the lines that's

          (*Gesture.*) definitely straight, straight drops. That's definitely irrigation activity, people are all stopping at the same time, starting at the same time depending on the conditions.

          It's dry for a while here. (*Gesture.*)

Karen made salient one of the vertical drops (jumps) in the graph. Because she could not assume that her audience knew how to read such drops across the seasons, she described it as one that has arisen from irrigation activity. (At other moments, vertical drops were related to clogged pipes, adding and removing boards, etc.) Whereas Karen had the tendency to move on after making such brief statements about the source of a graphical feature, her visitors often sought elaborations of the situation from which the graph had been abstracted. Here, Walter was not satisfied with a statement about irrigation starting and stopping at the same time. Rather, he began to address local farming practices (including the crops grown) and how these have undergone historical change.

Walter: Yeah, a lot of hay, people are into the hay and stuff.

Karen: Yeah, a lot of people cut it at the same time
Walter: Further, you go towards the Fox's farm. Down Henderson Creek. Because
            once you get past Fox's, it stops. There is corn. But of course, nowadays, there
            is late corn, too.
Karen: Yeah, they grow different varieties.
Walter: I think they grow mostly early corn (*Gestures toward earlier parts of the
            graph*) on the fields that are around Henderson Creek.
Karen: Corn definitely has a lot, requires lots of water, doesn't it? Compared to hay.

Karen did not expect Walter's interjection but her 'Pardon?' requests a restate-
ment. Walter therefore obtained a speaking turn in which he elaborated his
'passing theory' about the topic at hand. Karen then acknowledged his theory,
but continued with an explanation of the conventions (scale) rather than ad-
dressing the relation of the rainfall with the peak of the water level. As she was
in display mode, and these were her graphs, Karen talked. However, when a
graph is a public object, there is always the potential that other persons will
contribute their readings. In this case, a visitor to the open-house event (Walter,
the Mayor of Oceanside and principal of local middle school) did not just let her
continue, but contributed in an active way. Walter injected his hypothesis based
on his own reading of the graph. As a consequence, the episode does not just
constitute mere display, and is not only about the graph. It is an exchange that
takes the graph as its starting point and elaborates many other, related issues.

In this episode, Karen introduces the topic of irrigation, which goes with
particular vertical discontinuities (jumps) in the graph. But it is not just that
these jumps are signs that stand in a signifying relation to the water level
changes caused by irrigation. Rather, irrigation also stands in relation to the second, layered graphical information on the top border of the graph (*Fig. 6.8*). At
the moment, Karen and Walter stood in front of that part of the roll that was
recorded during the summer. There were no rainfall events marked on the top
part of the paper roll. Thus, the jumps attributed in this episode to irrigation existed in relation to the time of recording (summer), the number and size of the
(here lacking) rainfall events. Most importantly, the topic in the episode was not
some feature of the graph, but the farming practices and irrigation that goes on
in the valley at the present time. Yet all this is part of the thick layer of background understanding and experience that brings forth the extraordinary competence in the first place. As a seventeen-year inhabitant of the valley, Walter is
a knowledgeable conversation participant. He is as familiar as Karen is with the
extent of hay farming in the summer, which requires the dry conditions of (late)
summer in this part of the world. Karen then suggested that many farmers begin
and stop irrigation at about the same time, a fact again related to the weather

patterns in the valley and haying practices that require a dry period for each harvest.

Walter subsequently added that not all of the farms grow hay, but that they also grow (different types of) corn. He even provided a description of the specific farm where the corn crops begin to dominate the fields. Karen then made a statement—which can be heard as a question to Walter who had previously already talked considerably about strawberry farming practices—that corn takes more water to grow than hay. This, in turn, has significant impact on the irrigation practices (especially if there are different types of corn) with a resultant effect on the water levels and Karen's charts.

This episode shows that graphs are not just signs standing in a unique and unambiguous relationship to objects. They are not merely objects of knowledge in the way that cognitive developmental research in mathematics has treated them in the past. Rather, as in the previous analyses, the primary object of the interaction between Walter and Karen is the creek and the surrounding valley in which it is located. The graph is but an object that anchors the social space in which the two collectively engage in the construction of the watershed for the purposes at hand.

Walter: Well, say, I guess, as you know, the structure of the material of the soil material in the valley is- So, like they say, it is the best place in the world to have septic fields. From the point of view of a person putting one in, not necessarily for the rest, if all they wanted was go down the first number of feet, they don't necessarily think what else I have to have in.

Karen: Yeah, they don't think beyond that.

Walter: That's right, but it's sort of a lot of sand, and coarse soil so it's-

Karen: A lot of clay in the valley.

Walter: It drains well, yeah. So, that's probably why they have to pump so much water here compared to over on Graham Creek where Maber's Flats are. I don't think that they have to pull that much.

Karen: Because there are organic muck-type soils over there and they're wet.

Walter: Compared to the valley they are just pumping out the water in the valley up here. The water, I know, it goes straight through.

This episode exhibits a latent conflict in the claims as to the nature of the soil in the valley. Walter, who, at the time, had been living in the valley for seventeen years, is familiar with the people, their farming practices, and the resources on which they draw to support it. Further, as a mayor his is familiar with the wastewater disposal practices. Some families are connected to the town sewer system whereas others have septic fields. Septic fields require a stratification of different consistency to allow the wastewater to penetrate and absorb the solid matter. Walter knows that this is 'the best place in the world' for building septic fields because of the presence of the required strata. Karen claimed that the soil consists of clay. But Walter was less interested in making a claim about septic fields—which can nevertheless be a problem when they are close to the creek—but about the fact that wells have to be deep to reach into the aquifer that lies considerably below. He made a statement about the ideal nature of the soil for septic fields in order to support his subsequent statement about the farmers' needs to pump so much water. He contrasted this pumping activity (which actually occurs in his own neighborhood) with a much lower pumping activity in another part of the creek system that is situated in a former, now drained bog and therefore has sufficient water on the surface to warrant less pumping.

The conversation then moved into past farming practices, where water was pulled out of irrigation ponds, which were insufficient to supply enough water in more recent times when they were 'going out of style'.

Karen: I know, they had a lot of strawberry farms, there used to be a lot of strawberry farms in the valley, all of the properties used to be strawberry farms, Trentelman's that's where the road came from. The name of the road in that area and they said that they had just taken it from the ponds, they used to pump the water out of there until it was going out of style. So it'd be gone-

Walter: So it'd be gone very quickly.

Karen: Yeah, faster than they can recharge and push that into the ground.

Walter: So, they probably had a lot to do with prior damage to the creek. I would think and how it flows.

Karen: Yeah. See, we just don't know, we only have that base line.

Walter: That's right. So much you get through the verbal historical facts from people. I don't know that, I just I've only been here for seventeen years myself so.

Walter suggested that a lot of the damage to the creek has likely historical origins in these past farming practices, issues that he suggested could be found out more about in the form of 'historical facts' from the people living in the area. He felt that even his own (quite extensive) knowledge was limited given that he had been living for only seventeen years in the area.

This was one of those moments when Karen was speaking to a local who, in fact, was more familiar with and knowledgeable about the area. Walter remembered how people had farmed, the type plants that were grown, and the relative amounts of water that were required. Karen was no longer the sole expert concerning the meaning of the graph. While she remained the owner of the graph, Walter was more familiar with the referent of the graph. In other words, the graph was the object of the interaction but Walter had access to many tools that mediated *his* relation but that were not available at this point in time to Karen. In this sense, then, the significance and relevance of the graph were contested.

Presenting her work in public, continuously Karen navigated the tension between interacting with the audience and presenting it with what she might consider to be the foundational knowledge required for being able to read the graph. There is always a tension involved in such relations. Here it was one arising from the relation between the person who owns the interpretation of the graph and therefore who had control over the speaking turns versus the person who had come to open-house event to be informed. But there is an additional tension, which arises from the fact that the conversational topic can be shifted—here, the conversation articulates issues concerning the entire valley and its water resources. Karen began in display mode and attempted to retain the position as a presenter rather than engaging in a fully open interaction. This changed when the topic shifted from the sign that mediated the knowledge about it (at least in as far as Henderson Creek is concerned) to the watershed itself. A few minutes after the above exchange, Walter indicated that he also lives in the valley and that he has one of the water licenses. Subsequently, the interactions between the Karen and Walter changed in kind. They began to talk not only about the graph, but also about the issues for which the graph stands in a reflex-ive relation to the water, valley, history of settlement, and changes in farming. At this point, both owned the issues, and thereby allowed the respective other to be an equal contributor.

6.7. KNOWING GRAPHS IN CONTEXT

Throughout this chapter, I articulated Karen's graph-related competence. Rather than *interpreting* the graphs, which would have involved some form of inferen-tial process, she *read* the graph in a transparent way. Her way of reading is rep-resentative of the processes of reading graphs transparently which I recorded when scientists engaged with graphs from their own work. Each time, extensive situation-specific knowledge of the setting including tools and instruments was employed as resource to explain how their graphs should be read. Working in the community, knowing the farmers and their farming practices, physical, geo-graphic, geological, and meteorological characteristics of the valley, provides Karen with a rich tapestry of experiential and embodied knowing of the situa-tion. Through her work, she is also very familiar with how particular events and conditions will change the creek (which she visibly inspects daily) and therefore the recordings of the pen chart. When she looks at a graph, she experiences it as an aspect of an existential totality. Because graphs are an aspect of a meaningful whole, they have become transparent much in the same way that the walking cane becomes transparent to the actions of a blind person or that this keyboard becomes transparent to my actions of writing these pages. Karen is no longer concerned with the material details of the signs themselves, but her reading leaps beyond to the world she knows so well. The reverse movement from the world she knows to the expressive domain also continuously happens. Provided any situation description, Karen projects without hesitation what the graph will look like. Although she has not lived for twenty and more years in the area as some of the people that she interacts with, Karen knows from the older residents that the valley has experienced a housing and road-building boom. This allows her to make a reverse association to the graph and tell her audience what the graph might have looked like. She can also discuss the impact of a projected housing development which would increase the number of inhabitants by 5,000: the impact would be disastrous for the water flow patterns, which would increase even above the currently observed peaks, and the run-off would be even faster. These events then would be characterized by higher and narrower graphical peaks. Even 'non-natural events' are transparently read from the chart. We can imagine that the first time seen, however, Karen and her employer had to engage in a structuring process that allowed them to find a correspondence between some state in the world (here 'clogged pipe') and the sudden peak on the chart.

In the extensive web of signification that constitutes and is constituted by these relations, Karen navigates between the different representations of the creek (i.e., signifiers that characterize knowledge about the creek) and integrates over the tools and history. This integration is achieved in and through Karen's activity. The different representations (signs) Karen uses in her activity are nevertheless distinct and arbitrary. As she works in the creek, her familiarity with the setting increases. Furthermore, the outcomes of her activities provide additional mediating tools for understanding the creek. For example, Karen measures the water level in the creek using the pen chart recorder connected to the measurement device. As an outcome of this activity, she gets graphs that monitor the water level throughout the year. These graphs then become new tools that mediate her relationship with the object and in fact constitute Karen as a 'more knowledgeable' subject and the creek as a 'better-known' object.

In the foregoing sections, I showed different aspects of knowing graphs. These different episodes show that mathematics in everyday practice is constituted in a noisy field of practical action and discursive relations. In this practice, graphs do not exist as ideal Platonic objects with definite structure and elements. Rather, in the context of particular practices, graphs make available what is necessary in the situation at hand. Other structural aspects that a theorist may identify remain hidden—Karen neither reads individual data values nor identifies the slopes of the graphs. What matters to understand graphing are the significance and relevance of graphs within the larger context that constitutes Karen's lifeworld. It matters in the Karen-creek-graph relation that a particular graphical feature arises from a clogged pipe in the instrument or the lifting of a dam rather than some other event in the watershed.

The graphs do not just serve to express something about nature (in this instance the watershed). Rather there are very different, economic and personal matters in which these graphs are inscribed. For example, the proposal from which I had culled three graphs in this chapter (*Fig. 6.10, 6.12*) was written to seek funding for monitoring the water budget in the watershed. The grant would provide further funding for Karen and therefore financial security for another year. The graph that Karen read to different audiences is not just a representation, but it is lived as a relation within a range of practices and her life of being a water technician. This graph exists in relation to the ending contract that provides for her subsistence and a new contract with the prospects of continuing her work at a place that she has come to be very familiar with.

'Context' has been one of the focal points in the discussion of how to make mathematics more relevant. We may ask the question, 'Where is the context in contextual word problems?' The present chapter contributes answers, from my perspective, in new ways. It is evident that situated cognition does not mean that people think differently in different contexts. Even though the same physical sign may be used in different contexts it contributes to constituting the context in different ways because of the different relations that it mediates. In everyday practice, such as the one that Karen enacts, different relations contribute to an over-determination of any individual relation and the objects and tools involved. At the same time, changes in objects and tools also changes the way in which subjects are constituted. Thus, Karen's subjectivity changes over time, through interactions related to division of labor, and in individual interactions. Karen is a different person in her interaction with a beginning teacher new to ecology and the value than she is with Walter who, along a number of dimensions, is more familiar with and knowledgeable about the valley, community, creek, farming practices, or watershed. Without considering existing social and material relations, we draw inappropriate inferences about the nature and extend of someone's knowledge based on analyses of the data we have at hand.

From a phenomenological point of view, we need to consider the lifeworld of the person working with a graph, the world perceived and acted in by the person in activity. The concrete embeddedness and meaning of activity cannot be accounted for by analysis of the immediate situation. All concrete social institutions and relations are characterized by historically emerging contradictions. At the same time, objectively existing social structures do not have a determinate effect. Any meaning, as the background against which individual words and graphs become intelligible, is socially constituted in relations between activity systems and persons acting in the world; it always has a relational character. However, it does not take our analyses any further if we view context as a container that can be grafted intact onto cognition or cognitive development. The social is more than a container of the psychological, but each of the two arises from complex dynamics by means of which they are constituted in actual practice. Signs are produced and used within the dynamic intersection of actions, objects, and speech within a practice and therefore function as relations within the practice. Signification cannot therefore be reduced to representation. Participants themselves become in and through the relations in which they are embedded. Karen is who she is in relation to her employer-farmer, the water conservation and creek restoration efforts of her activist group, or a teacher who has requested her help for a school-related project. 'Karen' also emerges from the relations with the First Nations people, the creek that she so intimately knows, and the Mayor, Walter, who participated in the reading of graphs in relation to the community.

At this point, a troublesome question demands attention. How did Karen get there? How does transparent reading arise even though the same reader has faced graphs initially as unfamiliar and strange, abstract entities that are neither relevant nor significant? In the next chapter, I describe the results of a study in which I accompanied a doctoral student from the early stages of her research project, when many variables did not yet exist, to the later stages where she constructed graphs and presented them in a variety of different local, national, and international venues.

SEVEN

From Writhing Lizards to Graphs

The Development of Embodied Graphing Competence

In the previous chapter, I articulated some aspects of the intricate network of significations that the water technician Karen had developed and that provided the background against which the graph can be said to be meaningful. Because the graph was related to this background, which itself constitutes meaning, it also became associated with meaning. To Karen, the graphs had become part of the lived-in world, an integral part of the natural events that they refer to. How does such graph-related competency develop? What does it take for graphs to become (for particular individuals anyway) transparent windows onto the world that lies beyond? This chapter describes the results of an ethnographic study among ecologists, with a particular focus on Sam, a doctoral student who studied a certain type of lizard at the northern extension of its reaches. Previously, Sam had worked with another professor for a one-year period, but found the isolation of the secluded island where the research was being conducted too strenuous. Thereafter, she had changed to do her Ph.D. work in herpetology, which still required five months of fieldwork every year, but which allowed her to live in a town with regular access to all amenities and within a day's drive of two major urban centers. The chapter begins with a description of Sam's reading of the population graph, the one that gave her least trouble. Sam struggled and ultimate misread parts of the graph. This performance is contrasted with a description of the competence related to statistical analysis and graphs that came out of her own work. This contrast, then, constitutes the rhetorical starting point for the remainder of the chapter, in which a detailed description of the collection and transformation of data is provided. These activities constitute the background understanding against which her graphs that represented particular aspects of the lizard species become intelligible. This competence may be exemplary for the embodied competence that was observed among all the scientists in the interview study when they explained their graphs.

7.1. GRAPHS: INSIDE AND FROM OUTSIDE

By the time the study of Sam's work was conducted, she was in her fifth and sixth field season (two at the master's degree level) collecting data that she subsequently analyzed for inclusion in her theses, national and international conference presentations, and publications. Like many of her peers who did not have outside funding, Sam was working as a teaching assistant (TA)—as a TA she previously participated in teaching (population) ecology courses and a fourth-year statistics course for biologists. Among her peers and professors, she was known as being very competent in statistics, mastered a variety of statistical techniques including multivariate analysis. Despite this preparation and experience both in population ecology and statistics, Sam was among those scientists who experienced considerable difficulties when asked to read the graphs. (The session where she attempted to interpret the population graph was recorded during her field season in the laboratory where she kept the pregnant female lizards. The second session where she talked about her own graphs was recorded in my office where we met for a total of ten hours for interviews regarding her research, data analysis, and reading of a variety of population-ecology–related graphs.)

7.1.1. Graphs: Inside

The episode begins with a statement of despair about not knowing how to read the upper intersection of the death rate and birthrate curves (*Fig. 7.1*). (The deictic and iconic gestures are indicated in the text by verbs and Roman numerals are used as indices to link the text and the locations of the gestures in the figure.) Sam then articulated the different rates of change of the birthrate curve ('birthrate increases *more* steeply...' and the trend of the birthrate curve to decrease. She proposed to divide the birthrate curve and drew the corresponding line, while suggesting that the graph could be divided even more.

> I don't know what the hell's happening at this point (*Points to [1]*), birthrate increases more steeply, it decreases, it's is not clear. ... We can hack it up (*Draws [9]*), even more if you wanted to, since I'm not supposed to but I will do it anyway. Where is the peak... Where that peaks, so at this point (*Points to [10]*) that's increasing, so at this point population is... What am I talking here, death rate, so there's gonna be, population size, well actually this rate *is* increasing (*Points to [11]*).

The subsequent utterances focused on the maximum of the birthrate curve and Sam appeared to begin a statement about its effect on the population but then stopped, shifting her attention to the birthrate below (to the left of) the first in-

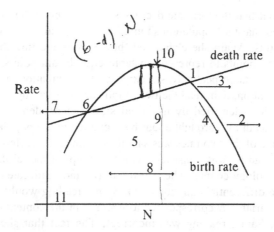

Figure 7.1. This population graph was marked up by Sam and, here, includes the locations pointed to and trajectories of waving gestures.

tersection. In this episode, we can see that it was not self-evident to Sam what to focus on. It was not clear where she should start articulating what the population will do given specific birthrate and death-rate values.

Sam's attention then returned to the upper intersection. She suggested that this might be an 'unstable equilibrium point' and elaborated (i.e. produced an interpretant) this description by stating that the population would 'just gonna crash eventually'. She made an evaluative statement about a population crash, which itself was further elaborated by several propositions that from a conservation perspective, it would be unwise to let the population reach a size larger than the point at which birthrates and death rates are equal to each other.

> I think they're probably, that's why I was wondering if that's (*Points to [1]*) an unstable equilibrium point, if it drift this way (*Gestures [2]*), then it's just gonna crash eventually, which is gonna be a bad thing. So for conservation of species (*Gestures [3]*), you potentially don't want to get- reach above this limit (*Points to [1]*) because the population will essentially crash. If the birth rate declines so rapidly (*Gestures [4]*) above that certain point, I expect that it will probably crash. So you'd want to keep it in this region right here (points to [5]), because I expect they're both (*Points to [6]*) (*Points to [1]*) Unstable equilibrium, in which case outside of these points (*Gestures [7]*) (*Gestures [3]*) the population essentially will crash very, very quickly. So, yeah, I'd keep it in here (*Gestures [8]*).

She pointed out that the birthrate decreases rapidly above the upper intersection and concluded that the population should be kept to values between the two intersection points. Again she elaborated 'both' by suggesting that she expected the intersection points to represent unstable equilibrium points. To the left and right of these points, respectively, she expected the population to crash.

In this situation, Sam read the two graphs as representing a stable state for a given population identified by its initial size, independent of the changes to birthrate and death rate brought about by a change in the population size. Here, the relative size of the two rates was salient to Sam and therefore mattered in her reading. Her reading was consistent with the simplest model that are taught in introductory ecology courses where birthrate (b) and death rate (d) are constants leading to the differential equation (3.1). Sam's reading would be entirely consistent if the variable N corresponded to N_0. Yet in the context of standard uses of the graph, Sam's reading was incorrect. The fact that she used the signs 'equilibrium point' and the qualifier 'unstable', both of which are elements of the standard population-ecology discourse shows that she is somewhat familiar with the domain. Like Ted (Chapter 3), she did not only perceive birthrate and death rate and their relative sizes but also found perceptual salience in the relative slopes. Like Ted and five other scientists (Table 2.1, p. 27), she inferred a collapsing population to the right of the second intersection from the relative sizes of birthrate and death rate.

7.1.2. Graphs: From Outside

Despite her background in ecology and her experience as a teaching assistant in population ecology and statistics, Sam did not do as one might expect in an expert-novice paradigm. Her difficulties surprise given that the graphs had been culled from an introductory textbook in her own domain. Furthermore, she was a willing participant and participated in my research for a period of two years so that one cannot argue that she was not motivated or unwillingly participated. This performance also contrasted information received from others in her department, which expressed that she was a competent ecologist and statistician. When observing her talk about the graphs that she produced as part of her doctoral research, one can also get a different impression of Sam's use and readings of graphs. When Sam talked about graphs and statistics from her familiar environment she did in fact display the competence that one might expect in the expert-novice paradigm and from someone with five years of independent research experience on topics of her choice.

As the scientists in the earlier part of this book, Sam was asked to talk about graphs and data from her own research. Repeatedly, I attended and sometimes

REPRODUCTIVE OUTPUT

	PC₁ (SVL)	PC₂ (Tail length)
Litter size	0.63	0.94
Percent live per litter	0.99	0.07*
Mean neonate SVL	0.83	0.61
Mean neonate tail length	0.99	0.02*
Mean neonate condition	0.52	0.03*

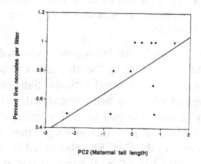

Maternal tail length positively affects
percent live per litter

Figure 7.2. Table of p-values and one graph for the linear models presented by Sam at an international ecology conference.

(video/audio) taped presentations that Sam gave at national and international conferences and seminars at her university. To be able to go more deeply into the details of her work, I asked Sam to talk me through a set of slides that she had used at an international ecology conference (e.g., *Fig. 7.2*). The data tables contained the *p*-values for correlations between a set of reproductive output measures and the residualized measures of snout-vent length and tail length. She then used three graphs to represent the relationships for the pairs of variables that showed a 'significant' correlation, which she had set at $p < .10$ because of the small number of data points ($N = 12$) that she had at that point in her study.[1] Because I had known Sam for over a year at the moment this interview was conducted, had spent almost two months in the field with her, had conducted interviews, and had met with her in informal setting, she went at this point to the specifics of the data analysis. Because of my familiarity with her work, she did not feel the need to provide me with the situation descriptions that scientists usually gave at the beginning of discussing graphs from their work (see Chapter 5).

> What I did was to reduce the mom characters using principal components analyses because these guys are correlated so I uncorrelated them so I could get some principal components analysis then I rotated them. So they're, these are entirely uncorrelated. [...] So these are five different regression analyses or general linear models, basically. Yeah, and so these are p-values and these are the ones that suggest- I've got twelve data points, so there is just an awful lot that is going in. So anyway, so I just presented the three graphs to sort of look at these relationships. So, the first one- and they all take the exact same point. In all cases it's the tail length, the mom's snout-vent length doesn't have any bearing on reproductive output which is, which is very unexpected, normally there's a very strong relationship.

In this first segment, Sam began by explaining what she had done to the data. Rather than talking about correlations in the abstract, and consistent with our analyses of graph use in scientific journal articles, Sam provided me with a context for understanding just what was being correlated. In other words, if one considers the correlation between the two variables 'mean neonate tail length' and 'mom's tail length' (reliable at $p < .02$) as a sign, then Sam's introduction constituted part of the context of the sign. She elaborated the table (itself a complex sign that has the collective 'lizard' Sam had captured as a referent) as representing five regression analyses or general linear models for which the corresponding p-values are displayed (*Fig. 7.2*). As she told what the different graphs represented, she interrupted herself to articulate a limitation of these analyses, which lay in the small number of data points. She then continued to summarize that significant correlations existed only in the column of tail length. She also read the table as indicating the lack of correlations of the five reproductive output variables with snout-vent length and immediately suggested that this result was unexpected, because there normally (in other species) are relationships.

We note that Sam began to talk about the context and the statistics, which constitute the very context in which the three graphs lying before us (one of which is reproduced in *Fig. 7.2*) existed. Furthermore, although the correlation was significant, Sam did not have confidence in it but already provided a first explanation for the missing but expected correlations: small sample size. This therefore was an elaboration in terms of the statistical procedure underlying her results. As she continued, she also provided contextual information from the field that contextualized the graphs.

> So the mean was about seventy-five percent, it was a huge range, so some, and in some litters fifty percent was stillborn. In fact, at this meeting, the one thing I

talked about was this and I got a variety of suggestions from people and I talked to one fellow and he suggested that perhaps this is not a real effect, perhaps it's an artifact of their housing conditions in the lab.

Sam suggested that the data from the 1996 season were perhaps artifacts of the housing conditions, which may have led to the high mortality rates among the offspring. She also referred to the conference where she presented these results for the first time and receiving feedback from another researcher, himself working with this lizard species but a more southern variety, who suggested that the mortality rates were unusually high. Thus, Sam drew on her familiarity with the conditions under which the data were collected and on things she had learned from others since, to provide a plausible reading of the result. When I asked her why the correlations between reproductive output measures and tail length were reasonable she articulated some of the understanding that intensive work with the lizard species has allowed her to develop.

The tail is essentially a big blob of fat. So, the best guess I've got so far is just that, the more fat you have the- Potentially that allows you to produce more young or bigger young. It's quite possible that in fact it's just that animals that have bigger tails are better quality because they've never had to drop their tail, they managed to stay away from predators and not ever had to drop their tail. So, maybe it's quite possible that it's just an indirect measure, that it just means you're good and if you're good enough to get away from a predator you're also good enough to have lots of kids. So, they're just excellent females. So [...] I'm trying to pull that apart... I haven't separated out females that have lost tails versus not lost tails but if you've not lost your tail, your original tail, you have a longer tail. A re-grown tail never gets as long and so, so those females that have never lost their tail tend to be just good.

Sam provided several, within her discipline reasonable, interpretations of the correlations between reproductive output measures and tail length. She knew from experience that the tails were essentially constituted by fat, which itself could be used during gestation as a resource or that some other factor made these 'better animals' allowing them to store fat and have a better output. In this case, the significant correlation is really a mediated one ('indirect measure'), that is, the cause for reproductive output is thought to lie not in the amount of fat stored in the tail but some in some other variable that also leads to longer and fatter tails. One sign for being a 'better-quality' animal would be the presence of the original rather than a re-grown tail, itself a sign that the lizard had been in a dangerous situation where it dropped the original tail in order to escape a predator.[2] Sam suggested that she wanted to get a handle on this question by conducting separate analyses for animals that had and had not lost their tails.

Even in these short excerpts, one notices the considerable competence that underlies Sam's reading of the data, which was often encapsulated in a simple statement such as 'there is a significant relationship between reproductive output and tail length or neonate tail length' during her presentations. Sam did not just 'plug numbers' but isolated the effects of various maternal characteristics and correlated these with other reproductive-output–related variables. She was familiar with the nature of the data that was correlated in the statistical tests and with the nature of other variables that might mediate the results, such as having lost versus not having lost the tail. On the biological side, Sam had also developed a deep familiarity with the animals so that she was able to assess whether a statistically significant relationship was plausible or whether the relationships possible were artifacts or indirect measures of some other variable. That is, Sam's understanding of the graphs was not abstract but deeply embedded in her familiar world that encompassed mathematical-statistical and biological processes and entities. This competence contrasted her problematic reading (from an analyst's perspective) of the unfamiliar graphs, which were not similarly embedded in her everyday world. I was able to ascertain that Sam indeed had considerable mathematical-statistical competence, shown not only in the type of analyses that she conducted but also in the fact that she was repeatedly employed as a teaching assistant in the fourth-year statistics course for biology students. Having been responsible for teaching the statistics courses for students in educational psychology and education in a previous employment, I know that a considerable portion of graduate students have difficulties understanding and conducting even simple analyses let alone do multivariate analyses such as those that Sam conducted. Many of her peers at the doctoral level in biology suggested that she had superior knowledge in the domain and that they had sought her assistance previously.

At this point, I want to leave Sam's readings of graphs and take the stark difference between Sam's readings of familiar and unfamiliar graphs as a given. One of the things I wanted to know more about was the question how scientists come to read their graphs (and other data) in the way they do. For this purpose, I conducted (with the help of my students) ethnographic work intended to follow one or more scientists and doctoral students over an extended period of time. Here, I present some findings pertaining to collection of data, data recording, and data transformation that constitute the background against which we have to see Sam's reading of her own graphs and data in the second part of this section. That is, I present (at least part of) the story about how Sam got to the point of deeply understanding graphs related to her work.

7.2. FROM WRITHING AND BITING LIZARDS TO DOCILE GRAPHS

7.2.1. Emergent Nature of Facts (Depicted in Graphs)

Although recent work in philosophy and social studies of science has critiqued traditional approaches to the sign-referent relationships, natural scientist unabatedly use graphs as a means to display 'facts' (rather than constructing them as many social scientists argue). In this chapter, I investigate the construction of such facts in field ecology. In this discipline, many members understand themselves as being engaged in an observational science and as operating without the grand theories one might find in physics. Although much of what members know is derived from naturalistic observation, the facts that they report in academic settings (posters, presentations, articles) are purely based on measurements. However, these observational data are diverse in their origin and error size and are collected over a considerable geographical area (here a twenty-kilometer stretch) and over long periods of time (three years). Considerable work is required to coordinate these diverse data in time and space before individual data can be converted into population statistics. Along the trajectory from data to reported fact, ecologists are not always certain whether they have observed something. The following quotes from different stages in the construction of one lizard species illustrate such uncertainties.

> *(Sam, seeking to catch lizards in one of her field sites:)* 'I usually find about five a day. I sort of am getting this feeling that they are more active later in the day. They can't tolerate, I think preferred temperature is about twenty, mid twenties or maybe high twenties- probably mid. So in the real heat of the day I don't look for the animals 'cause they're buried down too deep and then I go out again in the four to six kind of range and lately I've noticed I've had better luck'.

> *(Sam in the field laboratory, timing lizards as she chases them along a race track:)* 'I don't know if I will be able to use these speed measures, but I do it anyway. Maybe there is something, maybe not'.

> *(Sam presenting the results of her work in a colloquium:)* 'And it turns out the longer the lizards are kept in the lab, the slower they run. Which is kind of interesting, but I can statistically control for this effect and go on to look to see if there are other things that are important. And it turns out there are. One of the things that's important is what sex you are. Adult males are typically shorter than adult females, their body lengths are shorter. And adult males also have relatively longer back legs than adult females. And it turns out that this body length and back leg length is important for predicting how fast it runs'.

On the one hand, we see self-doubts in the field and field laboratory where Sam talked about the uncertainties of finding lizards and whether the sprint trials with the animals she captured and returned to the field laboratory were any good. On the other hand, far away from the field in a more formal academic setting, we observe the matter-of-factness of propositional knowledge about lizards such as the statistical significance of correlations between sprint speed (dependent variable) and body length and back leg length (independent variables). On the one hand we see factual statements and hard inscriptions, on the other hand there are uncertainties related to objects, instruments, and measurement processes in the field. From our participant observer perspective, there existed a sharp and seemingly irreconcilable contrast. We must wonder how such firm statements and claims are possible when they have emerged from myriad of decisions and uncertainties that are evident during the ecologist's fieldwork.

Working backwards and tracing (authoritative) statements and inscriptions, we would first find printouts from statistics software that had operated on a large database, which itself was imported from another, spreadsheet software package into which numbers had been entered during the ecologist's past field seasons. From here on, we would find a dizzying heap of proliferating inscriptions. There would be field notebooks; tables partially filled with records of widely varying origins; forms; printouts containing codes and coding schemes; numbered metal tags for field use; labeled and code-bearing vials, socks, plastic holding boxes and wooden enclosures all identifiable by a one- or two-digit painted number. Linked to these 'first' inscriptions we would notice an array of diverse instruments and associated measurement practices. Surprisingly, we would find little in terms of concepts, laws, and theories but, as our participants would tell us repeatedly, 'a lot of conceptual mayhem that lies beneath all of that'. Because the ecologists understand their practice as an observational science and because this particular project is concerned with correlations of phenotypic aspects of the lizards, much of the ecological fieldwork appears to be driven by what is do-able in terms of measurement. However, there are also considerable variations in measurement practices, scales of measurement, and measurement error. That is, our scientists' activities and the resulting data on which their claims are built arise from a variegated observational topology associated with considerable co-ordination work which allow the statistical and graphical correlation of quite unlike aspects of the study object. The distillation of the fieldwork, the lizard as it becomes visible to the scientists' audiences in Sam's graphs, is not just a natural object, not just an individual construction, and not just a social construction but arises out of the interaction of nature, individual, and culture. It is the results

of a dialectic involving Sam's agency and the social and material structures salient in her lifeworld.

7.2.2. The Lizard Project

More than thirty years ago, the author of the renowned Peterson's *Field Guide to
Western Reptiles and Amphibians* noted that

> There is much to be learned about the distribution, habits, and behavior of
> western reptiles and amphibians. The many question marks on the range maps
> should be a challenge to filling in gaps in our knowledge of distribution. We
> have not even found the eggs or young of some species, and much remains to
> be learned about time of breeding, courtship behavior, enemies, and other mat-
> ters.[3]

According to the informants in the ecological field camp, this description is as
viable today as it was more than thirty years ago when the field guide was first
published, especially for the lizard subspecies investigated by Sam. This lizard
subspecies is of particular interest because its range lies at the border of the animal's geographical distribution and little is known about its life and natural history. Sam was particularly interested in questions of reproduction. Some organisms have fewer offspring than physiologically possible; such decreases are
correlated with increases in survival rates and often paired with giving birth in
alternate years. According to Sam's research proposal, the ecological literature
suggests four methods of testing this hypothesis from which she chose that related to measurements of phenotypic correlations in the field. Consequently, she
had proposed to study life history traits (e.g., size at birth, growth rates, size at
maturity, length of life, mortality rates, weight), natural history of the animal
(e.g., temperature patterns, characteristics of den sites, reproductive habitat), and
the tradeoffs between current and future reproduction. These tradeoffs are to be
established for a variety of metric traits (distance moved seeking and defending
mates, mating duration, basking rates, length of gestation).

My observations and interviews suggest that Sam's work was little driven
by theoretical discourses in the field of ecology:

> There is only one underlying theory to biology and that's natural selection. But
> that's the only theory that we have. Physics has other theories like lots of laws.
> We don't have any laws. In biology there is but one unifying principle that
> works very well. And then, there's just a lot of conceptual mayhem that lies be-
> neath all of that. And so, a lot of uncertain things are going on.

Seen from the outside, Sam and her fellow ecologists were unaware of many
local theories about population dynamics or foraging theories that could have

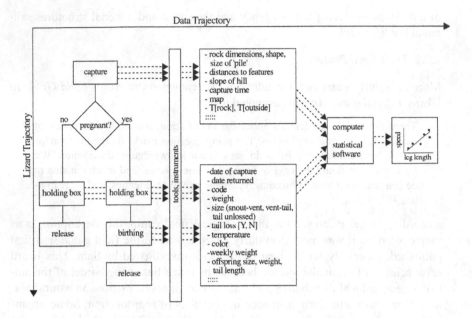

Figure 7.3. Trajectories of lizards and data during the ecological study observed.

been deemed relevant in the context of their work. Thus, the interviews with the ecologists in the field, which also included a variety of graphical representations of such theories (e.g., predator-prey relationships, habitat choice) uncovered the considerable effort that they spent on making any sense. The research was driven less by local or overarching global theories but by questions of whether something could be measured. If something could be measured, it was included in the data collection; conversely, if an instrument became available (such as the Munsell chart in this study), a new 'factor' became salient in the research.

For professional behavioral ecologists interested in one or more reptiles, the work of getting to know and representing their animals is much more laborious than the recommendations made in the field guide may appear to indicate. My ecologists' articles and presentations are, like most scientific articles, built around figures, graphs and tables; these inscriptions are the hinge that permits them to connect the intractable world of their field work with the docility of the page. Close inspection reveals that in this, figures, graphs, and tables are not unproblematic. Rather, they are themselves tied to the proliferation of inscriptions that begins in the field with the capture of lizards, moves to the laboratories where the inscriptions accumulate as the result of our ecologists' perceptual apparatus and practices, continues to the computer station where the digitized results are collected and processed, to ultimately result in the graphs published in journal papers and as conference presentations.

To make scientific 'perception' possible, lizards are quite literally extracted from the field and transported into the laboratory. There are many stages of a lizards' journey, inscriptions produced or used to co-ordinate other inscriptions, and the paper trails, which, despite different trajectories, end as digital electronic bits on some computer hard drive (*Fig. 7.3*). The site of capture is tagged, and many of the measurements that are used to construct the 'home range' are actually made at a later point. At this point, the lizard and its home range are literally separated until they are brought together again, as layer-able inscriptions, after many measurements, inscriptions, and transformations. The captured lizard is placed in a temporal holding area, a numbered cotton sock, to be brought to the field laboratory where it is positioned for perception. After undergoing several sets of procedures that bring the lizard physically together with measurement instruments, gravid females are placed in enclosures until they have given birth while male and non-gravid lizards are returned to the sites of their capture.

7.3. FIELDWORK AND EMBODIED UNDERSTANDING

Ecological fieldwork involves hard physical work often under exacting meteorological and geographical conditions that require physical, emotional, and mental discipline. In addition to the climatic conditions which are endured, dangerous situations and unpredictable circumstances can develop which frequently put the lives of fieldworkers into peril. The difficulties of fieldwork arise from the rigors of being exposed to the climatic elements and the demands of navigating difficult terrain and botanical coverage. During the fieldwork with Sam, the weather changed continuously even in mid-summer. There were days when the field notes stated that the 'weather is "not lizard weather" ... since it is cold and occasional splatters. Around 2 p.m. it worsens and is raining quite hard and is windy and cold'. Within days of those cold spells, scorching heat and high humidity could be experienced. Thunderstorms and torrential rains emerged from nowhere soaking us before we could return to take cover. Black flies and mosquitoes alternated ambushing us at different spots along the trails and at different times of the day; clothing did not seem to be a problem to them. The rocky slopes where lizards were found were often steep and we had to go up and down by means of switchbacks, while searching for the organisms in the jumble of loose rocks, gravel, and vegetation. During sunny July and August days, the temperature on top of the rocks was higher than the maximum of 60 °Celsius shown on our thermometer. Each person consumed several liters of water, if not to replenish completely at least enough to prevent complete dehydration. Despite those difficulties, Sam often talked about the fact that the physical aspects of lizard research is relatively easy compared to research with other organisms:

> With the lizards, you don't usually go out until eight or nine o'clock in the morning. And you are not going out when it is freezing cold and wet and freezing cold in rain, because that is when you find your things. You won't find them if you go out then. You have to go out when it is warm and sunny, but not when it is too warm and sunny. Because when it is too warm and sunny then the lizards are not out there either.

However, the difficulties posed by the weather and difficult terrain when conducting research with lizards are compounded by the repetitive nature of the fieldwork. Sam described the physical activity as being so exacting that she often felt mentally drained and incapable of continuing her work after she returned to the camp. She found it difficult to do necessary laboratory work such as taking various measurements and transcribing data into a computer database. She would indicate that she needed a rest and was unable to put in the extra hours required for immediate processing of the captured animals delaying these measurements until some break in field work (such as resulting from bad weather), or 'next week'.

Such delays may have unnoticed effects on the data produced by a scientist, which may then only become disclosed because of some fortuitous circumstance. For example, back on home campus during the winter, Sam ran a correlation between sprint speed and leg length. Yet the expected and reasonable correlation was not confirmed. Remembering a single instance of a lizard that apparently had lost some speed during a five-day stay in captivity, Sam used the statistical technique of 'partialing out' (controlling for) variation due to the number of days lizards had spent in the lab before she came to measure their sprint speed. This time, the correlation between leg length and sprint speed was significant.

Undoubtedly, ecological fieldwork leads to tremendous hardships taking both physical and mental tolls from the students. As such, fieldwork is not unlike other 'rites de passage' in which future members of some group exhibit the required prowess that allows them to enter the club. However, in ecology, fieldwork is more than just a 'rite de passage'. Fieldwork in fact leads to particular, embodied and holistic understandings that distinguish many field ecologists from other scientists (theoretical ecologists, management ecologists).

Fieldwork is an important source for their characteristic knowledge and understanding of the ecological interactions and biology of organisms in the field. Extended fieldwork leads to considerable knowledge of the field, but this information experiences a filtering as it is condensed for academic resources (i.e., journal papers, theses). This filtering often results in a formalizing of field knowledge in which knowledge claims typical to the literature—'scientific' evidence—are included in the reports and other experiences and knowledge of the field atypical to the literature are excluded. These excluded claims and understandings are often exchanged by ecologists in other forums and are referred to by them as 'anecdotal' knowledge. This anecdotal knowledge does not or seldom enters formal reports, but nevertheless is a central resource that field ecologists draw on when they read and interpret formal representations (e.g., graphs) in the course of their work. The informal aspects of ecologists' work are systematically deleted as they are considered unimportant or illegitimate in the context of 'real science'. Also arising from the fieldwork is a particular, animal-centered perspective (a 'feeling for the organism') of the ecosystems that ecologists are working with and attempting to understand. Field ecologists' understandings are much more holistic than the formal publications would lead one to believe. These publications, because they can only report 'scientific' evidence, do not communicate the embodied 'anecdotal' knowledge that is a prerequisite for the 'scientific' knowledge.

In the course of spending several seasons in the field collecting formal data, Sam also came to embody a substantial amount of what she calls 'anecdotal' knowledge and unarticulated understandings that constitute the foundation of her understanding of this lizard subspecies. There was so much more to her understanding than the few statistically significant relationships that were the results reported in her dissertation. So much of what she knew arose from the experience of 'being-there' and patterns she experienced in the field setting for which she did not have 'scientific' measures but for which she instead had a *sense* of what was happening. This *sense* was seen in the regularities that did not show up in the statistical analyses she conducted but which she had nevertheless constructed.

For example, it was this sense that led her to the construction of variables that had not been used before in the field. One of the things she reported to the scientific community was sprint speed and its correlates, such as relationship to leg length (gender and time in captivity statistically controlled). This variable did not exist a priori but emerged from her experience in the field. Even her operationalization of the variable for laboratory measurement was not an abstract operationalization of sprint speed but one keyed to her experiences of hunting lizards in the field. Asked for the origin of her variables, Sam suggested the fieldwork that gives rise to a sense of 'what might be important'.

> Some of that will give me some indication. I think in terms of the distance, the sprinting distance that they would typically do. Because they, if the rock is just flipped then they tend to bolt somewhere, they do tend to bolt to the nearest covered object so if it's close that tends to work well.

Even searching for the animals and knowing when and where to find them is largely based on the sense the scientists built up over the course of their field-work of under what conditions and in what locations the animals are likely to be found (see quote on p. 229).

When I followed up to find out more about her search strategy, Sam attrib-uted much to it to her 'anecdotal' knowledge of lizards and their particular con-text in the valley where she worked.

> I have some suspicion that they don't move around much but. You know, I go out, I look for them at the same time [of day]... They're in the same region but they could have been to town and back, and I don't know. It's quite possible they're moving a lot more than I think, a lot of it is totally anecdotal.

Yet, despite their use of this 'anecdotal' knowledge in finding their study ani-mals, making field decisions, constructing variables, and the frequent exchange of stories about these topics and other field experiences, ecologists clearly dis-tinguished between these understandings and 'scientific' ones. Beyond the field work itself, although embodied experience and a sense for what is going on was a crucial aspect of scientists' interpretation of formal representations, they often discounted their own and others' 'anecdotal' data and information.

> One of the things that I'd like, I have a little bit of an ulterior motive in that people have made some management decisions about these critters and as far as I am concerned it's not based on any solid scientific evidence... I mean, my stuff is somewhat anecdotal I mean, their stuff is anecdotal to the max.

My observations suggest that a *sense* of many regular patterns in the field were formed from Sam's many visits to the different sites over three field seasons from mid-April to mid-September. Over this time she built up a great *sense* of the relationship between lizards and their environment, which was so strong that she felt confident enough to seek grant money from an agency concerned with forest renewal—even though this evidence was not included in her dissertation. It was however noticeable that although ecologists may share stories of these regularities to others in person, and although this sense might inform their con-struction and analysis of variables, understandings that arise from this field sense but not from formal 'scientific' analyses do not enter into the formal writings. Generally, in the larger scheme of scientific work and governmental policies 'anecdotal' knowledge is discounted as a primary resource. From my perspective, of course, this anecdotal knowledge permeates all of the formal knowledge and representation use of those same scientists.

There were many instances when Sam elaborated a formal graph that resulted from her data in terms of her 'anecdotal' knowledge that resulted from her fieldwork. That is, Sam's knowledge of the site provides a justification of and a sense for the results of the statistical analysis.

> Site is significant in just about everything, which is not, to me, at all surprising knowing these sites. The only thing it's not different in, that is not important, is the nearest rock. But it is important, it's important in everything and to me it's not, it's not surprising knowing what the sites look like.

Sam talked about the statistically significant differences between dependent variables according to the geographical site where she had captured her lizards. Twice in this excerpt, Sam noted that the significant differences are not surprising given what she knows about the sites. That is, her knowledge about the sites, which she articulated in terms of what she elsewhere called 'anecdotal knowledge', made statistical differences not surprising. But of course, she acquired this anecdotal knowledge in the course of her fieldwork long before she ever conducted the statistical analyses during the winter season back on home campus.

In the case of the field ecologists, the primary knowledge was that of the biological systems. Whereas theoretical ecologists and physicists treated inscriptions as things that related to other inscriptions, field ecologists always talked about natural systems before, or instead of, talking about the representations. At best, graphical representations are used to 'quickly give a short synopsis of a biological system such that the example makes sense potentially without a long drawn out explanation'.

7.4. ECOLOGICAL FIELDWORK IS COORDINATION WORK

To track animals and co-ordinate capture sites (i.e., movements of the ecologists in the field), lizards, measurements, heaps of inscriptions and spreadsheets, ecologists use maps, tags, codes and tables. That is, articulation involves work and necessitates a variety of representations. I describe the following types of work: 'digitizing and tagging' as articulation work; tables as devices that prospectively describe and prescribe sequences of (measurement) activities and retrospectively provide accounts of these activities as completed; and maps as devices that relate lizards and their 'preferred habitat'.

7.4.1. 'Digitizing,' and Tagging: Articulation Work

The marking and recapturing of animals, associated with multiple measurements each time an animal is captured, constitutes a core practice in the ecologist community. It allows members to understand various aspects associated with the animals including changes in their behavioral characteristics along their life span. Without inscribing marks in/on the body of individual animals, therefore, re-capture would be meaningless, for it is by means of the inscribed codes that scientists relate previous measurement activities to subsequent measurements on the 'same' animal. From an ecologist's perspective, ascription of places, times and measurements to individual lizards has to be possible throughout their work to maintain at least the illusion of a lizard's identity throughout the scientific endeavor and to the reporting of its results. To allow such co-ordination, objects and inscriptions have to be marked, tagged, or otherwise labeled to allow an accumulation of layers into strata of facts said to have come from the 'same' animal. In the case of lizards, marking consists in removing individual or sets of toes such as to constitute a unique pattern (numerical code) for each animal. 'Digitectomy' constitutes an important aspect of electronically digitizing the lizard because it turns each animal into an identifiable individual. The animals' individualities seem to be preserved and constructed precisely through the loss of a body part.

Once an animal enters the laboratory in a sock, it is ready for 'receiving' a personal code by toe removal. However, the process does not begin with the animal but with an inscription—the 'Markings[location]' record where [location] takes some value such as D2 or PH and codes a particular field site—in which all codes currently in use are recorded. After selection of an unused code, 'Marking scheme' and 'Toe code' are used to determine the leg and position of the toe(s) to be clipped. Subsequent to the clipping procedure, lizards are returned to their holding box or enclosure (gravid females) for further 'processing', whereas the toes enter vials—themselves labeled with date and site (D2) of capture and the code—for later age analysis. The use of the specific code is recorded by means of an X in the 'G' column on the 'Markings D2' inscription.

The codes are literally inscribed in the lizards by clipping toes and the lizard's environment is tagged. This permits the researcher to return and do additional measurements at different points in time, which, as we observed, could occur even one year later. Thus, tags assure identities to lizard and capture site across time and space (such as when Sam relates information about a specific lizard to its specific site perusing the information in the spreadsheet). In the process of clipping the toe(s) from a lizard, the animal gains individuality in, and because of, the uniqueness of the missing body part. At the same time, the attribution of individuality is also the beginning of a corresponding loss of individuality, that is a co-product of the construction of universal truths about the lizard species as a whole. It is in the separation of the toe from the lizard that the animal obtains its individuality that we know the lizard as an individual. It is in the separation of the lizard from its habitat that lies at the origin of the construction of lizard knowledge. At the moment the toe cut is completed, the lizard has lost part of itself, but, at least as an actor in a set of human practices, gained individuality. As the scalpel's edge completes the cut, a first epistemological discontinuity occurs.

In an interesting twist of events, the digitectomy that marks the animal also marks the presence of the corresponding toes in the vials. Inscribing the code into the body of the animal associates the two. Thus, 200.0 indicates that the animal is missing the second toe on the front right foot. Ironically, the same code is the identification of a particular vial containing that toe. That is, in the case of the vial, the '200.0' marks the presence of the same toe that is marked as absence in the case of the animal. But whereas the missing toe uniquely specifies the animal, a toe lacking tight co-ordination with the associated inscription would be useless for it is no longer unique. In and as code, A200.0 links the contents of one vial to a particular animal and more importantly, to a proliferating set of inscriptions associated with the same animal.

One may ask, Why does a simple act such as 'digitectomy' involve several schemes and inscriptions? The not immediately evident answer is that the code is compact and minimizes the number of sign elements involved. As in computing, compression is work that in the case of 'digitectomy' is distributed over humans and inscription. For example, the *Field Guide to Western Reptiles and Amphibians* suggests to express a lizard as RF3-RH3-LH5 when its right for 3, right hind 3 and left hind 5 toe are removed; our herpetologist would use 474.0 to code the same marking. To achieve this compactness—474.0 vs. RF3-RH3-LH5—Sam's scheme makes use of three different properties of number including sign, ordinal and cardinal properties. Despite violating general mathematical practices, the coding scheme achieves its usefulness because it is locally sufficient and adequate for the intended task.

The 'marking scheme' codes legs by means of a number with two to four digits. Ones, tens, hundreds, and decimal position code the back left and right (BL, BR) and front left and right (FL, FR) legs, respectively. Sam codes the toes using the set {1, 2, 4, X, 7} where X corresponds to the long toe which she never cuts. Thus, the number 200.0 indicates the absence of the second toe on the front right foot of a lizard, whereas 200.1 indicates the lack of two toes, the second one on the front right and the first on the front left foot. What appears to be confusing, and where the coding sheet comes in, is the changed sequence for coding the feet in the number (e.g., 100.1) and the columns in the marking scheme (0 0 1 1).

In this coding practice, the researcher exploits numbers in their natural, ordinal, and cardinal (which allows arithmetic operations) form. Natural number means that there is a sign system that can be used to designate number; ordinal number exploits the culturally specific sequential nature of numerical signs within the entire sign system, that is, 2 as following 1 and 7 as following 4 (in the present case). Finally, for the system to work, the arithmetic nature of numbers, the possibility to add them to form new numbers, must be exploited. Thus, mixing arithmetic and ordinal properties, 1 and 2 can be added to create the sum of 3. Now '3' does not correspond to any one toe. But this missing correspondence, and the rules for composition, allows the researcher to decompose a 3 into a corresponding lack of Toes 1 and 2.

The code is neither complete nor consistent when viewed from a mathematical-semiotic perspective. First, as a set, {1, 2, 4, X, 7} and the operation {+} are incomplete in that they do not cover the results of the operations $7 + 4$, $7 + 2 + 1$, $7 + 4 + 2$, and $7 + 4 + 1$, and $7 + 4 + 2 + 1$. Second, the code is inconsistent in that three missing digits ($1 + 2 + 4 = 7$) are expressed in the same way as the missing 5th digit. That is, the code is degenerate in that '7' may refer to the missing fifth toe or to the lack of three toes 1, 2, and 4. However, the new set arising from the original set and the adding operation also changes the ordinal properties of the numbers. Despite the fact that some of the digits in a code indicate to the researcher that multiple toes have been removed (or lost) from the same foot, the size of the code is not indicative of the number of toes removed. Thus, the digit '7' in a lizard's code means one toe removed, whereas '5' means that two toes are missing. It is precisely because of the limited generality of the code that it contributes little to the adequation of lizards with mathematical structure, though the code does a lot of work along the process for achieving this adequation. Here, the use of mathematical properties in scientific research does not contribute in any direct way to the adequation of the natural world and mathematics; but it contributes indirectly, because of its contribution to the co-ordination work.

The coding scheme is not fully compact because Sam also has ethical concerns. Thus, the codes in the 'markings' inscription are sorted according to the number of toes missing so that our ecologist, checking used codes from the beginning of the table, always removes the minimum required number of toes. To further 'minimize damage to the animals', she reused the numerical codes for the different and distant sites and uses a letter to distinguish lizards (A) and skinks (S) leading to such codes as LPH A1.0, D2 A1.0, and D2 S1.0. But the coding system is sufficient for all practical purposes at hand and cover any *likely* occurrence of missing and clipped digits. This coding system, though inconsistent and incomplete in arithmetic domain covered by the numbers—sign, ordering, and arithmetic properties—is thoroughly and practically adequate. In this sense, the code is 'a very primitive holophrastic language' with a grammar that also involves inferential movements (in the case of recapture, for instance) and provides sets of instructions.

The work that goes into coding is correlated to the work of decoding. Thus, reading the code of a lizard is so cumbersome that in the actual laboratory processing of the animals, ID is seldom used. Even for the ecologist who is thoroughly familiar with the lizards as individuals, keeping track of the animals and inscriptions in the laboratory is much easier when she uses the arbitrary numbers on cages and socks. Thus, the day-to-day work at the capture sites and in the laboratory is characterized by much more easily remembered numbers on socks, plastic boxes and wooden enclosures, the latter two also being recorded in the tables that co-ordinate and record the various activities and their products.

> Melissa: We did N9? Or N1?
> Sam: That hasn't been returned yet. It must have got put in Box 1, but that one
>   was home already.

The animals are not referred to and tracked in terms of their individual codes, but in terms of the numbers on their boxes and enclosures. Thus, there are inscriptions such as the records from the color trials that only have the cage number recorded. Only when the color is entered into the coding sheet is it associated again with the individual code number inscribed into the lizard. In the field, too, numbers were employed for coding and therefore co-ordination work. For example, every time Sam caught an animal (lizard, skink, snake), she pulled a tag from her pocket. In the field notebook, the number of the sock and the arbitrary number of the tag marking the site of capture are entered side by side. Here, each lizard is assigned to an arbitrary inscription (on the sock) that is used temporarily to construct its identity: 'sock 3 corresponds to tag 166'.

Figure 7.4. A table in which Sam maintained the data collected in the laboratory. The data were subsequently transferred into a spreadsheet that was exported into a statistical package during the winter months.

7.4.2. Tables

Tables (e.g., *Fig. 7.4*) are the single most important representation in the ecologists' camp. A glance at the headings of the tables suggest that the construction of knowledge in ecological fieldwork involves many heterogeneous bits and pieces—materials, instruments, work practices and reptiles—which are selected, ordered, coordinated, adjusted and fitted together in layers. The object of the in situ work of ecologists is to coordinate these layers in a seamless way; this ordering activity is centrally tied up with maintaining tables, the contents of which can be transferred into a spreadsheet. In the intertwining of multiple purposes of tables (spreadsheets), as record of the representation activities and as a schedule determining future activities, the tables are intrinsic to the activity of our ecologists. Representational devices and represented phenomena stand in a reflexive relationship through their production as an intelligible ordered pair. In the present case, the pair involves a myriad of in situ accomplishments in constructing items that are plausible candidates for filling the cells in the spreadsheet.

Tables accomplish more than one type of work: they drive the work schedule in that empty cells are constant reminders of the work yet-to-be-done, and they record the fruits of the ecologists' labor. In this example, the table is reflexive: it suggests what activities have been completed, in its function as a recording device; and it is a plan for future activities that have as their goal the completion of an empty cell in the table. As both Sam and Bill (a postdoctoral fellow working in the field camp) noted, the tables were devices that 'reminded [them] to write everything down'. That is, each empty cell in a row already started is a reminder of a plan for observation-related activities that need to be completed; each empty row a constant reminder to collect additional specimens to increase statistical power and decrease Type I errors of prospective statistical analyses. Each cell entry therefore is reflexive with an activity, which, as part of the entry, is simultaneously being marked as completed; each completed cell is both a record of work done, and a piece of 'data' associated with a particular animal. We see this double role in the following episode emerging from the interaction between the field ecologist and an assistant (Mike).

> Mike: Percent black?
> Sam: I got it already. This is for N-one. Is it there?
> Mike: (Checks table) Oh! Sorry.

This episode hinges on Sam's question 'Is it there?' referring to the presence of an entry in the field at the intersection of the row which also contains 'N1' as entry, and the column for mottle ('percent black') to be determined with the Munsell chart. From her perspective, she was already further along in the activity and the mottle entry for 'N1' should have been completed. The absence of the entry would mean trouble. The field assistant's work was not mediated in the same way by the table; he asked for a determination of the mottle. Sam's question 'Is it there?' forced the assistant to perform an informal audit and check the status of the cell in question. His apology signals that the original request had been inappropriate. That is, when an intersection of a row bearing the digitized code and a particular column bearing the name of a 'variable' is empty, it calls for a series of actions that have as their result an inscription that fills this cell and thereby stops demanding to be filled. The intervening process involves a co-ordination of a particular animal with a particular device, such as A400.0 and a digital balance. From this co-ordination, which can involve multiple activities, emerges a new datum that can be related, as part of a vector containing the same datum from numerous and nameless other lizards, to another datum, equally part of another vector.

Tables not only record but they bring about a topology of observation, they project and record measurements; the 'everything' in our ecologists' statement that tables 'reminded [them] to write everything down' here means every item that the column headings specify as relevant to be perceived about lizards. Tables therefore constitute the framework that determines both what to look at, and whether an observation is complete. We can therefore say that perception is driven by the table in the sense that the current state of the table provides motives for regarding and disregarding activities, search the environment, and treat some measurement as a doubtful or normal claim. In fieldwork, ecologists may come to realize that their perceptual apparatus as embodied by the tables is incomplete. In this case, one or more new columns are added as Sam engages in her activities, or the labels (or content) for an existing column are redefined.

Here then, the spreadsheets in fieldwork are more flexible than those that have institutional histories, have co-emerged and therefore codetermine routine work practices of groups of people in the way tables (including spreadsheets) function in intensive care units. As the ecologist adds a column in the course of field-work, there are then some missing cells, or cells of which the content no longer corresponds to the information sought. In the latter case, Sam 're-measures' the previous items which means that such a change is prohibitive when there already exists a large number of previous items have been assessed in this way. On the other hand, when a new column is selected such as the color of the lizards, re-cords in this column start at the date this new variable emerged. Consequently, the empty cell is not unique in its meaning as request for work to be done, for a color determination of lizards already released or lizards from the previous year was impossible to conduct; color determination likely would have varied in the case of a recapture. Furthermore, an entry itself sometimes becomes a referent for future activities such as when the recorded time associated with one speed trial projects the earliest time at which a subsequent sprint trial can be per-formed. The entry does not achieve that on its own, but in the context of an ad-ditional entry at the top of the page that indicates an interval of forty-five min-utes between subsequent trials. This time interval was significant enough to become part of a presentation of her research results.

In the context of the tables, the role codes and tags play takes on additional dimensions. Codes and tags do more than inscribing the research into natural objects. They co-ordinate and align multiple and layered representations such as to preserve the structural integrity of phenomena. Codes and tags achieve this alignment in a way that is not unlike pushing a knitting needle through a refer-ence hole in a stack of punched cards so that columns become aligned and do joint work. Layered and coordinated representations are therefore useful and necessary in the domestication and control of natural complexity and disorder arising from the phenomena and researchers' activities which are heterogeneous and distributed across time and place.

When the field work is completed at the end of the season, the records of the work turns into a mere representation of the lizard, gathered in a neat spread-sheet (also printed on paper), summarized, simplified, juxtaposed. There is a reflexive relation between the table as a record (and literary phenomenon) and the practical performance of observation, measurement, and representation. Filled cells now constitute an audit trail of the summer's work completed, sometimes, as in the case of sprint speed and capture, specified to the nearest minute when the particular measurement was accomplished. Lizards therefore have been 'domesticated' to the point that they only exist in and as of the docile

records that are fitted into other literary arrays, scientific presentations at colloquia, conferences, and scientific articles. That is, activities are not *just* conducted, but they are conducted such as to produce items that can be inscribed in
the cells of a spreadsheet further translated and exported into a statistical program. Items that cannot be processed by a statistics program do not contribute to
the construction of lizard as a species. *This* is our ecologist's paradigm: if
something can be measured, digitized, and statistically analyzed, it constitutes
an observable; knowledge which she could not quantify (including best time of
the day to catch a lizard, weather conditions), despite its importance to her work,
was discounted as 'unscientific' and 'anecdotal'. The entries have no meaning
outside the particular activities. It is only after statistical processing that the
cells, in a highly transformed way, become objects that inscribe themselves in
and become part of a discourse that has an intelligibility beyond the local activities of the researcher: 'The speed of lizards is determined by their size and
length of the hind leg'.

7.4.3. Mapping Territories

The final inscriptions to be discussed here are the maps generated by our ecologist in which the capture sites are recorded (*Fig. 7.5*). These maps constitute an
interesting phenomenon because they do not conform to geographic-
mathematical conventions by preserving an exact metric in a global sense, but
do record distance measurement, sometimes with one-centimeter accuracy.
These maps are both mathematical and non-mathematical. As our analysis
shows, mapping the territorial preferences of lizards involves more than simply
driving a few stakes into the ground and unproblematically recapturing the animals within this area as the instructions of a leading field guide and an ethnomethodological analysis thereof make it seem to be.

For studying the 'home ranges' of lizards, the instructions of Peterson's
field guide recommend the establishment of a grid of stakes in the field. Lynch
argues that by establishing a grid, the naturalist impresses a graph into the natural terrain. Capture and recapture sites of animals are marked on a scaled drawing of the field with grid so that the natural 'habitat' of individual lizards can be
established as locations that are accumulated over time in and as of summaries
of recapture locations over time. Integrating over these sites, a territory can be
established for each lizard in the area.

Following the ecologists around, I soon realized that the capture process
and therefore mapping of territory and territorial preferences is not at all simple
at least here at what is thought to be the geographic boundary of this lizard species.

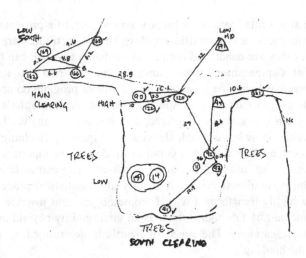

Figure 7.5. Excerpt from a map drawn to represent the location of capture sites. The map is to scale and the angle between any two lines are approximations. However, absolute distances and directions (at a later point) of each connection between two sites determined and recorded in a data table.

It's kind of a funny area. It's not actually one of my best sites. So it's down on the list. I have good luck here in the spring, last year, but when I came back throughout the summer I had very bad luck. So, I really would like to know where they go, because they are here at some point.

On some days, even working an area for several hours does not yield a single lizard sighting let alone re/capture. Considerable work is therefore involved in this search, hundreds and hundreds of rocks flipped and returned into their previous positions. Sam characterized most of her knowledge about the behavioral aspects of lizards as 'anecdotal', often no more than gut feelings; finding the animals appears to have much to do with luck. However, each time a lizard is captured, the exact site is marked by driving a tag into the ground—as if the confluence of lizard and researcher was beyond luck alone. At some later point, the distances between most tags, distances to adjacent rocks, bushes and trees, and the (approximate) distances to the forest edge are measured and recorded on the map and in a table where details of all capture sites are kept.

Finding the animals is a haphazard enterprise and large areas are searched. Sam's solution to sampling was therefore not to work with a fixed sampling grid, but to sample large areas and mark the sites of capture. Because lizards 'are usually found under rocks', sampling therefore potentially covers an entire area because the rocks of interest could be anywhere. On the other hand, the ecologists characteristically do not sample larger rock piles which, because of the noise involved in the flipping of the rocks would interfere with successful sighting and capture. Flipping rocks and capturing those lizards found underneath therefore does not constitute a simple mapping of lizard location onto a grid but is more like purchasing scratch and win tickets. Where scientists flipped rocks was not so much a systematic establishment of every unit area on the grid, but where the lizards are found is determined by the search strategy. Searching in rock piles, where lizards are likely to rest, is an inefficient search strategy for the noise of flipping rocks in one area of the pile would chase away all those lizards that rest anywhere else. The search strategy therefore focused mostly on rocks that are somewhat isolated from others, or in very small assemblages of rocks. Furthermore, on summer days, the temperature of the rock surface went beyond 60 °Celsius (the scale maximum of the digital thermometer) so that, depending on the thickness of the rock, it was too hot underneath for the most daring reptile. The tag numbers on the map which indicate a capture are therefore reflexive of the success in capturing animals but the white areas on the ecologist's maps (*Fig. 7.5*) do not constitute evidence that lizards cannot be found there in principle. Whereas the marked spots are indicative of a spot where lizards were found, the white areas are inherently underdetermined in their meaning: these locations could be those that lizards do not frequent, those that ecologists do not search, or those that make capture inherently difficult to impossible.

The map establishes both territories of lizards and territories of success for the ecologist. It takes shape as more and more lizards (as well as skinks and snakes) are captured; it is the activity of capturing that specifies the details of the map rather than the similarity with the activities of mapping a territory that we would observe in a geographical team. Through the sampling process and its (quite literal) 'findings', the map is reflexive not just to the activity, but also as a function of the success in finding those reptiles of interest. The map constitutes an inscription reflexive of 'success' in the sense that it records primarily the sites of capture, from which it is built up by imposing a metric on inter-site distances, and another series of distances to things that are relevant in human cognition, shrubs, forest, rock piles. The map also emerges and is constitutive of the inhabited space structure attributed to the lizards, the movements, the lifeworld, lived experience of those literally 'caught within it'. It is reflexive to the lizard's momentary position in its lifeworld and to the activities of those scientists who intrude in this lifeworld. The map therefore includes objects that are associated with the reptiles, capture sites, and objects that characterize the area surrounding the site in human cognitive and cultural terms. Whether these objects are relevant at all in the lifeworld as experienced by the reptile is not a matter of question though theoretical biologists raised this as an important issue long ago.

Although different from those described in the literature, Sam's maps, as exogenous format, still stand in a reflexive relationship to the 'endogenous' terrain of the lizard. The map that emerges from the activity does not preserve the geographic-mathematical properties associated with a two-dimensional expanse. If the tagged capture sites were located and mapped by surveyors, the resulting maps would not be congruent with that established by the ecologist. The maps, however, are very useful and achieve the practical purposes of our ecologist. They are used to construct and explain properties of the reptiles, their ranges and movements, and characteristics of their lifeworlds (from the ecologist's perspective). Capture sites as marked are treated as independent of the search process taken to be one space-time co-ordinate from the life history of the reptile. Although the maps were not literal impressions of the mathematical grid onto the natural world, there was a correspondence between the field and the map. The correspondence was then achieved by means of the tags that allowed a potential superposition of map and the world. In the same way as the digits, and other tags on socks, enclosures and boxes are elements of the co-ordination work that ensured that a particular measurement ends up in the appropriate cell of the spreadsheet.

7.5. TAMING NATURE: MEASURING

Representations are important devices to track the movements of lizards, toes, sheets of paper, measurements, boxes, enclosures, numbers and digits; they are also important in providing some assurance that the different sheets of paper can be articulated in subsequent layering, adding and correlating processes. These technologies also allow a co-ordination of measurements that come from different instruments with different ranges of value, accuracy, and practice even within what might be considered one and the same dimension such as 'length'. Thus, whereas philosophers and epistemologists may categorize length measurements into the same category, as pertaining to the same concept, the work practices associated with length measurement are heterogeneous in precision, concerns for replication, variations and so forth. Here, we examine several measurement practices which allow ecologists to digitize their animals and subsequently re-represent in mathematical space after starting with initially recalci-

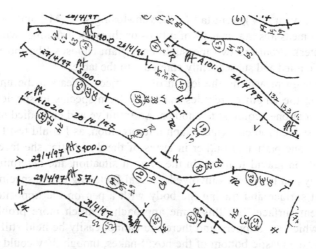

Figure 7.6. Section of a transparency where the centerline of lizards and major anatomic locations (snout, vent, tail end, break-off point of lost tails) are recorded. The transparency is subsequently used to determine various length measures including snout-vent length and tail length.

trant specimens which they work methodically to immobilize, expose, work with, and (literally in the case of color determination) perfect the animals' surface appearances.

7.5.1. Sizing Lizards

With a few rapid movements, Sam removes a lizard from its holding box or enclosure and places it into a clear plastic box. She covers the animal with foam and presses it against the bottom of the box thereby literally flattening the recalcitrant lizard and constraining it from further movement. After turning the plastic box over, she produces a trace along the center of the immobilized lizard on the other side of the clear plastic and adds a few orthogonal markers that later allow her to make measurements that will fill the 'mark sheet' table. This includes marking the positions of snout (H), vent (V) and tail end (T); in those cases where a lizard has previously lost part of its tail, the end of the part not lost is also marked (TR; see *Fig. 7.6*). Each trace is then identified using the same information that also marks the toe vial: date of capture and lizard code. After a days work, the marks on the plastic box are erased after they have been traced onto a partially translucent paper to produce a permanent inscription. Using a map measurer Sam traces the line twice between each pair of markers to produce snout-vent (SV), vent-tail (VT), vent-losing scar (M), and losing scar-tail end (TR) measures that consist of a two- or three-digit number without decimal (millimeter). From the two measurements, Sam chooses 'the most consistent', which is circled and subsequently entered into the table.

Although measuring the length of an object appears to be unproblematic, measuring the length of reptiles may vary from species to species and within species from one ecologist to another. An intuitive and embodied understanding of the possible problems enter graph interpretation, as I could testify, when Sam explained one particular outlier in terms of the difficulties she faced measuring leg length on gravid females. In the present situation, the determination of lizards' body and tail lengths involved a literal flattening of the specimen against a clear flat surface and tracing the body with a pen on the opposite side of the transparent surface. But this same approach is much more problematic with skinks, which are slippery and therefore cannot easily be held still between the foam and the plastic bottom of the box. Snakes, though they could be measured in the same way, are sometimes (especially by male ecologists) 'snapped'. Snapping consists in holding the animals behind their head and at their tails and stretching them along a ruler. Snakes resist this process which makes length measurement a show of strength ('macho thing') and, because the process often separates the vertebrae, make the snake longer providing ammunition for discussions of who captured 'the biggest' snake. Snapping would be a fruitless endeavor with lizards and skinks because they are capable of tail autotomy (shedding of parts of the tail).

The observed transformation is one from a physical, three-dimensional animal into a one-dimensional line that, because it curls in a second dimension needs another transformation (using the map measurer) before it is useful as a zero-dimensional number (point). Other practices produce numbers more directly by suitably bringing the lizard in contact with some instrument. To construct leg length, Sam aligns each leg, using the fingers of one hand to press the leg against a desktop ruler in the other hand. 'Counting on' or 'counting back' millimeter-indicating lines from the closest centimeter marker converts the analogue lengths into a two-digit number including a decimal (centimeters). To produce an item for the head-width (HW) column, Sam takes a vernier caliper, pushes the lizard's head between the legs of the caliper and closes it. A measurement is then produced by 'counting on' or 'counting back' twice—first to the end of the vernier, then to the closest match between a line on the vernier and one on the main instrument to result in a two- or three-digit number including a decimal (millimeter).

These descriptions illustrate that the perceptual topology is quite heterogeneous even concerning the measures along one dimension (length), from the 'same' object, of the same order, their frequency of replication, accuracy and units used (millimeter, centimeter). Furthermore, some instruments produce 'digital' read-outs and others analogue signals that are converted by the user into a number.

The analogue instruments such as ruler and vernier caliper, like many other instruments in the past, are based on reading from some form of ruler usually linear but occasionally circular. Thus, weights are read off at the position of a standard mass on the arms of balance beams or on a linear Newton scale; distances are measured by rulers and tapes; time and compass readings are on circular scales, but involve similar divisions of the circle into major and orthogonal minor lines; and thermometers convert an expansion of volume into a one-dimensional expansion of height. In such devices, measuring is fundamentally conducted by counting equidistant line elements. Ordinarily, line elements are grouped and labeled so that measurement is not a literal counting from the line at the beginning of the scale to that closest to the tip of the object, but a 'counting on' or 'counting back' from the closest number. Thus, these traditional devices involve some practice that converts the object into a linear analogue signal (length), and then a conversion by means of an appropriate count. Other instruments such as the map measurer (or scales) provide a digitized read-out. Here, cultural knowledge is embodied in the measurement instrument, which may decrease the number of errors in the conversion work that turns analogue signals into digital signals. The digital device does work that previously was done by the scientists. Several ecologists pointed out that the advancement of the field lies in the move from analogue to digital devices.

7.5.2. Sizing the Habitat

In the field, the heterogeneity of perception relative to 'length' measurement increases. Some distances are measured using a twenty-five-meter measuring tape to the nearest decimeter or centimeter (distance to next rock, bush, tree, forest, other tags); rock dimensions are measured with a small measuring tape to the nearest centimeter; other distances are determined by counting the number of paces; and the distances to the nearest forest are often estimated to the nearest five meters. That is, there is no single referent for measurement and accuracy for 'length'; and the practices distribute the work of arriving at numbers and digits in different ways. Using tape measures draws on embodied practices and 'counting on/back' described above; pacing is literally based on counting the number of steps and physically covering distances; and estimation draws on many embodied experiences and associated cognitive work to bring forth a reasonable measurement estimate by merely looking at the field context in its entirety. 'Appropriate accuracy' is a practical problem done in reference to the domain rather than in some abstract field of references; that is, the move into the field changed not only the size of the objects but the nature of the objects of interest was ill-determined as well. In the laboratory, lizard snout, vent, tail end, tail stub end, beginning and end of leg all appeared to be unproblematic locations that could be marked or against which a ruler could be held to make a reading. The objects of measurement in the field site, however, sometimes emerged from the snags encountered during the activity of measuring.

To construct some indices about the preferred habitats of lizards, Sam decided to measure 'distances to the nearest {rock, shrub, tree}' and the 'distance to the forest edge' because she expected these to be cover objects to which lizards run if they are disturbed at their site. But 'rock' did not literally mean rock, for Sam concluded that the rock had to be one 'that a lizard could run under' so it must be moveable and have visible areas that a lizard could run under. It also has to be 'at least ten centimeters wide so it can be a suitable cover object'. What it means to measure 'distance to the nearest shrub' or rather what 'nearest shrub' remained unspecified until Sam and her assistants 'sized' the second site. Here they had to deal with plants that are technically 'shrubs' (according to a field guide) but which do not offer any possibility of cover to a lizard. At that point, 'shrubs' were more clearly operationalized as those bushes that have a one-meter base diameter.

Here, in the variable 'distance to nearest {shrub, rock}' both distance and {shrub, rock} are problematic, for the objects have to be such that a lizard could hide from a predator. Nearest rock therefore was the nearest rock under which a lizard could hide given its length, that is ten to fifteen centimeters. If we think of the environment (home range) as a background of the lizard, it is only partially captured in the representations of the scientists, and curiously enough, in their terms rather than those of the animal. Even within ecology, measures such as 'Distance to the nearest {tree, rock, forest edge, rock pile}' are open to critique because they are decidedly human categories for which we do not know whether they play any role in the lifeworld of the animal. Objects such as bushes and forest edge, because of their extension, also render problematic the identification of a mark that identifies the point at which to read the tape measure and becomes a practical achievement in operationalization. In the case of the forest edge, the question of just where to mark the edge was solved with the decision to record the distance to the closest five meters.

To determine the size of rocks under which lizards are found, Sam uses two intermediary inscriptions: a sketch with the approximate shape of a rock in the form of a polygon, each side bearing the associated measurement in centimeters, and a drawing with the length of the sides to scale. Yet the drawings are not 'scale drawings' proper as the angles between the sides are preserved only approximately. Sam deconstructs the polygon into triangles, measures base and height using a ruler, and finds the area of each triangle by drawing on the school-taught algorithm (one-half base times height) finding the total area through summation over the areas of all component triangles. The size of the rock surface is not measured using an instrument for size, but, through a series of transformation that allow Sam to work with the resources she has at hand including measuring tape, paper and pencil, ruler and two mathematical algorithms (area of a triangle, summation). Although there are potentially many different ways for finding the area (including ways that are more accurate and more correct from a mathematical perspective), the ecologist's work of finding rock surfaces is taken as practically adequate way for constructing the quantity. The adequacy of the particular way chosen is a practical question rather than one that is usefully solved by reference of more accurate, rapid, etc. methods. Similar arguments hold for the accuracy of the measurements in the field. It is immaterial that a few lines of arithmetic might have yielded a formula for calculating the area of a triangle from the length of its sides. (A formula which can easily be embedded in a spreadsheet and therefore would reduce the work of finding areas and increase the precision of the result.) Though Sam has the required computer and tape measure, the associated practices and cognitive resources are not ready-to-hand. But her methods and techniques are the result of practical choices that obtain their efficacy, adequacy, and legitimacy in the embodied and socially organized praxis of her discipline.

7.5.3. S(e)izing Color

Producing scientific knowledge by working around and within the local contingencies of equipment and instruments places the scientist in the awkward position of having to 'make do' by balancing the local solutions with the anticipated, evaluative conventions of a wider scientific community and audience'. This is especially the case when a scientist has little familiarity and experience with some measurement instrument. During the first year of this study, Sam was just beginning to use the Munsell chart that provided an opportunity to view the construction and appropriation of color determination. Because Sam had not used the chart previously and had not seen the chart being employed by someone else, she had to discover in her own activities the use and details of practice. She learned to use the instrument through a protracted process of embodiment.

From the beginning of her work with the Munsell chart, Sam had particular concerns about the accuracy of color determination and color variability, and therefore set up conditions in the field laboratory that allow her to control variation. Sam constructed a special arena, a cardboard box closed on three sides by white walls, and installed a daylight incandescent lamp. In the movement of color determination into the laboratory and hence into an even more isolated and decontextualized situation, we observed a considerable change in the practices associated with the Munsell chart from those usually observed in anthropological and geological fieldwork where samples are investigated directly in the field. Each page of the Munsell chart consists of color plates in the center of which there is a viewing hole through which the object of interest is examined. In this way, each viewing hole of the Munsell chart already provides a context free reference and abstracts the object of inquiry from the surrounding context; moistening of soil samples brings about an additional uniformity among different objects. The double move into the laboratory and into the surroundings of the uniformly white box and the additional cleaning of the animals (a literal cleaning of the lizard's surface appearance), however, constitutes an increasing abstraction of lizards and their color from the conditions in their natural habitat. These moves have a transcendental flavor in their quest for constructing lizards 'as they really are' not as they appear to be in some context of their home range. In their natural surroundings, lizards are embedded in dense, complex visual contexts. Although change of color appears almost non-existent in this lizard species, determination of lizard color is transferred to the laboratory where contextual aspects can be controlled and are the same for all lizards. Thus, Sam creates a new physical space designed for the sole purpose of determining lizard color. Whether this has any relevance for the color in the field has not been determined, but the practice satisfies the scientific community norm of repeatability, consistency, and observer bias. It also satisfies the requirements of projected statistical analysis in which individuals and groups are compared and processed *ceteris paribus*. Comparing lizards on the basis of data collected in a special place irrespective of their native home range is the scientific practice related to implementing *ceteris paribus* conditions.

In the work with the Munsell chart, the ecologists found that they were rather consistent in determining the percentage dark coverage (mottle) but they have trouble 'doing the color'. A field assistant frames this trouble:

You have to be consistent about where you are looking both generally (center back behind the shoulders) and specifically (the outer edges of some scales were more lightly colored) and it seems that colors are sometimes between the ones available. The lizards keep moving around and stick there heads through the card holes too which makes it difficult to determine the color.

The ecologists' most difficult task is to 'find the closest match' which requires moving the lizards back and forth from hole to hole and switching cards back and forth three or four times until they finally announce a number followed by two more representing hue, value, and chroma based on the closest match to the lizard. However, what Sam wants, and what the chart was designed for, is not directly accessible because of the structure of the instrument. Because an exact match is a highly unlikely event (the text in the introduction of the Munsell chart book indicates about one in one hundred), the user has to determine the closest match between an existing color and the object at hand. To arrive at the closest match, the user therefore has to compare the results of two comparisons. That is, the user does not simply try to get a match, but the closest match from a range of matches. But the structure of the instrument, the viewing hole inside the color patch, only allows one match at a time, the next one being dislocated in time and space (often other page). If we take the distances of two matches with Color 1 and Color 2 as Δ_1 and Δ_2, respectively, the cognitive work to be accomplished is this. Seek a good match, determine and remember Δ_1, seek another match, determine and remember Δ_2, compare Δ_1 and Δ_2, and finally choose the smaller as the better match. In this, at most one of the distances Δ is presently visually available whereas the other one has to be held somewhere in memory.

Despite the multiple moves in decontextualization, the color measures do not appear to be 'repeatable' and 'consistent' without any 'observer bias on color measure'. Sam's concern is that of accountability: 'I want to get an idea about if it is consistent between observers, I have no idea'. What she needs is some kind of index and sense for the amount of variability, she needs 'information about the precision of the measure ... something like standard error'. At the time she does not know whether she can convert the Munsell classifications into numerical data. Sam decides to conduct a 'blind' test involving two field assistants, but her assessments agree on only two of eleven color categorizations with those of the first and in seven cases with the second field assistant. The mottle determination is somewhat better, but still not good enough so that Sam could be comfortable having somebody else do her color work. She decides that she will do all the color work herself, requiring her to determine her own consistency. In the end, Sam was discouraged. 'I am not sure I am measuring anything. I know I am measuring something, but I don't know if it really means anything in terms of whether it is repeatable'.

Although nobody holds Sam accountable then and there, and although it is unlikely anyone would ever engage in an audit, she works hard to make precision and with it the objectivity of their observation an accountable aspect of their work. Sam and her fellow ecologists treat 'replication' as repetition of process to check reproducibility; repetitive measurement contributes to the authority of the data and to the credibility that the research is replicable across sites. The results of 'replication' provide an index of variability, which is not a problem as long as it is accounted for and expressed in numerical terms. Ultimately, replication allows ecologists to make claims about the soundness of the research done and the quality of the data presented.

Although the ecologists claim that 'Doing things repeatedly is a habit', this is not universally the case in their own practices. 'Unproblematic' measurements like length or weight do not necessitate replication, but 'problematic' ones such as color determination make replication necessary. The trustworthy and objective character of the table cell entries emerge from the participants' detailed attention to the systematic practices used to constitute measurement, and their recognition of the difficulties involved in unambiguously measuring and classifying complex continuous phenomena. In the present case, the objective character of the descriptions that enter the computer emerges from the associated practices that make 'repeatability,' 'consistency,' and 'observer bias' rationally—that is, in field-specific ways—accountable. Here, 'repeatability,' 'consistency,' and 'observer bias' are enacted in the form of independent assessments. The objectivity of the work of measuring and coding is provided for by arrangements that encourage the emergence of an accountable practice. The products of these activities are inscriptions that can be transported to different places and worked on at different times. The definiteness provided by the schemes erases from subsequent documentation all the uncertainties involved in establishing the first step, that the researchers are grappling with, and their work practices within which their activities are embedded.

7.6. MEASUREMENT: ADEQUATIO REI ET INSTRUMENTA

In the previous section, I described the heterogeneous topology of measurement practices in the ecological field camp. In this section, I theorize the process of measurement as an operation in which the world is collapsed into an observable, that is, a process analogue to digital conversion characteristic of the epistemology of Western science. Lynch provides an analysis of the relationship between the world and mathematical structure that is topicalized in the isomorphism {world ↔ mathematics}. I extend this analysis by introducing a formal definition that includes the isomorphism between world and language {world ↔ language} and that shows how adequation centrally involves a physical relationship between object and instrument.

Traditionally, scientists assumed that 'nature' existed independently of human concepts and measurement. However, atomic physicists questioned this assumption in the 1920s. They suggested that we can speak of particular phenomena only in terms of the relation between some 'preparation' of the world and a measurement process; the phenomena, such as electrons, exist only in and as of the relation between the preparation and the measurement instruments. That is, the adequation '↔' involves a physically instantiated relation between 'world' and 'mathematics' provided that the instrument embodies structures that are consistent with mathematics. It is in the process of measurement—of physically bringing together preparation and measurement—that specific values of the mathematical representation and thereby the particles can be said to be realized (which in quantum mechanics is known as the 'collapse of the state function'). At this moment, an analogue world is collapsed into digital signals that are characteristic for Western cognition. We see a similar process occurring with macroscopic objects. The preparation involved extractions of animals not only from their natural habitats but also from their temporary habitats, the boxes and enclosures in the field laboratory. In the situated activity of color identification, the individuals physically isolate them and their perceptive apparatus from the larger setting, an embodied form of 'abstracting' the topics of research from the 'non-scientific' 'qualitative'. For length measurements, the lizards are transported into a special environment. Through the physical relationship between the lizard and a ruler (or some other intermediate process), some specific points of the animal (snout, vent, tail end, end of non-lost part of tail) are related to marks on the ruler.

To aid the tracing of the data construction and transformation, I introduce the following notations. Let 'x' stand for an (in principle unknowable) aspect of the lizard and A for some operation (associated with a mathematical operator). An observation arises from doing measurement A_i to the aspect of lizard j, $/A_i x = a_{i,j}x$. This 'doing' corresponds to the material activity of physically bringing together animal and instrument. The values $a_{i,j}$ produced in this way bring about what is observable and can be represented, and therefore what constitutes the knowable about each lizard. These $a_{i,j}$ are in fact the cell entries which together establish Sam's data table and spreadsheet. By doing the very physical process of approaching the lizard to different measurement instruments, an operation, the animal is associated with an ordered array (or rather vector) of properties. In this approach, the construct (e.g., 'length') and its metric are indexical to the data, they reflexively stabilize each other, but are only 'imaginably indexical' to worldly things. For example, the length operator L and values l_j it produces are reflexively related, and the l_js are only imaginably indexical to the phenomenon ($Lx_j = l_j x_j$). Constructs, theory and interpretation have relation only to the data but are not the worldly events themselves.

The notation makes it clear that from a potentially infinite number of states, only one is taken during the measurement process, a process of analogue to digital conversion. The epistemological rupture between the two ways of knowing occurs when natural objects (preparation in quantum mechanics) are physically associated with a measuring instrument. By physically holding the lizard to different measurement instruments, the animal is slowly associated with an increasing, ordered array of 'properties'. This formal definition embodies what quantum physicists and some social scientists have recognized for some time. Little is known about the properties of the underlying phenomena only how it looks through the imposition of the format whether this format is that of some instrument or a simple classification which is imposed on the data.

Measurement in science is often regarded as the process of assigning a number to fundamental attributes of an object or event; in the classic view, the correspondence between numbers and objective properties is used to reassign the outcome of mathematical operations on the numbers to the measured phenomena. Viewed from the present perspective, making a distinction between number and attribute is not useful. The operation determines both attribute and values that the measured object can take. Placing the lizard behind the viewing holes of a Munsell chart defines both the attribute color and the possible range of values this attribute can take. In the same way, placing the lizard on a digital scale both determines the attribute and the values, and holding it against a ruler determines length and values of the right front and back legs. My approach to measurement makes it literally impossible not to attend to the epistemological problems involved in the process. In the moment the animal and the instrument are brought together, an epistemological rupture occurs. The phenomenon collapses to a mark or number on some scale, which ultimately entails the construction of human knowledge. This process begins at the moment where the manifold and many-dimensional phenomenon, here the lizard, is brought to and held against, the instrument.

Scientific laboratory work is largely organized around the practical task of constituting and framing phenomena so that they are measurable and therefore can be described mathematically. The relationship also works the other way in that the available tools by and large do the framing as researchers search for ways of 'generating the numbers' on the basis of which they can construct claims. Tools and instruments are crucial implements in the ecologists' practices of constituting 'veritable' accounts of lizards. Although this may not be ecologists' predilection or work of choice, when they think of themselves as part of the community, doing as everyone else does—as they see it—is the central referent to their work. The tools and instruments are part of the perceptual apparatus of the discipline.

7.7. CALCULATING AT LAST

It is only after the spreadsheets have been completed, when the ecologist is back on home campus and begins the processing and analysis of the data that the spreadsheet becomes extrinsic and a mere representation of lizards and their home ranges. It is at the point of completing the last cell that the ontological status of the spreadsheet shifts to become mere representation. From all the paper sources, the data are gathered into one spreadsheet to be exported to the statistics programs for further processing. Lizards are now not only reduced to the digits in the tables, but are also literally digitized in electronic bits. At this point, the work of our ecologists has some resemblance with that of astronomers who receive electromagnetic waves which are converted into streams of digits which are then converted with the help of digital image processing software into colorful pictures and interesting phenomena. Sam also processes electronic digits but with tables and line graphs rather than colorful images as the end product. The image of lizards and related interesting phenomena are produced and subsequently presented to various scientific audiences by artful manipulation of electronic digits collectively addressed as 'the database'. In both cases, however, electronic blips are converted into new knowledge based on scientific 'prejudice' and through artful practice in handling the 'data'.

Another major (epistemological) rupture occurs as lizard individuality dissolves when the ecologist's attention moves from the rows to the columns of the spreadsheet. During fieldwork and during the laboratory data production, our ecologist is singularly preoccupied with the horizontal dimensions in the data table, the production of inscriptions by dealing with individual animals as identified by their codes. The individuality of each lizard, as indicated by lacking digits and associated codes in the appropriate spreadsheet column exists only in so far as it allows to trace the source of the records. It allows entering information into appropriate cells and coordinating it with other entries collected at different times and localities. The rupture occurs as Sam changes from a singular concern with 'horizontal' practice, completing all operations A_i that produce the cell entries $a_{i,j}$ for lizard x_j, to a singular concern for measurements as a class (vertical). At this point, the ecologist is no longer concerned with the properties of individual lizards (same row), but with the mathematical properties of a set of numbers (same column). This shift is facilitated, because mathematically the concern merely shifts from the first to the second index of each data entry ($a_{i,j}$). It is the zero-dimensional nature of each data entry $a_{i,j}$ that allow it to serve as a pivot in the move from rows to columns. Once the shift from rows to columns is completed, codes related to individuality are irrelevant for most purposes; lizards loose their identity and collapse into individual dots on graph paper. That is, associated with the discontinuity as the ecologist shifts attention from rows to columns, change of pace and change from one activity type to another, a second discontinuity occurs on an epistemological level. Occasionally do lizards as individuals re-emerge, for example in the cases of statistical 'outliers' such as 'Bertha', a particularly heavy gravid female who spent a particularly long time in the field laboratory, that turned up as a data point far removed from the other lizards. They also re-emerge when Sam is asked to talk about her graphs to individuals and groups not familiar with lizards and the kind of research she does. In the computer, the animals and the world are lost, only to re-emerge in constructed transformed terms of population averages, correlations, standard deviations, etc.

As lizards loose their individuality, statements about populations of lizards become possible. This seems to be the classical movement in the construction of science, paralleling the endeavor in physics, to construct law-like statements about particular objects of interest. These statements are about populations rather than individuals. Behaviors and phenomena arise at the level of the population rather than at the level of individuals. This does not have to be. As ethnomethodologists emphasize, structure is not there making individuals into cultural dopes by imposing behavior; rather, structure (such as 'power') emerges from the situated activities of people. Recent modeling in computational ecology shows that flight patterns of bird flocks can arise starting with the behavior of individuals that can be described by a few simple rules such as keeping proximate distance and speed to the nearest neighbor(s).

What is related during statistical analysis are not only heterogeneous aspects, but the members in each number pair were collected at different times and sometimes in different locations assuming that when and where measurements are made does not affect knowledge claims, unless there are contrary patterns emerging. For example, the number of days in captivity mediates sprint speed; there is a negative statistical correlation between sprint and days in captivity.

Then, so not to affect the claims about other relationships, the statistical apparatus is used to 'control for' or 'partial out' the variation due to this variable. Furthermore, there is the heterogeneity of tools and resulting measures which, on the surface, appear to have little to do with each other but which are associated in computer (hyper-) spaces that link the measures from these different devices. The processing of the lizards in the way done by the ecologists ahistoricizes the data—even with the time of measurement recorded—and leads to the semblance of ahistorical truths reported about the animals. Even with specifics of the measurement situation recorded, such as the temperature, days in captivity and so forth, the correlations bring together animals and records processed at different times. Despite the very time-consuming lived work of ecological research, the end product, propositions or graphical representations about the animals have transcended time.

Another separation quite significant to the work of ecology is that of data about the specifics of the capture location and the lizard itself. Ecology is often thought of as a discipline that emphasizes the organismic and holistic view of nature. However, in the present case, the animal was separated from its environment, as were the corresponding data. These were united only later in the final computer spreadsheet where the data cells with contents of different origin could be correlated again. Thus, statements about the lizards and their environment arose after a separation of the animal from its natural environment—which in the case of the females could last for several months. Scientists' work inverts the description 'Graphing a phenomenon identifies the thing or relationship with the analytic resources of mathematics' in the sense that the mathematical resources bring about the phenomenon. The very factual statements 'leg length determines sprint speed' and 'sprint speed is inversely related to body length' arise from a reading of the graph (or the corresponding statistical information on the computer printout). In this notation, the ecologists' data analysis relates the observable outcomes of two or more measurement operations. That is, if sprint speed is the outcome of completing some time measurement (t_j x $_j$ = /T x $_j$) and body length the outcome of the associated measurement operation (l_j x_j = /L x_j), the assumption implicit in the ecologist's statement about the correlation is that it is not only a statement about the covariation of two arrays of numbers, but by retro-projection, also about lizards (x) as a class. Here, the outcomes of numerical operations are projected into the world, a similar to digital to analogue conversion. In the graphs used to illustrate significant covariations between coordinated sets of numbers, individual lizards have been reduced to single points in a hyperspace. Now, even their codes are gone and their individuality reduced to an infinitesimal point in a computer and paper world. Yet in this singularity, it contributes to the construction of factual propositions such as 'The speed of lizards is determined by their size and length of the hind leg'.

7.8. MATHEMATIZATION OF PROFESSIONAL VISION

Lizards are not unitary things, and, in the way scientific audiences experience them in journal articles, colloquia and conference presentations, are not living things at all. A heterogeneous conglomerate of inscriptions, linked to objects of different scale, type, and nature; data are collected at different places and time scales. To bring together these heterogeneous things, and at one place, things have to be inscribed and transformed. The moment lizards and their home ranges are brought to and mapped onto the topology of the scientific instrument, the multifaceted complexity of real biting and defecating lizards is transformed into the phenomenological categories of the behavioral ecologist. At this point, this rapprochement of cultural and natural objects that the world is domesticated, made compliant and transformed into culture and that the lizards are disciplined. They are made to comply with the disciplinary perceptual machinery and they are made part of the disciplinary perceptual machinery and knowledge structures. Lizards, unknown and wild, are domesticated (and 'disciplined' in multiple senses of the word) into that which we know. The captured animals constitute boundary objects in that they mediate between the natural world from which they were literally abstracted and the scientific world of the laboratory. In the world of the field, rocks are flipped and returned, lizards caught and observed, plants identified, distances measured, rocks drawn and so forth. In the world of the laboratory, lizards are 'digitized,' weighed, measured, run (speed trials), fed, sexed, color s(e)ized and so forth. All these activities lead to a dizzily proliferating collection of various kinds of inscriptions. Only later further up the ladder, once the local details are removed and forgotten, can these inscriptions with heterogeneous origins and transformation trajectories be brought together and related such as in a graph or table.

Although the research I observed is generally descriptive, the perceptual apparatus involved goes beyond and transcends the metaphorical retina and has a more complex and heterogeneous topology—i.e., it is not a plane or spherical surface. Though the retina is still involved, the recording devices are instruments of different size, quantity, error rates, etc. and the practices associated with them. The field laboratory constitutes the observational machinery of the ecologist, and the associated practices constitute the practices of perceiving. But the topology of the perceptual apparatus is neither homogenous nor homomaterial, but quite heterogeneous and heteromaterial and the associated practices have many of the local and ad hoc characteristics observed in other studies of scientific laboratories. This topology has a cultural history and constitutes a disciplinary frame for what can be seen and the particular metric than can be employed. In a similar way, the column headings determined the dimensions along which the ecologist could perceive; these dimensions are historically determined by the cultural knowledge and the instruments, which themselves have long histories and embedded assumptions about the world. The selection of these dimensions is an inevitable accompaniment to the crafting of inscriptions relevant to the purposes at hand. What makes the selection problematic is when they are presented as matters of necessity rather than as a practical, political or economic choice. What remains are depictions that are naturalized as obvious and disinterested perspectives. Perception in this way is also consensual in that the topology of the perceptual apparatus is preserved. Sam makes her practices accountable and credible to the extent that others will take her description as a legitimate perception of the lizard in the foreground of the unstated naturalistic assumptions that take for granted the existence of the species, identifiability of the species members as separate and nameable, and their indexability through various forms of evidence. My topology of the observational machinery is part of the project to establish an *alternative geography of cognition*. Crucial phenomena are not located exclusively in the brain but in historically and physically situated activity systems that make up the lifeworld of the individual. Activity systems, by definition, are always situationally contingent and therefore require accounts that permit the ecological and evolutionary nature of the system to become salient.

The proliferation of inscriptions and use of numbers was noted throughout this study. However, numbers alone do not indicate a 'mathematization' of the lizard. Some of the number used draw on indexical properties, other numbers draw on their ordinal properties; some number used draw on arithmetic (cardinal) properties of numbers but only to produce efficient and 'least-damage-causing' ways of coding. Number use then is not equivalent with mathematization that contributes to statistical and mathematical modeling of nature. Furthermore, the fact that measurement exploits the arithmetical properties of numbers is deeply embedded in the cultural history of the field, for measurement does not involve cardinality by necessity. Similarly, the cardinal property of time is an invention of Western culture but was largely treated in an ordinal dimension by most cultures. How a lizard becomes a mathematical (statistical) object is partly due to the historical processes that produce instruments in the first place. When such instruments are part of the perceptual apparatus, the mathematization is partially made invisible. But using numbers to indicate measurement is only part of the story of mathematization. An epistemological shift occurs, and with it the full impact of mathematization, when the measures on individuals that fall into the same column are treated and processed as members of a class of numbers which have statistical properties.

Lizards are literally visible to those in the field and field laboratory who catch and 'process' the animals. In this sense, Sam has a relationship with the lizards as living beings that move around, are difficult or easy to catch, and show specific behaviors allowing her to identify each individual—some of whom she gives names. She describes changes in observations on female lizards by taking the perspective of the animals. Her depiction of lizards shares a lot in common with naturalist portrayals of animal life. But, from her perspective, all she can collect by means of literal seeing is 'anecdotal' information; in the parlance of her field, even the color descriptions one might find in field guides are anecdotal. In the context of her domain, behavioral ecology, Sam has to make lizards visible to others by means of a different perceptual machinery that allows her to arrive at 'scientific' descriptions of the animals of interest. This process requires animals to go through the process of digitization, a conversion that involves 'hard' numbers and electronic bits, before other ecologists can 'believe their eyes'. But as part of this process of converting nature to electronic bits, Sam's field activities are open to an indeterminable horizon of contingencies that arise in the course of the scientific work that results in 'knowledge about the lizard'. Through Sam's work, the lizard is constructed and thereby made visible, involving a proliferation of inscriptions, conversions of nature into numbers, and electronic digits. Lizards are reflexive of endlessly awkward processes and objects left behind in the field laboratory and terrain. That is, literal seeing while important to the individual ecologist is not as important as the 'observations' possible once the animals and their environment have undergone multiple transformations. Instruments, field laboratory, inscriptions, and associated practices therefore constitute pieces of an observational machinery that does not have the even, spherical surface of the human retina but has its own multi-dimensional, heterogeneous, and heteromaterial topology. This chapter is about the topology of this observational machinery that turns nature into (afferent) digital signals that are the data for subsequent processing in centers of computing and ultimately into the graph that are the topic of this book. These signals are not pure in any sense but are always and inevitably formed by the enacted disciplinary practices. That is, any relevant aspect about the lizard's life history or natural history emerges from the interplay between the domain of inquiry and existing discursive practices and material configurations. In the present study, these discursive practices were more related to instruments and measurement and less to unifying concepts and theories regarding ecology and evolution.

To be represented in graphs, lizards have to be observed using the whole observational machinery available to a field ecologist, including analogue and digital scales, analogue and digital length-measuring devices (e.g., ruler, tape, map measurer, paces), Munsell charts, and digital stop watches. As a result of physically bringing together natural phenomena and the instrumental topology, numbers composed of digits begin to fill rows and columns of a spreadsheet. The entire laboratory machinery is used to produce digits which are then summarized, compared, correlated and otherwise processed by means of statistical software producing tables and graphs (i.e., series of non-trivial and arbitrary transformations of subsequent stages). These tables and graphs, in the context of some form of text, originate, support, and legitimize scientific propositions about the lizard such as 'sprint speed is inversely related to lizard length' or 'maternal tail length predicts number of offspring'. In the process of the scientific work, lizards as biting, writhing, running, feeding, and defecating creatures are turned into series of numbers composed of one or more digits—first on paper and subsequently into the electronic digits of spreadsheet and statistical software.

Ethnomethodologically specified, measurement is a hopelessly vulgar competence. My interest lies in the local, in situ practices by means of which ecologists conduct their measurement-related activities which includes producing local judgements related to the practical adequacy, accuracy, and appropriate correspondence between measuring device and measured phenomena. My interest lies further in the transformations that the original measures undergo until the final graphs emerge. This chapter shows that there is a lot of accountable work involved in the collection and transformation of data that ultimately are turned into work. This chapter provides documentary evidence of this work, which is ultimately indexed by the graph in a metonymic way (see Chapter 5). There exist the potential dangers of 'going native' arising from my own training as natural scientist. But I did not treat scientific measurement as a homogeneous set of methods and standards but as an ecology of heterogeneous techniques that allow scientists to attain locally recognizable and locally adequate measures. I provide documentary evidence here that measurement, though a familiar term of the ecology trade, is neither a coherent interdisciplinary practice nor coherent for the same scientist across situations, even along physically similar dimensions (e.g., 'length' and 'distance'). Although the measurements generated by our ecologists may have had referents in the practices of the discipline, what we observed was how each of the scientists locally elaborated and enacted for him/herself the meaning of the cultural referents in *this* setting.

Mathematization that underlies graphing has been described as one aspect of the process of scientific seeing. In this chapter I show *just how* mathematization is achieved in the context of physically demanding ecological field work and describe some of the discovering practices in ecological fieldwork to disclose the order of the local contingencies of the day's work in an ecology field research camp. I show how the scientists in the field made their work accountable in, of, and as instances of measurement practices. However, what mathematization achieves is more than just seeing, for it is no longer individuals that are seen but a conglomerate of indistinguishable replaceable individuals. It is the move from 'this lizard is doing X' to 'lizards do X', from the psychology of the individual to the sociology of the masses. Because sweeping and 'fuzzy' generalizations describes no one individual in specific, statements such as 'leg length determines speed' can stand as a factual statement about this animal in general. This epistemological rupture between the description of individual animals and statements about all the animals as a class is associated with scientists' shift from the field laboratory practices to data processing in computers. There is a similar shift from filling rows of a spreadsheet associated with individual animals to processing columns of numbers that summarize individuals into classes.

Tables and graphs, which abound in scientific publications, are visual documents that 'integrate the substantive, mathematical, and literary resources of scientific investigation, and create the impression that the objects or relations they represent are *inherently* mathematical'. At the same time, the real mystery is the adequation of mathematics with the empirical world not the superimposition of one mathematical form with another. This isomorphic relationship between mathematics and the empirical world is a given for many scientists. However, rather than being a simple relation or adequation, there is a possibly infinite chain of signs linked together by the embodied practices of their users. Reference is therefore a quality of a chain of signification that mediates between the purer elements of empirical world and mathematical form toward the extremities of the chain. Scientists who use numbers in their research do not necessarily mathematize the world though they still draw exploit the mathematical properties of mathematics.

EIGHT

Fusion of Sign and Referent

From Interpreting to Reading of Graphs

In this chapter, I focus on the interpretation of (unknown) graphs as they come off the research apparatus in an advanced research laboratory in biology; based on this initial interpretation, the scientists discard the data or save them for more detailed analysis conducted when they are no longer under the time constraints imposed by the ongoing experiment. These interpretations have not been practiced and rationalized; rather they are the results of analyzing the graphs in a first-time-through manner. But in this first time through, the scientists exhibit the methods by which what they do becomes naturally accountable. The scientists also use their interpretations to make decisions about whether their experimental apparatus is properly functioning and whether the sample is still in sufficiently good state to permit an analysis. My analysis renders evident that scientists build up complex schemes for reading graph through extensive experience with all parts of their apparatus. This apparatus is a form of optical device that allows the scientists to see the natural object of their interest—here salmonids, or rather, pieces of their retina. I use a cultural-historical analysis to show how the laboratory and its inhabitants come to internalize processes that are initially mediated by tool use.

8.1. INTRODUCTION

Throughout this book, there are examples that show how scientists do more than making inductive inferences that lead from graphs to corresponding 'natural objects'. What is a salient aspect of a graph is not known a priori, that is, without knowledge of the natural system. All evidence shows that graphs, tools that produce them, and personal familiarity with the natural system are mutually constitutive and are convergently stabilized by the scientist in praxis. This became quite clear to me when I first joined the biological laboratory that is the focus of the chapter. Despite my background in physics and statistics, and de-

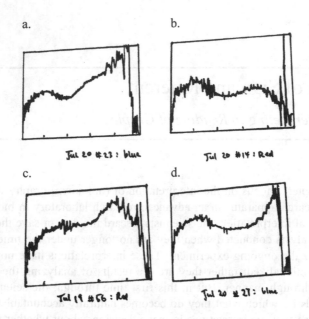

a.

b.

c.

d.

Jul 20 #23 : blue

Jul 20 #14 : Red

Jul 19 #56 : Red

Jul 20 #27 : blue

Figure 8.1. Facsimile drawings of four graphs on a computer monitor as they were produced online during experiments in a biology laboratory. Whereas the scientists familiar with the apparatus and the biological system at hand make clear identifications, a person unfamiliar with biology and physics of instrumentation does not know what to interpret and how to interpret it.

spite years of research on the topic of graphing, I could not correctly interpret the graphs that were the result of a data run. In fact, it was not clear to me which graphical feature was to be salient and therefore to be interpreted. Initially, I was frequently attuned to features not at all salient to the resident scientists. Before taking a deeper look at the processes of working with graphs in this laboratory, we take a look at four graphs and the corresponding final conclusions reached by the team of scientists.

In the first two graphs of Figure 8.1, we might be tempted to identify two features each that appear at about the same locations in the wavelength spectrum (plotted on the abscissa with increasing values from left to right). However, the analysis of the laboratory interactions between the two scientists who collected and interpreted these graphs left little doubt that in first graph (*Fig. 8.1.a*) one had to see the left hump whereas in the second graph (*Fig. 8.1.b*) one had to focus on the right hump. The first graph was said to signify a blue retinal cone whereas the second graph signified to a red retinal cone. That is, in the first graph what was salient to the scientists is the first, smaller peak whereas in the second, the right (smaller) peak predominated in their perception and interpretation. Yet, a quick inspection of the two graphs shows that the left peaks in both are almost identical. How then do the scientist come to look at the second graph (*Fig. 8.1.b*) and apparently its left peak, not paying any attention to it as if it was transparent or even invisible to the eye? We might ask, how do scientists come to the point of articulating the left hump in the first graph (*Fig. 8.1.a*) and the right hump in the second graph (*Fig. 8.1.b*) as the signal and the other hump as noise, respectively? When scientists begin researching a particular phenomenon they do not and cannot know what the 'right' peaks are, that is, how to 'correctly' interpret the graphs that they have at hand. And yet, with some time they develop a competence that allows them to make assessments, among others, increasingly quickly and without hesitation.

These decisions are all the more remarkable given that a third curve (*Fig. 8.1.c*) was regarded as consisting entirely of noise and not referring to a cell at all. Here, the potentially misleading second peak in the blue part of the spectrum (left) has disappeared and the peak in the red part of the spectrum predominates. It is also much more pronounced than the red peak in the comparison situation. In the fourth situation (*Fig. 8.1.d*), the graph was discarded because it is 'too wide.' If we overlay this blue line with the other blue line (*Fig. 8.1.a*), they are virtually identical. Yet the scientists retain only the first but not the second graphs for further analysis. What mediates the situation is the fact that the final analysis (and therefore retention for publication) will ultimately be made in a subsequent analyses. These are largely conducted by one of the two scientists who collected the data using a variety of graphical analysis tools such as curve fitting, smoothing (Fourier transformation), etc.

8.2. EXPERIMENTING AND VISUAL TOPOLOGY

In this chapter, I continue to use the operator notion developed in the previous chapter for conceptualizing scientific actions, which includes manipulations and observations. This notation is conceptually identical to the graphical notation developed by David Gooding to describe Faraday's discovery of the electric motor but, because it is not diagrammatic, is more parsimonious in terms of the amount of space it takes up.[1] Each performative act is articulated by an operator A_i and each observation signified by an operator O_i. An experiment is therefore represented by a sequence of operations that abstract and transform a material object 'x' (cones from salmonid retina) into graphs. Scientists may inspect the object in its various transformed stages, thereby interspersing direct observations in the sequence of operations that ultimately produce the graph. This sequence is expressed as $O_m \cdot A_n \cdots A_2 \cdot A_1 \cdot x$ and leads to the observation 'a' on the natural object 'x' (which I write, consistent with the quantum mechanical operation notation as 'a·x'). One observation is therefore identified by a series of actions and observations is therefore represented as

$$O_m \cdot A_n \cdots A_2 \cdot A_1 \cdot x = a \cdot x$$

where 'a' is an 'observable' aspect of the natural phenomenon. It is evident that if any of the elements in the sequence to the left of the equal sign are altered, then a change may occur in what is being observed. Furthermore, because of the particular (random) influences due to the individual operations A_i, there are variations in what is actually observed. These variations are not all attributable to the object x. They are considered to be noise in the system. Separating a signal, which is attributed to x, from the noise, which is attributed to aspects of the instrumentation and to random influences on the natural object, is an important aspect of scientists' work. Doing this separation in a competent manner is part of the practical expertise that scientists develop in the course of their laboratory work.

We saw in Chapters 6 and 7 that scientists and the technician Karen read their graphs in a transparent way. That is, I observed a complete equivalence between a graphical feature and some aspect of the world such that pointing to a feature Karen said, 'This is a water from the north creek'. Such remarks show the transparent nature of the sign and the entire process by means of which it has become connected to the natural phenomenon (here, water levels). In terms of the notation I developed, the entire sequence $A_n \cdots A_2 \cdot A_1 \cdot x$ has become transparent to Karen so that her observation O pertaining to the graph at the same time pertains to an observable fact in the creek. A feature of the graph, isolated in the observation operation 'O' has been directly attributed to an observable aspect 'a' of the creek:

$$O_m \cdot A_n \cdots A_2 \cdot A_1 \cdot x = a \cdot x$$

On the other hand, her remark, 'This is a clogged pipe' still shows that some of the operations that intervene between the intake pipe and the graph have become transparent. Now, however, Karen attributes a feature to the pipe (A_2), which mediates the measurement process, rather than to the creek.

$$O_m \cdot A_n \cdots A_2 \cdot A_1 \cdot x = a \cdot x$$

In this vision laboratory, the scientists actually make two observations of different type. They optically observe the preparation and look at the different cells in their medium and select promising objects, rods or cones depending on the current focus of their research project. They also measure the amount of light being absorbed in a rod or cone, which requires many additional operations before a graph that ideally signifies the amount of light absorbed in a cell is displayed. Because a specific cell is associated with a graph appearing in a particular region of the spectrum, a first check whether the data are correct is based on the visual inspection of the cell through the microscope and of the graphical features. This process may therefore be expressed in the form

$$O_{m+1} \cdots A_{n+1} \cdot O_m \cdot A_n \cdots A_8 \cdots A_2 \cdot A_1 \cdot x$$

where two observations are set into relation. From the outcome of this comparison, the scientists make inferences about the object under study, the apparatus, or their actions associated with the preparation and measurement process.

8.3. LABORATORY AND PEOPLE

The research group featured in this chapter is interested in better understanding various aspects of the life history of salmonid fishes on the Pacific Coast of Canada. The group has specialized in research on the visual system of these fishes, its uses, and the changes it undergoes throughout the life cycle of an individual. An important aspect in salmonids is their migration from the rivers where they hatched to the salt water feeding grounds and back to their spawning grounds in the same river systems that they earlier had left. Researchers have found that just prior to seaward migration, salmonids change their physiology, including apparently a change in the eye pigment composition from a freshwater to a seawater form. The researchers are interested in measuring the temporal changes in visual pigment conditions in the photoreceptors (rods and cones), along with changes in other physiological characteristics such as silvering of the scales, saltwater adaptation, or modification of gill epithelia.

As part of the research pertaining to the fishes, the research group also designs new apparatus. For example, it designed an apparatus to increase the speed with which the data are collected or to improve the quality of the signals that they measure from the natural object (here retina). During the initial three months of my stay in the laboratory, the team was working on the data to be published together with a description of the apparatus. These processes of improving instrumentation cannot be separated from the research concerning fishes. Both lead to the same kinds of laboratory activities, though the publications will focus on one issue or the other (instrument, fish). These mutually dependent processes also influence the competencies related to graph interpretation and the developmental trajectories of these competencies.

The videotapes on which I draw in this chapter were recorded while the research group was collecting the data for a specific article. In this article, the scientists presented their apparatus, which decreased the time of measurement from approximately twenty seconds down to about one second (500–1,200 milliseconds).

The entire experiment is conceptually presented in Figure 8.2. In order to be able to work with active retinal tissue, the fish is kept in a dark container for a minimum of two hours. Because of the sensitivity of retinal tissue to light, the experiment has to be conducted at very low intensities of red light—which requires the researchers to dark-adapt themselves for a period of thirty to sixty minutes. The fish is anaesthetized and, immediately before removal of the eyes, sacrificed by severing the spinal cord. After removal, the eyes are bisectioned and the retina removed. Under the microscope, the researcher cuts one piece, which he mounts on a slide whereas the remainder of the retina is stored on ice in a saline solution. The retinal piece on the microscope slide is then macerated using surgical knives and under an infrared microscope, the image of which appears on a small monitor. After completion of this operation, a drop of saline solution is placed over the macerated retina; the preparation is then covered with a cover slip, which is sealed to the slide with liquid paraffin to prevent evaporation of the solution. The slide is then mounted in a the main microscope fitted with two light sources, one for providing the stimulus beam the other, an infrared lamp, for providing the background illumination to search for the objects of interest (*Fig. 8.2*).

Conceptually, the measurement unfolds as follows. In order to obtain information about the photoreceptors in the retina, two measurements have to be made. In the first, a light pulse is made to traverse the slide at a spot where there are no cells ('reference'). In the second, the pulse is made to go through the cell ('scan'). Because more light is (normally) absorbed in the cell than in the surrounding saline solution, the difference in the two light pulses is attributed to absorption in the cell. When a blue cone, for example, is active, then light in the blue part of the spectrum is absorbed. On the other hand, if the cell was

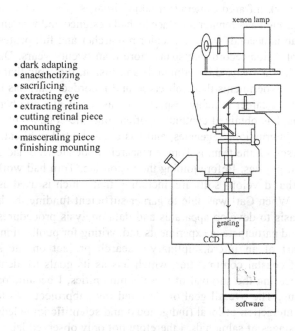

- dark adapting
- anaesthetizing
- sacrificing
- extracting eye
- extracting retina
- cutting retinal piece
- mounting
- mascerating piece
- finishing mounting

Figure 8.2. Processes in the preparation of the microscopic slide of salmonid retina and experimental apparatus that produces the graphs based on rods and cones identified and aligned through the ocular or CCD-based optical system. A second CCD acquires the light dispersed by the grating (bottom), thereby allowing a comparison the light spectrum transmitted through the object of interest and a reference.

bleached—previously deactivated by chemical transformation of light—a different absorption spectrum would be observed.

The experiment is set up such that one person could both align the preparation with the beam of light and operate the computer (counter and data recording); but the scientists practice a division of labor. This division does not only increase the speed with which the data are collected—excised retina is viable for only one hour—but also contributes to collective interpretations made on line. Thus, one scientist is responsible for searching suitable cells in the preparation, to align the slide for experimental and control measurement, and to manipulate the various shutters and light sources (background, stimulus) involved. The other scientist operates the data acquisition device (charge-coupled device or

CCD detector), infrared camera (for taking images of the cells), and the plotting software through a computer interface he had designed and written.

The lab is headed by Carl, a senior researcher and full professor in biology, with a publication record that spans more than twenty years. During this time, his research was focused on salmonids and vision, the topic that was also at the heart of the experiments that I observed and recorded. Carl has been successful throughout his career in many respects: he has received a number of awards and fellowships, has obtained continuous often multiple-concurrent grants from national and international agencies, and has a substantial publication record. The second person on the team is Tom, a research associate who had done his graduate work in physics. Before joining the vision lab, Tom had worked for a small company that develops software, including that which is used as part of Carl's apparatus. When Carl was able to garner sufficient funding, he hired Tom on a fulltime basis to develop apparatus and data analysis procedures, conduct data analysis, and participate in experiments and writing for publication.

As part of an interdisciplinary research project on the socioeconomic changes of coastal communities, which has as its goals to identify alternative ways of securing the survival of these communities, I became part of the laboratory group. The overall goal of Carl and my subproject was to better understand the interaction of local (indigenous) and scientific knowledge with respect to the life stages of salmonids. I therefore not only observed laboratory activities but also underwent something like an apprenticeship that enabled me to participate in the biological side of the research. As a physicist I was familiar with the concepts of absorption spectra and how to obtain them and with many of the data transformation techniques used in the laboratory, including polynomial curve fitting, Fast Fourier Transforms (FFT) and its inverse, curve smoothing, and so forth. Throughout the initial part of the research, I videotaped laboratory sessions in real time without stopping the camera and kept fieldnotes. These videotapes were transcribed verbatim, enhanced by images of the graphs presently being talked about that I copied from the videotape. Subsequently, I analyzed the tapes, annotating the transcripts where appropriate, and attempting to model mathematical procedures when it was unclear just what was going on. For example, the Carl and Tom talked about rotating a particular type of graph. However, whereas I expected a clockwise rotation of the graph to lead to a movement of a graphical feature to the right, the actual movement observed on the monitor was to the left. By modeling the 'rotation' both as mathematical rotation and as a eliminating a linear trend, I was able to ascertain that the scientists 'meant' to talk about 'detrending' as the subtraction of a linear function from the data displayed.

8.4. ONE DATA RUN

Experiments are complex events that are difficult to get a sense of unless one has participated in a series of them. To allow readers a sense for the phenomenon of interest, the first-time-through interpretation of graphs as these are produced during the experiment, I present and analyze one data run, that is, the collection of an absorption spectrum from one cell.

Data runs begin with the measurement of the control spectrum taken several micrometers away from the target cell. The senior scientist, Carl, who has prepared the slide, manipulates it under the microscope, and operates the shutters and light sources, requests a reference spectrum.

01   C: Ref!
02   T: Ref done!
03   C: Scan (*Pause*) Looks like a single cone.

Tom announces that the 'reference' (control) has been recorded, which allows Carl to move the slide and thereby bring the cell into place (with beam turned off). Once the cell is aligned, Carl requests a scan and describes the cell as a 'single cone' based on his visual inspection. Readers should note the hypothetical nature of the description ('looks like...') and the anticipation of the possibility that the object might turn out as something else (e.g., double cone, broken rod). The software is set to display immediately the absorption spectrum, that is, the difference between the two spectra taken through and next to the cell. Tom, who observes the monitor, announces a 'first problem' (Line 04).

04   T: It's very wide, that's the first problem. (*Continues transforming the curve*)

As Tom manipulates the display, which involves limiting the horizontal wavelength axis below 300 and above 800 nanometers and adjusting the vertical scale to magnify the displayed graph, he provides a description 'it's very wide'. For the non-initiate in the laboratory, it is probably difficult to ascertain the referent of Tom's description. If there had been a feature that noticeably stood out, the non-initiate might attribute the statement to this salient feature. Although the display shows some features near the left and right boundaries, these do not appear to be 'very wide'. There seem to be more problems ('first problem') but these are neither articulated nor evident without knowing more about the retina itself.

So far, the scientists have made three observations. The first was a visual inspection of the object O_1, with the result photoreceptor $a_1 \cdot x$. The second was a recording of the amount of light transmitted through the slide, cover slip, and the liquid surrounding the cell. Thus, the operation O_2 yielded an observation a_1 of everything but the photoreceptor, x', that is, $O_2 \cdot x' = a_2 \cdot x'$. Equivalently, the second operation yielded an observation of everything plus the photoreceptor, $O_3 \cdot x = a_3 \cdot (x' + x)$. The difference $a_3 - a_2$ is ascribed to what has been seen as cell, a_1. However, the exact nature of both a_1 and $a_2 - a_3$ is uncertain. It is only when the two converge and therefore stabilize each other that the data are good enough to be kept for further analysis.

In response to Tom's assessment that there are problems in the display, Carl raises the possibility of a relative signal ('little peak out there').

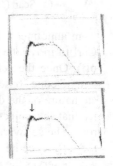

05   C: What's that little peak out there in the...

06   T: This one right here (*Points with cursor*)?
07   C: Too short?
08   M: About three-forty.
09   C: Yeah, too short.

10   T: Yeah, too short.
11   C: See where I am here. (*Looks through microscope*)
12   Off!
13   T: Off, Right.

Tom is uncertain about the referent of 'that little peak' and by pointing the cursor to a small feature in the UV part of the spectrum requests confirmation that both of them are aligned to the same peak (line 06). Carl asks whether it is 'too short'. Although the nature of 'it' would not be clear to a casual visitor, those present know that Carl is talking about the location of the peak along the wavelength scale on the horizontal axis. Sitting close to the monitor, I respond 'about three-forty', the laboratory shorthand for '340 nanometers' on the wavelength scale. Carl both responds to his own earlier question and thereby disconfirms the hypothesis that it implied (line 09). After Tom's confirmation, Carl moves to assess the position of the cell with respect to the stimulus. Looking through the ocular, he suggests that cell and path of light are no longer aligned ('Off!').

This small episode already allows us to get a sense for the fact that reading graphs in a first-time-through manner during scientific experiments is not straightforward. We observe the scientists engaged in assessing various (yet-to-be-disclosed) features rather than discarding the graph out of hand as not at all portraying the target object. The problematic nature of the graph becomes more certain only after Carl checks the position of the stimulus and notices that it had moved off the cell. What cannot be known is how long the cell had stayed in place, absorbing some of the light pulse, and therefore leaving its characteristic signature on the graph. After noticing that the cell and light were out of alignment, the data and its interpretation are abandoned without leaving open the possibility to do more the data. Without further deliberation, Carl requests another scan.

14   C: Scan... Definitely looks like a single cone.
15   M: Hmm?

16   T: It's rather high, in the four-sixty again.

17   M: Yeah.

In this case, we notice that Carl does not request a reference. Rather, the scientists take the reference from the previous trial. Carl does not request a new reference, which indicates that he either attempts to take a measurement on the same cell or on a cell in the vicinity, so that the previous reference remains appropriate. His affirmation that the cell 'definitively looks like a single cone' suggests to the other participants that Carl has realigned the earlier cell. Both assessments of the resulting spectrum (lines 15, 16) raise doubts. My own utterance indicates that I do not recognize an identifiable feature and Tom describes the feature salient to him as 'rather high, in the four-sixty again'. From this utterance, we can infer that the feature salient to Tom lies near the elbow in the second quarter of the graph, that is, near 460 nanometers on the wavelength scale. His 'it's rather high' suggests the existence of a reference—a peak at 460 nanometers is rather high relative to the 'single cone' announced by Carl.

In order to understand this exchange, readers need to know that Carl identifies single and double cones. The single cones absorb in the UV (near 370 nanometers) and blue (near 430 nanometers) parts of the spectrum. Each double cone consists of a green- (near 520 nanometers) and a red-absorbing photoreceptor cell (near 575 or 623 nanometers, depending on life stage). Thus, when Tom perceives a possible feature around 460 nanometers, it is incompatible with Carl's assessment of a single cone, which should be considerably lower on the wavelength scale. Tom's assessment of 'four-sixty again' suggests that one of the other, here not articulated problems related to the position of the feature salient in his perception. The 'four-sixty again' also shows, to paraphrase Garfinkel and his colleagues, the exhibitable-analyzability-of-the-graphical-feature-again, which is a property achieved in, and which consists of, the competent, in situ interpretive practices of laboratory science.[2]

The next turn shows that this graph is more complex than it first appears and that it requires further inquiry. Carl asks Tom to 'do a pixel shift' perhaps because he thinks that the peak is not where Tom presumed it to be.

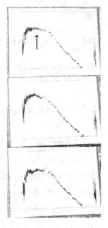

18	C: Well, do a pixel shift, because I think that one is the
19	peak.
20	M: This one, yeah.
21	C: Again.
22	T: That doesn't look like it.

Carl's gesture indicates that he sees a feature near the top part of the graph. His request to do a pixel shift indicates that a transformation is required before this feature can be seen more easily. My own agreement can be read as an indication that I could identify a possible peak at the location where Carl had pointed.

What a 'pixel shift' involves, why it works, and why it is a (culturally) legitimate way of modifying the graph requires substantial familiarity with the experiment, in particular with the data acquisition devices. After the beam of light passes through the slide, it hits a fine grating that disperses the different wavelengths in the way a prism would do. Being spread across an area, the photons corresponding to different wavelength enter different 'bins' on the CCD, slowly filling them up; each bin directly corresponds to a point in the graph. For a variety of reasons, the reference spectrum and the sample spectrum may be shifted relative to each other. The light corresponding to a particular wavelength does not fall into the same bin in consecutive reference and sample measurements. A 'pixel shift' realigns the two spectra, which then leads to a change in the absorption spectrum. Of course, the scientists do not know how much, if any, shift occurred—were it not for a particularity of the spectrum that gives their assessment some certainty. I will return to this issue later.

Without responding verbally, Tom does a 'pixel shift', then another, and yet another. He then comments 'This doesn't look like it'. We do not know at this time whether the comment pertains to the process or the product of the modification. Carl responds 'No, in the other direction' In the context of his earlier request for a pixel shift and the product of Tom's actions, this response can be heard as requesting a pixel shift in the other direction. It also suggests that Carl made an observation that confirmed Tom's assessment that the outcome 'doesn't look like it'. Because of the subsequent shifts, a peak seems to emerge more clearly in the area that Carl made salient earlier. Carl now appears to be satisfied ('OK' line 28) but immediately requests a different action ('now detrend it').

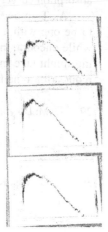

23   C: No, in the other direction.
24   T: Yeah

25   M: This one?

26   C: May be one more. No, the other way. So two that
27   way.

28   OK. Now, detrend it.

This request is an indication that Carl is not yet satisfied. The peak apparently does not correspond to his expectation. Again, considerable familiarity with the laboratory environment and with the phenomena of interest is required. In the publication issuing from this experiment, the scientists will suggest that the microspectrophotometric data are inherently noisy because of the small size of the phenomenon and weak absorbance. This corresponds to the cultural knowledge among experimental scientists that the difference between two (large) signals, each associated with small errors (noise), will be associated with relatively large errors; in other words, the signal (difference) to noise ratio decreases dramatically.

In reference to an earlier publication, the article will suggest that 'if [a curve] showed a clear linear trend 'tilt' it was linearly detrended'. In the present situation, we can imagine the peak earlier indicated by Carl to sit on a linear curve from top left to bottom right. However, the article also suggested that the 'long-wavelength limb baseline was used as the first criterion' for the suitability of a curve. That is, any detrending had to preserve a horizontal base line of the absorption curve.

As Tom begins to detrend the curve, I ask whether they believed the curve to be one with a maximum around 380 nanometers. Carl confirmed this all the while encouraging Tom to detrend until suggesting that the process of detrending might take more time than they have at their hands. He then makes another request for a modification of the graph 'Put a three-eighty up'.

29   M: Do you think it is a three-eighty?

30   C: Yap. Keep on detrending.

31   OK, well, this one can go on forever. (*Tom keeps on*
32   *'detrending'*)
33   But put a three-eighty up to see what it looks.

Both the utterances in lines 29 and 33 suggest that the participants in the laboratory are aware of something that is to be found at 380 nanometers—the UV
peak that turned out to be determined in this experiment as having its maximum
at 371 ± 8.2 nanometers. But what is the meaning of 'putting up a three-eighty'?
An interested anthropologist would eventually find out that in the scientific literature there exist measurements of absorption spectra on other species. Also
available in the scientific literature are the coefficients for polynomial functions
of degree seven that model absorption spectra. The special-purpose software
written by Tom makes it possible to plot spectra with maxima at various points
along the wavelength axis. In his request (line 33), Carl is asking to overlay their
data with a curve that uses the published parameters. After some problems of
finding the appropriate keys in the dark of the laboratory, Tom succeeds in 'putting up the three-eighty' and suggests that it in fact fits (line 35). Carl is pleased
but checks to make sure that it was in fact the 'three-eighty' that had been plotted as a reference, which Tom confirms.

34   T: Yeah. (*Works on getting three-eighty, seeking flash-*
35   *light*) It does fit under three-eighty.
36   C: There you go. Three-eighty?

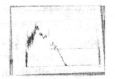

37   T: Yep.

Although this might have appeared to be the end of the process, Carl's next turn
indicates that they are not done measuring the cell under investigation. He suggests that they are now 'going to bleach it' and asks Tom to count.

38   C: So now we are going to bleach it. Ready to count?
39   T: Counting.
40   (*Pause*)

Again, the understanding the particular request requires some familiarity with
the biochemistry of pigments. When light falls onto a photoreceptor, it is absorbed and its energy provokes a change in the chemical composition. Because

of this chemical change, which is not reversed in the excised tissue, the cell is no longer photosensitive. The cell has been 'bleached'. The resulting 'photoproducts' are sensitive in the UV region of the light spectrum. That is, if the scientists do another measurement, the previously visible peak should have disappeared. 'Bleaching' can therefore be regarded as an experimental technique to confirm that a useable peak in fact had been observed, which disappeared as a (causal) result of the bleaching. Because the signals are often weak and noise, bleaching is not frequently done. But its existence in the practices of these scientists requires some further analysis.

Earlier I described that the scientists take a reference beam, leading to the observation a_2, from which they subtract the second observation a_3 taken with the photoreceptor in place. This procedure, however, is based on the assumption that the 'everything else' denoted above as x' is the same when the light travels through the photoreceptor and the preparation next to the photoreceptor. However, in a strong sense, this cannot be the case, as, for example, the walls of the photoreceptor and any medium inside will not be the same as the saline solution within which the photoreceptors are suspended. Thus, the 'everything else' in the two operations O_2 and O_3 are assumed to be the same and the scientists both hope and have some arguments for why they can make this assumption. However, when they bleach the cell, the third measurement $O_4 \cdot (x + x')$ taken after the destruction of the light absorbing pigment involves light that has traveled through the same materials (e.g., cell walls) now only lacking the absorbing molecule. Therefore, the difference $a_4 - a_3$ is taken to be a better estimate of the absorption because the extra materials x' in observations O_3 and O_4 is the same.

While the participants are waiting for the bleaching to take place, they continue to talk about the graph visible at the display. First, the utterance 'It's both half-width and lambda max' reflects two observations. The reference curve correctly models not only the position of the maximum along the wavelength axis (λ_{max} [lambda max] is the signifier used in different scientific communities) but also the width of the curve at half the height of the maximum. Carl confirms this observation but states a final problem in this episode. The detrending actually needs to be such that the base line comes to be parallel to that of the reference.

41	M: It's both half-width and lambda max.
42	C: Both fit.
43	C: The only problem we are having right now is the de-
44	trending we have got to get this (*points, indicated by*
45	*arrow*)

46   up there (*points, indicated by arrow*). I think that that
47   is possible. Don't you, Tom?
48   T: Yeah, I think it is possible.

Carl suggests that it should be possible to detrend in such a way that the long-wavelength baseline of the measured curve comes to be parallel to the reference. Being requested to voice his opinion, Tom, who will be conducting all analyses at a later point, agrees that an alignment of the two baselines should be possible. The scientists subsequently do another scan of the 'bleached' cell and, without inspecting it, move on to the search for the next cell to be measured.

In this one run, we can see a host of issues related to the first-time-through interpretation of graphs in experimental science. First, what is salient in a graph differs between scientists. Such differences exist not just between insiders and outsiders (such as myself during the initial stages of the research) but also among more experienced members of the lab. It is not a priori self-evident from a graph which of its features pertains to the natural object under investigation. Whereas the bent around 460 nanometers was initially salient to Tom, a little hump around 380 nanometers was perceptually salient to Carl. The difference is the following: in the first instance, scientist have to conclude that the object of interest is absent whereas in the second case they may see the signature of a UV cone. Whether the graph shows something of interest arises out of a process of stabilization in which apparatus, object, graph, and existing graphs become aligned. This alignment involves multiple observations of the object (through microscope or CCD and in graph), transformation of the acquired graph so that it begins to resemble another graph already accepted in the community (and therefore a cultural object). These scientists use multiple forms of transformations on the data such as pixel shifting, detrending, and comparison with normative curves; they 'bleach' the object to acquire yet another graph against which the acquired graph can be compared. Finally, there are other transformation processes involved not evident from this transcript. For example, the identification of λ_{max} and half-bandwidth is not made from the modified graph. Rather, the resulting graph (the one with appropriate baseline) is approximated by a 'Fourier transform fit', from which the relevant values of λ_{max} and half-bandwidth are determined.

8.5. CONTINUITY OF THE OBJECT

In the classical paradigm on graphing, people are asked questions about graphs that they frequently have no ideas about how these came into being. Research participants, such as students, have never had the opportunity to produce graphs of the speed of an object and relate it to its appearance in graphical form. They have little or no experience with how modifications in the object (speed) and characteristics (troubles, noise) of the recording mechanism change different aspects of the corresponding graph. They often have little or no experience with how graphs change when plotting parameters are changed, with the transformation of data, or with the fitting of data. In the few reported cases where students do have extensive experience with the transformation of natural objects into graphical representations and with the natural objects themselves they also become quite competent in interpreting new graphs pertaining to the domain. This is no different from what I observe in following scientists in their daily work. Their competence in reading a graph is related to their familiarity with the object, the transformation it undergoes during the experimental process, and with the particulars of the entire chain of translations that produce and change the data. As their experience and familiarity increase, the assessments of the graphs becomes increasingly routine; even those hardly noticeable features that take a lot of effort in the early parts of their work, including recourse to all the mediating tools at hand, are correctly identified on sight and without discussion.

8.5.1. Contingency of Troubles

Knowing the apparatus and being familiar with each action taken in the preparation of the object is important because from one day to the next there can be variations that initially remain unnoticed. These changes are subsequently 'discovered' when the graphs (or images under the microscope) are different from what is expected and from what had been observed on the previous day. Once the scientists have the sense that something is different, in the graphical representation or in the preparation, they will begin searching for clues to understand what might cause these differences. In the present situation, they had been able to use the new CCD camera connected into the computer monitor to replace 'direct' viewing through the ocular on one day. On the next day, however, they could not find their image in the way they wanted.

Carl: Now, that doesn't seem right. Hmm, Hmm.

Tom: You can change it to a different (*Muttering*) then (?) here then we should be back.

Carl: Hmm, do you remember what it was yesterday?
Tom: It was this- we didn't change anything here.

The transcript makes evident that there is not only something that 'doesn't seem right', but that neither scientist knew what was different from the day before. The absence of the image was, of course, troublesome in a double sense: they could not line up their object of interest and, even if they had been possible to do so, they missed a crucial piece in the stabilization of their reading of the graph. They engaged in an extended sequence of activities in order to bring about the visual image on the monitor to correspond to their expectations. That is, the experiment turned out to be troublesome after the preparation had been completed and at observation O_{vis} making it impossible to proceed.

$$O_{m+1}\ldots A_{n+1}\cdot O_{m,vis}\cdot A_n\ldots\cdot A_2\cdot A_1\cdot x = a\cdot x$$

$$\uparrow$$

In some instances, the visual inspection (occurring during the search for cells) of the preparation suggests some problem with an earlier part of the preparation or with the general state of the cells. In these situations, the visual inspection serves to stabilize or interrogate a particular transformation earlier in the experiment.

$$O_{m+1}\ldots A_{n+1}\cdot O_{m,vis}\cdot A_n\ldots A_8\ldots\cdot A_2\cdot A_1\cdot x$$

In the following excerpt, Carl links the absence of 'fringes' to the way he had prepared the retina—which was actually the accepted way.

Carl: I didn't use a razor blade in the macerating process and I notice that I don't get those fringes of photoreceptors that- that I have grown to like somewhat.

At this point, he critically interrogated a change in the practices, preparation, which appeared to have brought about a change in what he saw. He did not see the fringes from which cones stick out on a successful day. When these cones are attached to a larger piece of retina, they do not float around and Carl can easily take a micro-spectro-photography reading (MSP). In this case, the cells also stick out horizontally into the fluid and therefore can be identified and used. That is, he changed one of the operations A_i and when he did not find as many cones as he normally did, he associated this finding with the change in procedure. The change in the experimental sequence of actions was associated with a changing visual field ($O_{m,vis}$) and therefore was rendered problematic. In the course of collecting the data, the scientists talked about many different factors that could have mediated their success at finding suitable cells. Tom could have prepared the MEM solution differently, the age of the MEM solution ('When did you make up this solution?'), the way he had dissected the eye ('This dissection really bothered me'), the cells could have begun to decompose in their solution ('Or it goes into solution, too'), the cell has been changed by transmitted light ('It's broadened and looks pretty bleached already'), or just go bad ('viable for only one hour after being excised', 'double cones clearly exhibited signs of degeneration'). In the following episode, Carl inquires about the pH of the MEM solution, which is dropped on the retinal pieces before the cover slip is placed to fix the preparation.

Carl: Was it seven point two last time?
Tom: No, seven point four two, hmm the same kind, about two hundredths different. It was only the time before that I had seven point two.
Carl: And you made up a full liter, right?
Tom: Yeah. It's always one of the jars of MEM.
Carl: Hmm.
Michael: Do you see trouble?
Carl: Well- it's just that I am not seeing cones and I see rods shrivel up pretty quickly, which means that the osmolarity of the medium is possibly suspect. It's hard to draw conclusion just now. I just don't see the number of photoreceptors that I normally see.

Carl perceived the photoreceptor cells in a particular way, which he articulated as shriveling up. Furthermore, the number of photoreceptors was much smaller than the number that he normally observed. He used this observation as the starting point for an interrogation to ascertain that previous steps in the preparation of the specimen may have changed. Here, the investigation does not identify a possible candidate, for Tom appears to have done everything to specification.

8.5.2 Finding and Modeling the Source of Structure: Signal or Noise?

This is what I like doing too, looking through the data, too, trying to find reasons for such artifacts (*Pulling finger along 'baseline'*). This is an artifact again, but not as easily explained.

The graphs that come off the experiment share little with the graphs that ultimately get published, and even less with the smooth lines that enter textbooks. Even a graph that turns out to be 'a beauty', 'a textbook example', or 'as good as

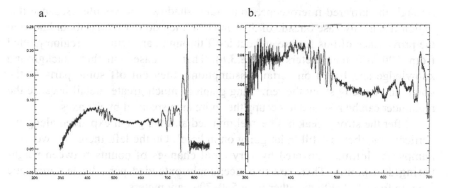

Figure 8.3. There is much structure in the graphs as they come off the experiment. (a) The unmodified absorption spectrum includes strong peaks in areas not of interest, here, in the long-wavelength part. (b) The same graph as in (a) but with only the 'region of interest' displayed still displays structure to be accounted for by the scientists.

it gets' exhibits features that scientists would have to account for in their publications. The scientists therefore have to find not only the 'real' signal of interest but also come to understand the nature of unwanted features, which allows them to eventually eliminate these features by changing certain pieces of equipment or by manipulating the raw data in particular ways. Let us examine some specific examples.

A graph as it comes off the experiment (*Fig. 8.3.a*) displays much structure but the real signal of interest is invisible to the untutored eye. The wavelength axis is scaled from 200–900 nanometers. There are strong features in the far-red and infrared region of the spectrum (660–780 nanometers). The scientists do not know the exact origin of the tall peaks between 740–780 nanometers. In one instance, I inquired about what appeared to be a signal in the long-wavelength part of the spectrum.

Tom: Well, it is in the deep red.
Carl: It could be something with the pigment illuminated by the light.
Tom: The shadows that you see on the infrared scope show that there is some I-R absorption not by our pigments, but by something else. But you don't see it.

Carl and Tom assume that these peaks come from material that is not visible to the eye because it absorbs in the infrared. On the other hand, when inspected through the infrared microscope, some faint shadows are visible. Because this region lies in any case outside of their interest, they do not worry about knowing the provenance of this feature and cut it off through narrowing the region plotted from 300–700 nanometers (*Fig. 8.3.b*). That is, based on their background knowledge and based on certain assumptions, they cut off some parts of the graph and thereby show the remaining graph in much greater detail because the remainder can be rescaled to occupy the entire area spanned by the axes.

After the strong peak has been eliminated and the graph expanded along the vertical axis, there is still 'a lot going on in here'. On the left, there are two little humps, the leftmost covered by very rapid changes of counts between neighboring data points. There are two, seemingly extended periodic features, one ranging from 430–490 the other from 560–700 nanometers.

In the course of their work, these scientists have learned that the charge-coupled device (CCD) that measures the light intensity along the spectrum in five-nanometer interval produces a certain amount of noise even if there is no light. To reduce this noise, they are held to expose the CCD until its bins are almost filled maximizing the signal to noise ratio. However, because the intensity of the light pulses is not constant across the spectrum, the noise will be high where the source output intensity is low—this noise is observed on the left.

It is not that scientists just know about the existence of this noise. They have direct experience with it in the sense that they see its signature changes in response to actions that they take. If they let the CCD warm up, the noise increases relative to the signal; when they decrease the temperature of the device, the signal to noise ratio improves. There is therefore a direct, causal relation between their actions, change of temperature, and their observations.

Another feature visible in the graph (*Fig. 8.3.b*) was more resistant to explanation and the scientists spent some time to understand its source and nature. In the region from 430–490 nanometers, each graph taken by the scientists shows a strong 'ringing', that is, a nearly regular oscillating feature. (A similar periodic that constantly shows up is visible in the right part.) This ringing became an important aspect that these scientists attempted to understand and eliminate. Because the scientists see this feature as 'ringing', they conceptualize it in terms of the superposition of two nearly identical signals. Tom therefore modeled what would happen if the spectrum of the light transmitted through the cell was slightly shifted. Independently, I also modeled what would happen if a periodic feature such as a group of three peaks were slightly shifted or if the signal were changed, for some reason, in its intensity. (According to the anthropological approach, my own modeling efforts turn up the same or similar 'ethnomethods' for dealing with the problems at hand.) In both situations the 'ring-

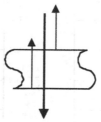

Figure 8.4. A part of the main beam of light reflects at each surface traversed, thereby decreasing the light intensity. Furthermore, an interference phenomenon is created between the beam and its reflected parts, thereby further changing the spectrum of the light.

ing' feature becomes visible. However, the two situations differ conceptually and therefore what the scientists can do about them.

In the first instance, the light pulse could be changed through superposition. Each time the light traverses a surface, some of it is reflected. If there are two parallel surfaces near each other, a light beam may experience superposition with that part of itself that has been reflected (*Fig. 8.4*). The reflection is minimized if the light falls perpendicularly onto the surface and increases with the angle. If the pulse and its reflection are slightly out of phase, that is, shifted with respect to each other, an interference pattern would be observed (*Fig. 8.5.a*).

The second assumption pursued is that the 'ringing' phenomenon derives from a variation in the intensity of the light. The scientists assumed that the xenon source may not be produce a stable output so that the light intensity varies slightly. In this case, subsequent light pulses have different intensity distribution so that—with or without cell in the path—the ringing can be observed even without a wavelength shift in the light. Such a change in the source intensity would give rise to the ringing (*Fig. 8.5.b*). The reflection on the surfaces of the cell would also decrease the intensity and give rise to a ringing due to changes in intensity.

At the time that the research was conducted, the scientists thought they understood or had narrowed down the phenomenon. In their research article about the methodology submitted at the time, they later wrote that a 'light intensity controller was being added 'to reduce variability of the signal amplitude across the spectrum'. However, they also suggested that if it turned out that the feature remains despite the addition of the controller, the phenomenon was likely caused by reflection and interference phenomena at the surfaces of the cells of interest.

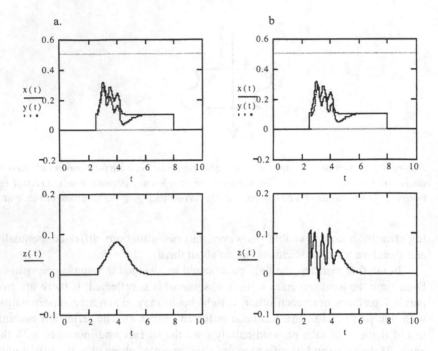

Figure 8.5. a. If there was only absorption due to the photoreceptors, the difference between the reference and scan signals (each in top diagram) would show the characteristic peak (bottom). b. If the source signal is varied (variation of source intensity, reflection on the cell itself), a ringing becomes visible.

In this case, they would not be able to remove the feature because it is intrinsic to the interaction of the light with the phenomenon under study.

01	T:	OK. Hard to interpret.
02	C:	The (head?) set noise?

03	T:	I dun know. Is it still in position? I have a look in a
04		moment, it doesn't look like anything.
05	C:	Anything in there?

06   M: I haven't seen anything.
07   T: It doesn't look like it.
08   C: Flat-liner.
09   T: Yeah, flat-liner.

10   M: Are you still on?
11   C: I was on. I move on to another.

This run begins with the request to scan. This run immediately succeeded a double-cone measurement that showed a clear signal. The absence of a request for a reference therefore means that this measurement is done on the second member of the double cone. (Only in very few isolated instances were two different cones measured using the same reference.) This other member, if the first graph was read appropriately, should bear the signature of a red cell. Although a novice member in the laboratory may raise the issue that there is a broader peak in the center and a narrow clearly defined peak on the far left, Tom looked at the graph in terms the expected curve corresponding to a red cone. Tom found the graph 'hard to interpret'. All three did not find the slightest evidence of a peak, which Tom and Carl characterize as a 'flat-liner'.

In this case, an object of potential interest can be observed at one stage (O_m). Carl confirms that he is still on the cell (line 11). However, the expected signal (O_{m+1}) corresponding to the object is missing. The object therefore fails to stabilize. The scientists abandon their effort regarding this cell and 'move on to another' one.

12   C: It looks like a U-V- To me- at least (*Looks through*
13   *microscope*) I could try to get in there better. It
14   could be low amplitude, but I think- I don't know.
15   T: I save it as a U-V but may be we should rescan. The
16   problem is only, yeah, if you could- If you have
17   these photoreceptors in the clear fluid around. Then
18   this is a problem. The problem is what you have
19   below and above it- Or it goes into solution, too.

In this situation, the scientists do not know what is going on and how to interpret the graph. They save the data to be able to take another look when they do not take measurements and therefore have more time. That is, they defer a definitive evaluation until some later point. Here, not knowing what the graph means or whether the object under the microscope is actually a UV cone does not have the consequences that such a behavior would have 'in captivity', where abandonment, deferral, and not knowing are not counted as successful strategies of problem solving.

8.5.3 In-Situ Experiment

Doing an experiment in situ is one of the options that these scientists have in ascertaining which part of the graph can be attributed to the photoreceptor. As we have seen earlier, scientists may decide to 'bleach' a photoreceptor. The difference between the first and second measurements can then be directly attributed to the photoreceptors, which have disappeared in the process. That is, they can attribute changes in the graphs to a manipulation of the object, thereby enacting a classical experimental design of the type

$$\frac{O_{m+1} \cdot A_{n+1} \cdot O_m \cdot A_n \ldots \cdot A_2 \cdot A_1 \cdot x}{O_c \cdot A_n \ldots \cdot A_2 \cdot A_1 \cdot x}$$

where the temporal sequence of actions in the activity is from right to left. First, the scientists take a 'reference', which they acknowledge as observation O_c. After moving the photoreceptor into the path of light, they take their sample (O_1), which leads to the display of the absorption curve O_c - O_1. In those cases that they decide to bleach, they enact the operation A (bleaching) followed by another measurement O_2. Here, the control observation (O_c) is assumed not to be changing until O_2 is being taken. However, the scientists are not even interested in O_2 - O_c, though they may occasionally plot it (*Fig. 8.6.b*) but in documenting the change between O_1 and O_2 (*Fig. 8.6.a*), which is attributed to the experimental manipulation (bleaching).

The scientists have a concrete model of the object and its surrounding that allows them to navigate their experiment and to establish coherence between the animal and the graph that constructs an aspect of knowledge about it. The model assures the constancy of the phenomenon through the many transformations that occur. In the present case, although the cell itself cannot be seen with the naked eye or even under the microscope that was used to prepare the tissue on the mounting slide, the scientist have a physical model of what happens to a piece of the retina. They 'know' where it is located, what particular operations do to it, how this affects the tissue, and so on. This coherence is further enhanced by the gestural expressions that they use to show what happens to the tissue when it is pulled or cut, and how the cells are aligned or found when under the microscope. Furthermore, they have models for what is happening on the slide, for example, if there is a larger piece of tissue that separates the cover slide more than normal

a.   b.

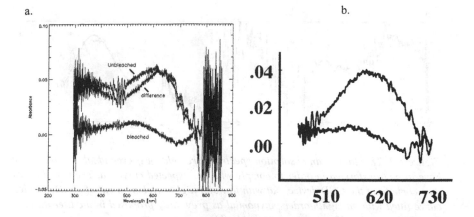

Figure 8.6. Bleaching changes the signal from a photoreceptor. (a) The difference of the absorption spectra from an unbleached and a bleached gives 'clear evidence' for the presence of a red cone. (b) For the publication resulting from the observed data collection sessions, the scientists present a best case for the absorption spectra from an unbleached and bleached photoreceptor. Noticeable is the absence of the noise in the short- and long-wavelength part of the spectrum still present in the left panel.

from the carrier slide so that there is a greater potential for moving about. The movement of the images under the high-powered microscope and the CCD image are therefore explained in macroscopic terms, or in terms of microscopic theories such as Brownian motion or convection. When these scientists interpret a graph in first-time-through mode, they bring their experience and familiarity with the equipment as interpretive resources. Already the nature of salient perceptual features depends on this familiarity. Other features that may stand out to the untutored eye have become all but transparent to the perception of the scientist familiar with equipment, experiment, and objects of inquiry.

8.6. TRANSFORMATIONS

The data collected by scientists seldom show the smoothness that characterizes functions in introductory mathematics course or the graphs that are featured in textbooks. Through their work with the phenomena and producing smooth curves, scientists gain substantial familiarity with the phenomenon, transformation of data, conceptual surroundings of phenomena, and mathematical transformations and their relations to the scientific phenomena. There is therefore a

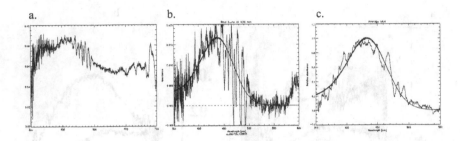

Figure 8.7. The data of an absorption spectrum are seldom if ever ideal and therefore have to be transformed in order to comply with the expected curves. a. The raw data as they come from the CCD device, but with low and high wavelength cut off. b. Detrended curve fitted with an eighth-order polynomial as previously proposed in the literature. c. Result of averaging 29 curves such as that in (b).

mutually constitutive relation of the scientific phenomena, the conceptual underpinnings explaining them, the raw data, and the transformations that turn them into publishable data (in support of the conceptual underpinnings).

The raw data from an experimental run (*Fig. 8.7.a*) not only make for a very noisy graph but also this graph does not resemble an absorption curve, which, when 'picture-perfect', are nearly Gaussian. After the scientists in this study have done a series of manipulations (to be explained below), their graph now shows a much greater resemblance with a Gaussian. This curve continues to be associated with a substantial amount of noise, which is of the order of magnitude of the signal itself (*Fig. 8.7.b*). The data show a lot of noise. The scientists therefore add the data from several runs in order to average out the noise and increase the signal to noise ratio (*Fig. 8.7.c*). In these last two curves, scientific readers can recognize an underlying similarity with the solid curve, a 'template' that has been accepted as the scientific norm for such data.

It is this template that scientists see even when looking at the graph (*Fig. 8.7.a*). All the transformations necessary to go from this graph to the cleaned up version (*Fig. 8.7.c*) appear to become transparent so that scientists actually see that a peak is suitable, provided some detrending, pixel shifting and so on has been completed. In those instances (red sensitive A_2 cones), the scientists use a particular technique that actually gets rid of the noise. Rather than fitting a curve to the data, they transform and re-transform their data, in the process cleaning them of the (high-frequency) transients (noise) visible in the raw data and remaining after the first few transformations. It becomes evident that scientists' understanding of phenomena, graphs, transformations, and instrumentation are all highly inter-related and mediate each other.

In the following, I take readers through some of the transformations and the mathematical and scientific conceptions that underlie them. Scientists' understanding of the graphs that they produce are inherently mediated by these understandings such that understanding scientists' understanding requires, in turn, anthropologists of science.

8.6.1 Fitting a Template

Existing graphs, such as those published in the literature, are cultural tools that strongly mediate the work in this laboratory. These graphs assist scientists in the vision laboratory to tease 'real' graphical features from spurious ones. Whether a particular graph is a likely candidate for becoming part of the published data sometimes is determined by drawing on existing 'templates'. In an already-cited excerpt, Carl asked Tom to 'put a three eighty up' so that they could 'see what it looks like'.

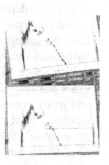

31	C:	OK, well, this one can go on forever. (*Tom keeps on*
32		*'detrending'*)
33		But, put a three-eighty up to see what it looks.
34	T:	Yeah. (*Works on getting three-eighty, seeking flash-*
35		*light*) It does fit under three-eighty.
36	C:	There you go. Three-eighty?

Carl explicitly requests a second graph to be superposed over their already somewhat detrended data. Tom puts up the three-eighty and confirms that their graph 'does fit under three-eighty'. The sample graph with a maximum at 380 nanometers on the wavelength scale functions as a template that, having been published more than a decade earlier, has become the norm in this scientific domain. It is therefore the outcome of an earlier experiment that has become accepted by the scientific community, to become a tool for subsequent research on the absorption spectra in visual pigment. What matters in the complete data analysis are not only the location of the maximum but also the width of the spectrum at half of the maximum height of the curve ('half-max bandwidth'). However, as part of the first-time-through interpretation of graphs, these scientists were initially more concerned with the location of the maximum rather than with 'half-max bandwidth'.

In the present instance, the computer plotted the comparison graph. That is, an outcome of a previous experiment, accepted by the scientific community, was implemented in the custom software. When the scientists see usable data in a noisy graph (*Fig. 8.7.a*), it is because they recognize in it the similarity with a template that could be plotted (visible as smooth line in *Fig. 8.7.b, 8.7.c*). That they have internalized the template to some degree can be seen from the following assessment of a curve that is 'very wide'—which in fact is too wide when compared with the template around 500 nanometers.

04	T: It's very wide, that's the first problem. (*Continues transforming the curve*)	

When the identification of a peak is more tenuous, the scientists call up the stored templates and plot them over the top of their own data. This template allows the scientist not only to compare their graph with the appropriate λ_{max} and half-bandwidth but also serves as a criterion for the other transformations that they conduct. These templates therefore are used in a double way: to find the exact position of lambda-max and subsequently to assess whether lambda-max and half-bandwidth are consistent with the cultural knowledge on the topic. For example, in the opening episode, Carl exhibited concern regarding 'baseline compliance', that is, the fact that a detrending procedure had to make the baseline of their data consistent with that of the template.

43	C: The only problem we are having right now is the de-	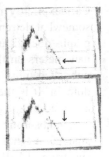
44	trending we have got to get this (*Points, indicated by*	
45	*arrow*)	
46	up there (*Points, indicated by arrow*). I think that that	
47	is possible. Don't you, Tom?	
48	T: Yeah, I think it is possible.	

Furthermore, detrending was also constrained by the slope of their curve, which had to be consistent with the slope of the template. If any of these constraints was not met, or if differences could not be explained such as in the widening of the spectrum with a shift from freshwater to saltwater pigment, 'the spectrum was rejected'.

> If it was well defined, the spectrum was given further consideration. If it showed a clear linear trend (tilt) it was linearly detrended (Hárosi, 1987). Then a template was fitted to the data. For A_1 based visual pigment cones an eighth order polynomial (Bernard, 1987) was fitted to the absorbance spectrum.

Because there were no published templates for A_2-based red-absorbing cones, the scientists use an A_1 template and then use a Fourier transform of their data to show the expected widening and decreasing of the peak with respect to the A_1 pattern. Here, new scientific knowledge is constructed out of the difference between accepted pattern and the new patterns created by this research team. At present, the publication of this work has to await further investigation, not the least because the (biological) mechanisms underlying these shifts are not understood and therefore require substantial investment.

8.6.2. Fourier Analysis

In the following excerpt, I am talking to Tom about the results of his analysis pertaining to the red cones (*Fig. 8.8*).

Tom: The better the shape (*Gestures along shape of curve*) the better F-F-T works. So I actually have for the broad, where it has the A-two pigment, where I do not have such a template, not for the A-two, for the red I have the F-F-T going through it. And it fits perfectly. That works rather well the better the shape is but it definitely works well for the average [of all reds]. You see this is a single red measurement (*Fig. 8.8*). Because A-two has a larger bandwidth and the red (bold smooth line) is the A-one template. And this is the difference– (*Points to the difference between narrow and bold smooth lines*)

Michael: So you probably have some mixture?

Tom: This is an A-two more or less. This is what we are trying to get to. So it is slightly lower and slightly wider and this is exactly the effect we are looking for.

Here, templates for the widened A_2 peaks do not yet exist. Doing a Fourier transformation and its inverse (FFT and iFFT) removes the transients and gives the scientists a first idea about how the curve will look like. However, because FFT only removes the transients, the shape of the curve already has to correspond to what scientists take to be the true shape.

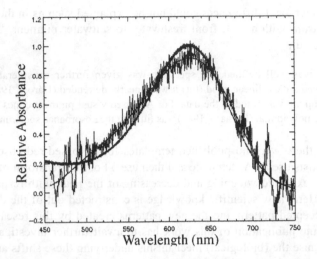

Figure 8.8. Scientists do not just fit their data using the published eighth-order polynomial template, but also apply a FFT procedure to get rid of the noise in their data.

In this situation, the data averaged over about eight red cells appear to be described quite well by the published eighth-order polynomial (bold smooth line). When they used the FFT smoothing and fitting technique, they arrived at the narrow smooth line. Visual inspection shows that the narrow line is wider at half-height. That is, the half-bandwidth is larger. This may be interpreted as a problem of the methodology that Carl presents to his peer. However, because he knows that rods change not only λ_{max} but also the half-bandwidth of the signal when the salmonid changes from the freshwater to saltwater form, this change in half-bandwidth is taken as a positive sign that similar changes exist for cones. In some situations, information such as the estimates of the mean half-maximum bandwidth is actually taken from the Fourier transform fit.

The scientists' understanding of the relationship between the smooth curve and the original data is mediated by the transformations, including the Fast Fourier Transform and its inverse. Utilizing the Fourier transform they clean the signal of undesirable noise but this does not fundamentally change the data other than eliminating 'transients'. Fitting the curves using some higher order polynomial does not have the same equivalent, unless the curve conceptually, that is, theoretically, should have the form of a higher order polynomial.

Conceptually, Fourier transformation works like this. Any function can be thought of as the result of summing a series of sine waves, each modulated by a parameter a_i, yielding a Fourier series:

$$x(t) = a_0 + a_1 \cdot \sin(1 \cdot t) + a_2 \cdot \sin(2 \cdot t) + \dots$$

Fourier transformation does the inverse: for a given function $x(t)$ or set of data, it produces the coefficients a_i required for expanding the curve in terms of a Fourier series. Inverse Fourier transformation is a re-composition of the original curve based on a subset of the coefficients a_i derived through the initial transformation. Setting a coefficient equal to zero corresponds to leaving out one of the sine functions in the equation. In research practice, those sine curves with high frequencies, thought to describe noise, are thereby lopped off by setting their coefficients to zero.

Consider a true signal with a Gaussian shape similar to the one our scientists expect that is riddled with random ('shot') noise (*Fig. 8.9.a*). A Fast Fourier Transform (used with discrete data points) provides a set of coefficients (*Fig. 8.9.b*). If these coefficients were used in the inverse FFT, the original data would be reproduced. If, however, the coefficients corresponding to high-frequency sine curves are set to zero (here all coefficients but c_0 to c_4), a smooth curve approximating the original data is produced (*Fig. 8.9.c*).

In this situation, the data averaged over about eight red cells appear to be described quite well by the published eighth-order polynomial (bold smooth line). When they used the FFT smoothing technique, they arrived at the narrow smooth line. In some situations, information such as the estimates of the mean half-maximum bandwidth is actually taken from the Fourier transform fit.

The scientists' understanding of the relationship between the smooth curve and the original data is mediated by the transformations, including the Fast Fourier Transform and its inverse. Utilizing the Fourier transform therefore cleans the signal of undesirable noise but does not fundamentally change the data other than eliminating transients. Fitting the curves using some higher order polynomial does not have the same equivalent, unless the curve conceptually, that is, theoretically, should have the form of a higher order polynomial.

The graphs exist both in the context of the natural phenomenon and as mathematical objects that can be transformed, smoothed, translated, and so on. However, in the understanding of these scientists, all transformations have an equivalent in the natural world. For example, the raw data are thought to be the result of the true absorption curve plus noise, fast transients. By doing a Fourier transformation, the scientists lop off the fast transients that do not belong to the

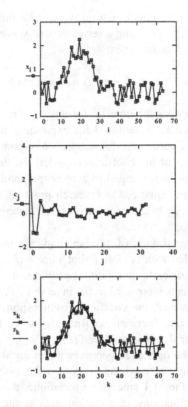

Figure 8.9. The Fast Fourier Transform is used to smooth and fit a noisy curve. a. Gaussian with white (random) noise. b. The coefficients of the Fourier-transformed original curve. c. All but the first five coefficients are 'lopped off' (set to zero) and an inverse Fourier transform is performed, a smooth curve results, here plotted together with the original graph.

phenomenon. Removing the transients is not just a mathematical technique to make the data less noisy but it also has a physical analogue. In the present situation, the true signal is nearly Gaussian (actually, a higher one for the α–band and a lower one for the associated β–band, which partially overlap). The rapid oscillations are part white (shot) noise from the CCD, in part response of refraction properties of the cell (oscillations around 450–500 nanometers). Both types of oscillation are of rather high frequency compared to the body of the Gaussian.

Given that these oscillations are not part of the signal from the photoreceptor they can legitimately be eliminated.

With experience and familiarity, the scientists can see already in the initial graph what the likely result of transformation would be. More, they do not have to mentally perform a transformation but their perception has developed to the point that they see what will be a good graph. The perceptual processes that were earlier mediated by the additional analytic tools, including transformations, Fast Fourier Transform, pixel shifting, and detrending, now have become embodied in their perception.

8.7. INTO THE COMMUNITY

In the preceding part of this chapter, I show how the graphs are stabilized together with the phenomenon that they come to represent and the instruments and practices that led to their existence. But unless graphs become accepted in the scientific community, and therefore stabilized at a social level, they cannot be used to stabilize subsequent objects. Thus, presenting their graphs with captions in the context of a scientific text so that it cannot be called into question involves itself considerable stabilization work.

In this section, I analyze the manuscript that the scientists Carl, Tom, and a postdoctoral associate produced based on the research that I observed. This analysis shows how the graphs are stabilized in textual form and by the articulation of cross-links between different semiotic devices, including different forms of verbal text (caption, main text) and tables. In Figure 8.10, several excerpts pertaining to one aspect of the research—photoreceptor absorbing in the red part of the spectrum—are provided.

Each panel provides a title that immediately makes salient the nature of the different curves visible. In the left panel, the title includes the signifiers 'rhodopsin' and 'porphyropsin' that parallel two pairs of (smooth and noisy) graphs; in the right panel, the signifiers 'unbleached' and 'bleached' parallel the distinction between a black and blue graphs. In the caption, readers are alerted that what might be taken as the same type of curves, the two smooth (red) lines approximating the associated noisy curve are in fact of different type. In the first instance, the red line signifies an eighth-order polynomial template whereas in the second instance, the red line signifies the result of a Fast Fourier Transform fit. Rather than simply intimating that the red lines resulted from some curve fitting procedure, the scientists provide detailed information as to the nature of the function. The caption also provides further assistance in how to read the graph by linking the black [upper] curve to an unbleached photoreceptor and the blue

MAIN TEXT (relative to red-absorbing member of double cones): All double cones, recorded in this study, were non-identical pairs having dissimilar spectral absorption properties: one red-sensitive and the other green-sensitive (see Fig. 2f). . . . The mean λ_{max} for the α-band of the red-sensitive outer segments differed for the sample of 14 double cones examined: 574 ± 13.9 nm (n=6) and 623 ± 17.9 nm (n = 8) (See Table 1 for spectral data and Figs 3d and 3h for the absorbance spectra of the red-sensitive outer segments of the double cones . . .

Table 1 Spectral data for rainbow trout photoreceptors examined in this study

Photoreceptor Type	Number of Cells	Mean λ_{max} (nm) ± SD	Mean Half Maximum Bandwidth (HBW) (cm^{-1}) ±SD	Mean A_{max} (Optical Density) ± SD	Mean Outer Segment Diameter (μm) ± SD[*] (n)	Specific Optical Density (μm^{-1})
Red-sensitive cones (A_1 based)	6	574±13.9	3256±79.4	0.012±0.009	3.26±0.48 (9)	0.00368
Red-sensitive cones (A_2 based)	8	623±17.8	3709[**]±89	0.022±0.012	3.26±0.48 (9)	0.00675

[*] one standard deviation   [**] estimates derived from Fourier transform fit

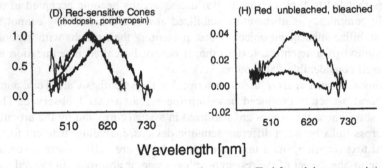

Figure 3. Absorbance spectra of rainbow trout cone photoreceptors. The left panel shows mean relative absorbance for the . . . and red-sensitive cones (Fig. D, two spectra represent A_1 and A_2 visual pigment dominated curves).
The solid black line shows the mean absorbance spectrum and the red line [smooth curve] shows the eighth order polynomial template. Note that the two spectra (red lines) in Fig. D represent the A_1 based visual pigment red-sensitive cones fitted by an eighth order template and the A_2 based visual pigment red-sensitive cones fitted by a Fourier transform function. The right panel shows the unbleached (black line) and bleached (blue line) [lower] of a single cone recording for the . . . red-sensitive cones (Fig. 3.h). See Table 1 for additional details on spectral data of cones.

Figure 8.10. Excerpt from the manuscript prepared for publication.

[lower] curve as resulting from the same ('a single cone recording') but bleached photoreceptor. At the end of the caption, the scientists explicitly point readers to the table where they provided specific readings that could be taken from the graphs such as λ_{max} and mean half-maximum bandwidth.

We can conceptualize what the authors have done in terms of providing links between different resources. The caption refers to specific panels (e.g., 'Fig D' and 'Fig. 3.h') and to a table located elsewhere in the article ('Table 1'). The main text refers to the table and to the figures. By the time the reader has advanced to the position in the text where table and figure are referred to, they have already read about the different fitting procedures. Therefore, the appearance of these signifiers in the caption constitutes a re-occurrence, which it can only be if the sign has occurred before. The scientists stabilize each of these representations in terms of the other forms of text (networks of signifiers).

The main text draws on this and other graphs as the material bases for the claims authors make; graphs form rhetorical resources in the demonstrative practices of science. Because graphs can be used to express continuous variations, they are more economical ways of reporting relationships than text alone. Furthermore, it is quite clear that the scientists are the authors of the text, although they have by and large written themselves out of the narrative. However, the author of the graph is the natural phenomenon itself. That is, it is possible that other scientists question the textual interpretation while accepting the graphs as the signature of a real phenomenon.

The scientists do not leave it up to the reader to extract values for mean λ_{max} and mean half maximum bandwidth, which could be taken from the graph, but supply another source, a data table, from which these values can be taken by the reader.

We can think of the relation between graph, caption, and main text in terms of the relation between sign, referent, and interpretants. In this case, caption and main text provide interpretants, that is, elaborations on the original sign-referent relations. In this, the interpretant text co-articulates the rules (though often implicitly) how to read. (This is the sense of the c in $R = f_r(S,c)$).

Although a reader might be tempted to consider the caption and main text *merely* verbal descriptions of what is there, and therefore providing *merely* redundant information, this, as any scientific article, would never get published as such. Rather, 'literal description' not only describes what is there but also (and inherently) prescribes how to read the graphs and their relation to the natural object. The literal description embodies a pedagogy that tells readers how to read the tables and graphs provided.

As its title 'Microspectrophotometry measurements of vertebrate photoreceptors using CCD-based detection technology' intimates, the scientists' article resulting from the observed laboratory activities had the detection as its major focus. That is, by showing that their scientific results compared favorably with previously published similar data, they supported their argument that the CCD-based detection is a viable alternative. However, the CCD-based technology does not only provide comparable data, but comes with a number of advantages, which makes the device particularly interesting. At the same time, the scientists publish a set of graphs and associated estimates of λ_{max} and half-bandwidth.

In themselves, the graphs and tools do not sufficiently stabilize one another. It is the comparison to previously existing and accepted methodology—among others used by the lead researcher himself—that provides substantial stabilizing elements. 'Data derived from the two different techniques compare favorably in λ_{max}, A_{max} and HBW'. Differences between the two techniques are attributed to 'variable A_1/A_2 visual pigment ratios of the fish used in the two studies'. (Readers could still argue that the instrumentation produces artifacts [graphs] and that graphs and methodology [including instrumentation] simply reify each other.)

In the context of my earlier research on the use of graphs in ecology journals, this example is rather typical in the tight integration of graph, caption, table, and main text. The integration is tight, in part, because from an informational point of view, the four parts of a scientific article provide a certain level of redundancy. From a semiotic perspective, they provide signifiers that are linked thereby constraining the significations a reader can construct. This linkage and redundancy decreases the interpretive flexibility of any single aspect and assures a greater likelihood of intersubjectivity between authors and their readers. The present analysis is consistent with the more extensive analysis in that it shows that scientific authors construct this graph-caption-table-text such that they mutually stabilize each other and permit but one reading. The scientific illustration constitutes an 'informational ambush' that leaves no escape for alternative interpretations. Rather than letting readers interpret the graph, the caption directs them to particular aspects: the unit of analysis (mean, single cone), the nature of the different smooth lines modeling the experimentally acquired curves, the different objects (UV, blue, green and red cones), and the different chemical composition of cones (A_1, A_2). The main text provides a particular reading and therefore instructions about how to read the graph. Thus, although the audience of this article is definitively scientific, descriptions of (instructions for reading) graphs are provided, and attention is called to units and scale. This is a practice that is not the rule in high school textbooks or in research on students' competencies related to graphing.

Biology provides ways of looking at nature. Biological texts, therefore, both in research papers and textbooks, serve as demonstrations in which readers are shown, and learn to see, the order reflexively specified by the textual arrangement of words and graphics. The text, then, is a work site where knowledge is constructed not through the passage of information from the book or article to the reader, but through the recovery of the order encoded by the author in text and accompanying graphs. It is through the noticeable work of reading that the knowledge of biological objects is constructed. The convincing text engages readers in a process of authentication so that readers recover what the biologist previously saw. Texts that permit readers to achieve this authentication, that is, texts that scaffold readers in achieving the demonstrable fit between graphs and text, display a compelling pedagogy in teaching readers something about the world they inhabit.

8.8. MUTUAL STABILIZATION OF GRAPH AND 'NATURAL OBJECT'

Graphs as objects do not exist as independent entities but are part of complex networks that integrate entities and processes. The competence of a scientific reader of a graph (relative to its 'standard' reading) depends to a large extend on the reader's familiarity with the entire network of entities. The producers of a graph, therefore, always exhibit great competence because of their intimate familiarity with the network. Other scientific readers also need to be at least familiar with the phenomenon and know how such graphs are produced and transformed. That is, the reader of the graph has to be already familiar with how such places (laboratories) function, and therefore, how graphs and the phenomenon it is said to stand for come to be associated. The understanding of graphs therefore depends on the stabilization both in the laboratory and its subsequent adoption and distribution in the scientific community (journal), from where they can then be internalized by other scientific laboratories, their instruments and people.

8.8.1. Signs and Referents, Tools and Natural Objects

The interpretations are the results of historical processes in which a system of signifiers and processes stabilizes itself. Results of earlier research by this and other research groups are internalized by individuals or are built into the instruments and calculation devices (computers, software), mediating future interpretations. As a result, what my research shows is a transforming activity system in progress. Some of its historical transformations can be reconstructed or are visible when there is trouble.

During the scientists' laboratory work, the graphs that they interpret in a first-time-through manner are both tools and object in the inquiry (*Fig. 8.11*). On the one hand, they use graphs as tools to interrogate and interpret the 'natural object' and the entire chain of translation by means of which the 'natural object' is made present again. On the other hand, the graphs also become objects of inquiry in their own rights that are interpreted in terms of the scientists'

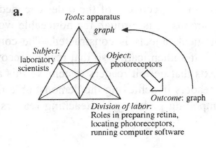

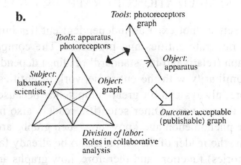

Figure 8.11. a. As soon as a graph has been produced, it becomes tool or object in its own right. b. The graph then participates in the stabilization of natural object, instrumentation, processes, and representation. The outcome of this stabilization work is a publishable graph.

understanding of and familiarity with the natural phenomenon under investigation and the contributions of the investigative process on the representation. The graphs therefore exist in a dense network of mutually stabilizing entities that also include natural objects, researcher skill in preparation, mathematical transformations, instrumental contingencies, and so forth. Scientists do not just make inferences that lead them from the sign to the referent. Rather, they always move back and forth. Based on the graph, they imagine possible scenarios that could have given rise to the representation; based on their understanding of the natural world, they imagine what a graph would look like given the graphical (representational) space at hand. It is out of the degree of congruence in the results of both movements that scientists develop a sense that they understand. From the activity theoretic perspective used here, we say that the scientists interchange tools (instrumentation, concepts of natural world and instrumentation) and objects. Their sense of understanding arises when tools and objects mutually stabilize each other.

The scientists in this laboratory did not discover the retinal absorption and the underlying scientific theory by making inferences from controlled experiments. Rather, the phenomenon, ability to produce it, and conceptual understanding were involved in a process of mutually stabilization. In the earlier stages of this inquiry, the scientists were not yet able to make the distinction between the representation and its object'. It is only at the end of his inquiry that they came to separate phenomenon, process of inquiry, and scientific theory. The correspondence of representations (signs) to their natural objects is the result of a process in which experiment and narratives are made to converge. In this process, the distinction between signs and their referents in the world are reified while the traces of the work that enabled the distinction are being removed. In scientific experimentation, gestures, skills, and so on—all of these aspects of disciplined human agency come together with the machines that they set in motion and exploit. Disciplined human agency and natural phenomena become intertwined in mutually constitutive ways and therefore interactively stabilize each other.

8.8.2 Movement of Objects (Graphs) through Activity Systems

After the scientists have stabilized graphs, natural objects, and instrumentation (and associated practices), they use textual and graphical resources to stabilize their graphs (and associated methodology) so that they will fare well in the community and become, eventually, accepted among the tools available to researchers. The scientists articulate how they drew on legitimized (accepted in peer review and subsequently published) procedures for preparing the natural object (photoreceptor) and for transforming the raw graphs (e.g., templates, detrending, Fourier transformation, and pixel shifting). Explicitly locating their own activity in the concerns and practices of the community guarantees a certain level of success to the prospects of their own work becoming an integral feature of the community. That is, the prospect of bringing about changes not only in their own activity system but also in all the aspects that are also part of the activity system of other research groups. When the scientists believe that they have sufficiently stabilized their tools and outcomes and that they have produced a narrative that is not only reflective of the stabilization process but also as unassailable as they can possibly consider, they submit their work to a journal and thereby to a peer review process (*Fig. 8.12*).

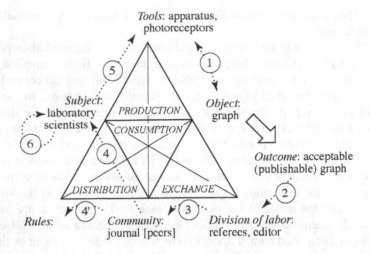

Figure 8.12. After the graph has been stabilized in the context of instrumentation and natural object (1), it moves via the review process (2) to become an aspect of the community (3). From there, it can be appropriated into a another (or the same) laboratory (4) where scientists begin to use it as a tool (5), build it into another tool (5), or internalize it themselves (6). In their publication scientists also formulate rules for reading graphs or using tools in specific ways (4').

Once the article has been accepted (which involves editors and reviewers serving for the community as an aspect of the division of labor) and published, the tools it describes and the graphs and data it presents are ready to become used by other researchers. Being used really means that these research tools, graphs, and data are available and can be built into other tools or be internalized by other researchers (subjects in activity systems). This process of being taken up was evident for the outcomes of other previously published research in the present situation. For example, the eighth-order polynomial curves, outcomes of earlier research, were built into the custom software to be used during the first-time-through interpretation as templates against which to judge the graphs produced by the researchers. These templates therefore constitute built-in tools that stabilize or destabilize a given graph as a candidate for further consideration. However, these eighth-order polynomials were not just built into the software but, having used them extensively, the researchers have become so familiar with them so that they can see in a noisy graph those features that make them match to a template. That is, by looking at a graph they can see it as a candidate—which, however, they have to clean up (detrend, pixel-shift, enter into averaging procedure, etc.) before their peers will be able to see in the published graph what they can see in the messy one.

Scientific tools and graphs are never just presented but are always published with detailed instructions of how to use or read them, respectively. That is, scientists do not just present a table with the values of λ_{max} and mean half-maximum bandwidth and the corresponding graph but use main text and caption to constrain the ways in which these can be read. Graphs therefore do not simply display relationships, adjuncts to text, or proofs for something stated in the text that readers can 'take in' in an unproblematic way. To understand the biological phenomena of which the graph is a part, readers must engage in a reflexive elaboration in which the main text and caption provide iterable instructions for where to look and how to read the graphical display; they provide organizational resources to the reader's gaze and interpretation. But this is not sufficient. To understand the graph, the reader already has to bring an understanding of the biological phenomena and the ways these come to be presented in something like a graph.

Appendix

The Tasks

Originally, my students (G. Michael Bowen and Michelle K. McGinn) and I conducted a research project on representation use among second-year ecology students. Because of the difficulties that these students experienced and to have a better idea for the graph-related practices of experienced scientists, I decided to use graphs from this ecology course. It was expected that scientists would represent a norm group featuring the kind of expertise that some of our university students would eventually develop.

We had established an extensive database on representation use in the ecology course, which was comprised of three weekly fifty-minute lectures and one weekly fifty-minute seminar for 13 weeks. For the seminars the class was divided into three groups so that there were about 15 students per seminar meeting, although some students did not attend the same seminar consistently. Whenever possible, the written artifacts produced during students' collective work were obtained. Copies were made of student responses to quizzes at the beginning of seminars and field notes of student interaction and participation made. Copies of student exams and the instructor's class notes, lecture notes, and his seminar instruction notes for the teaching assistant constituted additional data sources. These data and those constructed for a comparison of representation use in high school and university textbooks and scientific journals[1] allowed us to identify 'typical' graphs that second-year university students were expected to interpret. I finally chose three graphs for use with the scientists.

The three graphs we ultimately chose for our work with the scientists differ in type and complexity. The distribution graph (*Fig. A.1*) features the relative frequency of occurrence of three plant types according to published data. The population dynamics graph (*Fig. A.2*) represents a theoretical model that plots the functional relationship of birth rates and death rate on the population size, which itself depends on the rates. The isocline graphs (*Fig. A.3*) constitute a theoretical model of three pairs of resources which affect some third, growth variable; the size of the third variable is not directly available (as in a three-dimensional graph), but in the form of isoclines. All three graphs are common to the literature in ecology and in textbooks on the topic. For example, one widely used ecology textbook features 41 distributions, 19 graphical models with functional dependencies, and 65 isographs over a total of about 800 pages.[2] In several instances, we used additional graph interpretation tasks to understand the competencies of individual scientists. Thus, in the case of the field ecologist featured in Chapter 7 I used additional graphs that represented population dynamics; I also used mathematical and statistical models closely related to her work were introduced and discussed in a series of interviews. In other instances, I probed scientists' understanding of population dynamics by posing a task including several interrelated graphs.

A.1. PLANT DISTRIBUTIONS

The distribution graph (*Fig. A.1*) is similar to that originally published in a scientific article which used it to plot data. These data confirmed a model, according to which different metabolic pathways afforded differential adaptation to the climate. The graph portrays the relative distributions of three different kinds of plants along an elevation gradient; the data being recorded in the Big Bend National Park, Texas, this elevation gradient is also associated with temperature and moisture gradients indicated above the graph. CAM and C_3 distributions were on the same scale, and in fact complement each other; the C_4 distribution used a different scale in the original research.

Such graphs are central to ecology of niches, for they relate biological activity to a single environmental gradient and therefore represent the distribution of a species' activity along one niche dimension.[3] They often display the level of activity (e.g., oxygen metabolism) as a function of temperature or consumption rate as function of prey size to convey the ability of an individual to exploit resources in a particular part of the niche space. Conversely, such graphs thereby also represent the degree to which the environment can support the population of that species.

The graph used in this research was modeled on that used in the lecture course and represented a simplified equivalent of the original, published graph. The caption was similar to that in the course, but formulated the task such that the participants were instructed to talk about the inferences that could be drawn from the graph. This task therefore closely resembles the activity of a person opening a journal and trying to make inferences without referring to the main text. Here, the graph allows the reader to make an inference about ecological

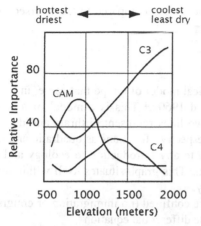

Distribution of C_3, C_4 and CAM (succulent) plants in the desert and semi-desert vegetation of Big Bend National Park, Texas, along a moisture and temperature gradient due to differences in elevation. CAM plants with nocturnal gas exchange for water conservation predominate in the hottest, driest environment, C_4 plants are maximally important under intermediate temperature and moisture conditions, and C_3 plants predominate at the cooler, least dry end of the gradient. (After data of W.B. Eickmeier [1978], Photosynthetica, 12, 290-297). What implications can you draw from this graph?

Figure A.1. In this task, the different distributions of three types of plants is used to make inferences about the differential adaptation to environmental conditions (here climate).

niches and the adaptation of the different types of plants to climate. For example, the caption provides a potential starting point for making an inference about adaptation: 'CAM plants with nocturnal gas exchange for water conservation predominate in the hottest, driest environment'. To make the (by the original author) intended inference, the relative positions of the three distributions along the elevation or climate variables have to be compared and attributed to the three photosynthetic (metabolic) processes.

Rather than presenting a graph with individual data points, I followed the practices in textbooks and lectures to use smooth graphs; that is, I eliminated variance to arrive at relatively simple distributions. However, in other respects, the task presented interpretative resources that are common in the scientific literature. It turned out that the abscissa label /relative importance/ was interpretively very flexible leading some of our scientists to abandon the task. However, the sign /distribution/ in the caption, which constitutes articulated context, should allow a reading of the graph as a relative distribution. The caption provides a reading of the three curves, especially of the relative location of the distribution maxima. Although the three curves are labeled (C_3, C_4, CAM), and although the caption suggests that CAM plants use the process of nocturnal gas exchange, the task does not make evident that the labels distinguish plants according to their photosynthetic mechanism. There are several intersections of the three lines, a fact that becomes relevant when compared to the intersections in the population graph.

A.2. POPULATION DYNAMICS

The second graph constitutes a graphical model of a type that came, in the field of ecology, into use in the 1950s and 1960s.[4] This is a model of a density-dependent population in which the two lines represent birthrate and death rate (*Fig. A.2*); the caption specifies the respective functions as quadratic and linear, respectively. Such models are central to any introduction to ecology and often appear in the first chapter of textbooks. This graph situates itself within the discipline of ecology in the following way.

Single species populations that are confined (no immigration or emigration) are modeled in ecology by means of the differential equation

$$\frac{dN}{dt} = (b - d)N \qquad (1)$$

where birthrate b and death rate d are some function of the population size N, i.e., $b = b(N)$ and $d = d(N)$. In the simplest case, b and d are constant so that the differential equation has solution in the form of an exponential growth rate

$$N(t) = N_0 e^{(b-d)t}. \qquad (2)$$

Exponential growth curves are biologically unrealistic. A more realistic model is obtained when birth and death rates are linear functions of population size, $b(N) = b_0 + b_1 N$ and $d(N) = d_0 + d_1 N$, respectively (*Fig. A.3*). In this form, these models have been the subject of other research on graphing. For example, Hermina Tabachnick-Schijf and her colleagues used the model in an economy context, where birthrate and death rate are replaced by supply and demand.[5]

In this case, the differential equation is

$$\frac{dN}{dt} = \big((b_0 - d_0) + (b_1 - d_1)N\big)N, \qquad (3)$$

which is usually expressed as the logistic growth

$$\frac{dN}{dt} = rN(1 - \frac{N}{K}) \qquad (4)$$

In the derivation of the logistic model, we assume that, as N increased, birthrates declined linearly and death rates increased linearly. Now, let's assume that the birth rates follow a quadratic function (e.g., $b = b_0 + (k_b)N - (k_c)N^2$), such that the birth- and death rates look like the figure. Such a function is biologically realistic if, for example, individuals have trouble finding mates when they are at very low density. Discuss the implication of the birth- and death rates in the figure, as regards conservation of such a species. Focus on the birth and death rates at the two intersection points of the lines, and on what happens to population sizes in the zones of population size below, between, and above the intersection points.

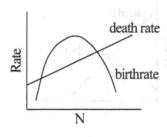

Figure A.2. In this task, competent ecologists infer that there are stable and unstable equilibrium points for the size of a population.

with positive constants r and K defined as

$$r = (b_0 - d_0) \text{ and } K = -\frac{(b_0 - d_0)}{(b_1 - d_1)}. \tag{5}$$

The constant K is the equilibrium capacity or carrying capacity of the environment. With $N(0) = N_0$, the solution of (4) is

$$N(t) = \frac{N_0 K e^{rt}}{\left(K + N_0 (e^{rt} - 1) \right)} \tag{6}$$

which, for different starting populations values $r > 0$ and $r < 0$, is illustrated in Figure A.4.

Apart from the fact that the birthrate is a quadratic function of population size, the behavior of the population near the two intersections is very similar to the two situations displayed in Figure A.3. The lower intersection can be approximated by a linearly increasing birthrate and a linearly decreasing death rate. In this case, $(b_0 - d_0) < 0$, giving rise to a situation as in Figure A.3.b where the population moves away from the equilibrium value ($N = 10$). The intersection of birthrate and death rate on the right-hand side (*Fig. A.2*) leads to a stable equilibrium, for $(b_0 - d_0) > 0$.

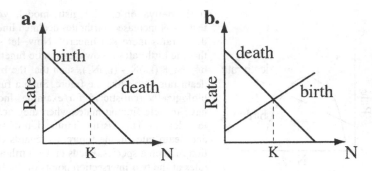

Figure A.3. Linear models of birthrates and death rates. a. The value of r = (b₀ − d₀) > 0. b. The value of r = (b₀ − d₀) < 0. At K, because birthrate and death rate are equal, the population size remains constant and the population is said to be in an equilibrium state.

In the present situation, the task as displayed in Figure A.2 consists in establishing the dynamics of a population, from the circular relations between population size, $N = N(b,d)$ and the two rates, $b = b(N)$ and $d = d(N)$. In this task, participants are specifically asked to focus on the two intersections and the resultant three sections along the abscissa. The target response identifies the two intersections as constituting an unstable and a stable equilibrium, or alternatively, provides a model in which the temporal development of the population size has been characterized as approaching the upper equilibrium or crashing below the lower equilibrium.

The task is typical for many that might be found in the context of teaching and research on graphing; there are no scales or units that might link this graph to any particular situation. Whereas the theoretical ecologists and physicists in my sample had had no qualms about the population graph, field ecologists and non-university researchers generally described graphs of this type as little helpful, falsifying the real issues that face ecology today. There are no indications as to the temporal variations in birthrates and death rates through the year and no adjustments are being made for migration. Thus, one researcher who had a successful career including several fellowship throughout his graduate work remarked 'I kind of get a bit bitter about behavioral ecology by assuming graphs like this because they're really idealized, abstractions'. Another scientist with large national and international funding of his research suggested, 'You're never gonna find a data set that looks like this. This is a theoretical model; it's based

a.   b.

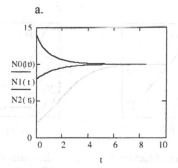   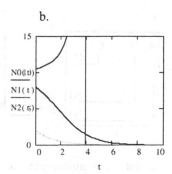

Figure A.4. Logistic population growth for different initial populations N_0 and stable (a) and unstable equilibrium (b) at the value for carrying capacity (K = 10).

on, you know, nice mathematics and equations... I don't know, any data set where you ever find this and you can ever point out there are probably two steady states'.

A.3. ISOCLINES

The third type of graph selected makes use of 'isoclines'. The word isocline is employed in a special way in ecology. Any interested student who looked up dictionaries for the word isocline would probably find brief one-sentence definitions, none of which related to the use in ecology. Thus, as the lecture text stated, the term isocline derives from the Greek isoklinês, iso- (same) and klinein, to slope or incline. Across a variety of dictionaries college students may use, isocline is defined as: (a) An anticline or a syncline so tightly folded that the rock beds of the two sides are nearly parallel; and (b) a line on a map connecting points of equal magnetic dip.[6]

Isoclines are an important means, for example, in toxicology, for representing the interaction of two independent variables (e.g., two heavy metals) on a third variable. In the present case, the term isocline is appropriate for cases where a derivative g (e.g., growth rate g) of some function (e.g., height h) is expressed:

$$g(t, R_1, R_2) = \frac{\partial}{\partial t} h(t, R_1, R_2)$$

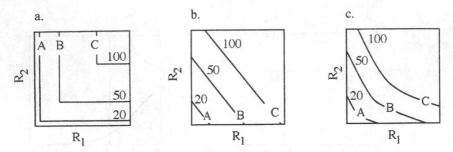

The amount a plant grows depends on a number of factors, for instance, the availability of nutrients (R). A shortage of any single nutrient can limit plant growth. Sometimes scientists study the effect of pairs of nutrients. The graphs depict three different biologically realistic scenarios of how two nutrients (R_1 and R_2) might combine to affect plant growth. Discuss the effects of different levels of the two nutrients on each amount of plant growth (20, 50, 100) in each scenario (i, ii, iii).

Figure A.5. In the isographs, two independent variables are related to a third, dependent variable which is represented only in the form of isoclines, lines of equal value.

Graphing g as a function of R_1 and R_2 then gives a surface plot in three-dimension. Any point on the surface is given by the values of R_1 and R_2, the co-ordinates in the horizontal plane, and $g = g(R_1, R_2)$ the value of the z-coordinate. If the graph is sectioned parallel to the x-y plane at different values g, one yields the curves in Figure A.5, where each line represents a constant growth rate. Isoclines allow scientists to analyze the interaction of two variables such as R_1 and R_2 on a third variable and therefore play important roles in the construction of scientific knowledge.

The three graphs (*Fig. A.5*) represent the conjoint effects of two variables on a third, where the third is represented by lines of equal effect. Figure A.5.a represents 'essential resources' which can mathematically be represented, for example, as (using convention of x representing horizontal, y the vertical axis, and z the axis coming out of the plane).

$$z(x, y) = \begin{cases} ax + b \text{ above} \\ cy + d \text{ below} \end{cases} \text{line connecting vertices} \}$$

The graph in Figure A.5.b represents 'substitutable resources' which can be represented mathematically as $z(x,y) = ax + by + c$. The two resources are therefore substitutable in the proportion of a and b. Simply saying that the model is

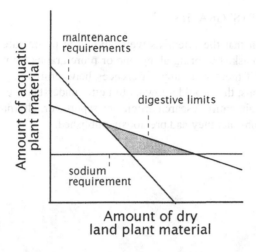

Figure A.6. Sample use of isoclines in combination to model the dietary choices of moose. (For detailed example see Stephens and Krebs op. cit. note 7.)

linear does not distinguish it from the first graph which is also linear in one or the other variable. Finally, the third graph (*Fig. A.5.c*) represents 'complementary resources' because there is some optimum combination that minimizes the sum of the individual components. Mathematically, the equation $z(x,y) = (a{\cdot}x + b){\cdot}(c{\cdot}y + d)$ would yield isoclines of the way presented.

Isoclines such as those presented here are central tools for understanding phenomena in behavioral ecology.[7] Take the following example, which I constructed to show how ecologists might predict the feeding choices of moose whose diet consists of aquatic and dry land plants (*Fig. A.6*). Moose require a certain amount of salt, which is modeled by a horizontal line, because it can be obtained only by eating wet plants. Then, there are both minimum quantities of plant material moose need to fulfill their dietary requirements and digestive limits on the amount of materials that can be processed. That is, three isoclines similar to those in Figures A.5.a and A.5.b establish the boundaries within which we should find the mix of wet and dry land plants that moose choose to feed on. Despite the prevalence of isocline graphs in ecology, this series of graph turned out to be the most difficult to interpret for our scientists.

A.4. SCIENTISTS' GRAPHS

At the moment that the scientists were invited to participate in the interviews, they were also asked to bring along one or more graphs from their work. It was expected that if there were any differences between reading our graphs versus their own graphs, this would allow us to better understand 'expertise'. Scientists normally brought entire research articles or reports, or, alternatively, printouts of individual graphs that they had previously published.

Notes

PREFACE

1. See Wolff-Michael Roth and G. Michael Bowen, 'Mathematization of experience in a grade 8 open-inquiry environment: An introduction to the representational practices of science', *Journal of Research in Science Teaching*, Vol. 31, No. 3 (1994), 293–318; Wolff-Michael Roth, 'Where is the context in contextual word problems?: Mathematical practices and products in Grade 8 students' answers to story problems', *Cognition and Instruction*, Vol. 14, No. 4 (1996), 487–527.

2. With respect to the comparison between high school textbooks and scientific journals, see, for example, Wolff-Michael Roth, G. Michael Bowen and Michelle K. McGinn, 'Differences in graph-related practices between high school biology textbooks and scientific ecology journals', Journal of Research in Science Teaching, Vol. 36 (1999), 977–1019. Concerning the study of scientists' use and interpretation of graphs, see, for example, Wolff-Michael Roth and G. Michael Bowen, 'Of cannibals, missionaries, and converts: graphing competencies from grade 8 to professional science inside (classrooms) and outside (field/laboratory)', *Science, Technology, & Human Values*, Vol. 24, No. 2 (1999), 179–212; Wolff-Michael Roth, G. Michael Bowen and Domenico Masciotra, 'From thing to sign and "natural object": Toward a genetic phenomenology of graph interpretation', *Science, Technology, & Human Values*, Vol. 27, No. 4 (2002), 327–356.

CHAPTER ONE

1. See Michael Lynch, 'Extending Wittgenstein: The Pivotal Move from Epistemology to the Sociology of Science', in Andrew Pickering (ed.), *Science as Practice and Culture* (Chicago: University of Chicago Press, 1992), 215–300, at 230.

2. Although psychologists such as Jill Larkin and Herb Simon differ in many ways from sociologists, anthropologists and other 'students of science' such as Latour about the nature of cognition related to inscriptions they agree on the fact that these are central to scientific activity. See Jill H. Larkin and Herbert A. Simon, 'Why a diagram is (sometimes) worth ten thousand words', *Cognitive Science*, Vol. 11 (1987), 65–99; Bruno Latour, *Science in Action: How to Follow Scientists and Engineers through Society* (Milton Keynes: Open University Press, 1987); Bruno Latour, *La clef de Berlin et autres leçons d'un amateur de sciences [The key to Berlin and other lessons of a science lover]* (Paris: Éditions la Découverte, 1993).

3. See for example Hermina J. M. Tabachneck-Schijf, Anthony M. Leonardo and Herbert A. Simon, 'CaMeRa: A computational model for multiple representations', *Cognitive Science*, Vol. 21, No. 3 (1997), 305–350; Karin D. Knorr-Cetina and Klaus Amann, 'Image dissection in natural scientific inquiry', *Science, Technology, & Human Values*, Vol. 15, No. 3 (1990), 259–283; Kathryn Henderson, 'Flexible sketches and inflexible databases: Visual communication, conscription devices, and boundary objects in design engineering', *Science, Technology, & Human Values*, Vol. 16, No. 4 (1991), 448–473.

4. Wolff-Michael Roth, G. Michael Bowen and Michelle K. McGinn, 'Differences in graph-related practices between high school biology textbooks and scientific ecology journals', *Journal of Research in Science Teaching*, Vol. 36, No. 9 (1999), 977–1019.

5. See Jay L. Lemke, 'Multiplying meaning: Visual and verbal semiotics in scientific text', in J. R. Martin & R. Veel (eds), *Reading science* (London: Routledge, 1998), 87–113.

6. Gaea Leinhardt, O. Zaslavsky and Mary K. Stein, 'Functions, graphs, and graphing: Tasks, learning, and teaching', *Review of Educational Research*, Vol. 60, No. 1 (1990), 1–64.

7. Wolff-Michael Roth, Michelle K. McGinn and G. Michael Bowen, 'How prepared are preservice teachers to teach scientific inquiry? Levels of performance in scientific representation practices', *Journal of Science Teacher Education*, Vol. 9, No. 1 (1998), 25–48.

8. For examples of this approach see Phil E. Agre, *Computation and Human Experience* (Cambridge: Cambridge University Press, 1997); Dana H. Ballard, Mary M. Hayhoe, Polly K. Pook and Rajesh P.N. Rao, 'Deictic codes for the embodiment of cognition', *Behavioral and Brain Sciences*, Vol. 20 (1997), 723–767.

9. Jean Lave, *Cognition in Practice: Mind, Mathematics and Culture in Everyday Life* (Cambridge: Cambridge University Press, 1988); Lucy A. Suchman, *Plans and Situated Actions: The Problem of Human-Machine Communication* (Cambridge: Cambridge University Press, 1987). Donald Davidson makes the point that we should abandon the notion of language as a third thing intervening between an individual and realityor of languages as barriers between persons or cultures. He encourages us to erase the boundary between knowing a language and knowing once way around the world more generally. See Donald Davidson, 'A nice derangement of epitaphs', in Ernest Lepore (ed), *Truth and Interpretation: Perspectives on the Philosophy of Donald Davidson* (Oxford: Blackwell, 1986), 433–446.

10. Bourdieu and many other practice theorists are made this point quite forcefully. See, for example, Pierre Bourdieu, *Le sens pratique* (Paris: Les Éditions de Minuit, 1980), which was published in English as Pierre Bourdieu, *The Logic of Practice* (Cambridge, UK: Polity Press, 1990).

11. Craig Berg and Philip Smith designed a study to find out the variation in responses with the type of instrument used. One of my own studies showed not only that students' responses to lever problems changed with format but also (for certain formats the responses) that the responses did not overlap. Craig A. Berg, and Phillip Smith, 'Assessing students' abilities to construct and interpret line graphs: Disparities between multiple-choice and free-response instruments', *Science Education*, Vol. 78, No. 5 (1994), 527–554. Michelle K. McGinn and Wolff-Michael Roth, 'Assessing students' understandings about levers: better test instruments are not enough', *International Journal of Science Education*, Vol. 20 (1998), 813–832; Wolff-Michael Roth, 'Situated cognition and assessment of competence in science', *Evaluation and Program Planning*, Vol. 21, 155–169.

12. See Latour 1987, op. cit. note 2, at 258.

13. Examples of cognitive anthropologists are Ed Hutchins and Jean Lave; the sociologists include Pierre Bourdieu and Michael Lynch; among the cognitive scientists we find Ed Hutchins (1995). Ed Hutchins, *Cognition in the Wild* (Cambridge, MA: MIT Press, 1995); Lave op. cit. note 9;

and Michael Lynch, *Art and Artifact in Laboratory Science: A Study of Shop Work and Shop Talk in a Laboratory* (London: Routledge and Kegan Paul, 1985).

14. See Paul Ricœur, *Oneself as Another* (Chicago: University of Chicago Press, 1992).

15. See Pierre Bourdieu, *Méditations pascaliennes* [Pascalian meditations] (Paris: Seuil, 1997), at 170.

16 See for example Valerie Walkerdine, *The Mastery of Reason* (London: Routledge, 1988).

17. Martin Heidegger, *Sein und Zeit* [Being and time] (Tübingen, Germany: Max Niemeyer, 1977); I also consulted the English translation by Joan Stambaugh (Albany: State University of New York Press, 1996).

18. See John Seely Brown, Alan Collins and Paul Duguid, 'Situated cognition and the culture of learning', *Educational Researcher*, Vol. 18, No. 1 (1989), 32–42.

19. See Jean Lave, 'The practice of learning', in Seth Chaiklin and Jean Lave (eds), *Understanding Practice: Perspectives on Activity and Context* (Cambridge: Cambridge University Press), 3–32.

20. Sharon Traweek, *Beamtimes and Lifetimes: The World of High Energy Physicists* (Cambridge, MA: MIT Press, 1988).

21. See Jean Lave and Etienne Wenger, *Situated Learning: Legitimate Peripheral Participation* (Cambridge: Cambridge University Press, 1991).

22. On this point see Michael Lynch, 'Method: measurement--ordinary and scientific measurement as ethnomethodological phenomena', in Graham Button (ed), *Ethnomethodology and the Human Sciences* (Cambridge: Cambridge University Press, 1991), 77–108.

23. James G. Greeno, 'Situated activities of learning and knowing in mathematics', in M. Behr, C. Lacampagne, and M. M. Wheeler (Eds.), *Proceedings of the 10th Annual Meeting of the PME-NA* (DeKalb, IL: IGPME, 1998), 481–521, at 482.

24. This point was quite apparent in the studies of the proof of the four-color theorem Donald Mackenzie, 'Slaying the kraken: The sociohistory of a mathematical proof', *Social Studies of Science*, Vol. 29, No. 1 (February 1999), 7–60. Among theoretical physicists, there is therefore a cultural divide between those who seek 'analytical' solutions and those who seek 'numerical' solutions to complex problems. Daniel Kennefick, 'Star crushing: Theoretical practice and the theoreticians' regress', *Social Studies of Science*, Vol. 30, No. 1 (February 2000), 5–40. There therefore exists a preference for doing theoretical physics in collaborative teams, where members are both familiar with the method and willing to do an aspect of a calculation independently. See Martina Merz and Karin Knorr-Cetina, 'Deconstruction in a 'thinking' science: Theoretical physicists at work', *Social Studies of Science*, Vol. 27, No. 1 (February 1997), 73–111.

25. This motivated the noted semiotician Umberto Eco to write the novel *The name of the rose*, in which a monk solves, in Sherlock Holmes like fashion to solve a murder by establishing a network of signification. Similarly, Bruno Latour's report on the failure of the French ARAMIS project—an individualized rapid transport system—was framed as a semiotic project in which the investigating sociologist established networks of signification based on signifying artifacts found during the investigation (Latour, 1992). Umberto Eco, *The Name of the Rose* (Boston, MA: G. K. Hall, 1984); Bruno Latour, *Aramis ou l'amour des techniques* (Paris: Éditions de la Découverte, 1992).

26. Among the philosophers and general semioticians, I count Umberto Eco and Paul Ricœur. See Umberto Eco, *Semiotics and the Philosophy of Language* (Bloomington: Indiana University Press, 1984); Paul Ricœur, *From Text to Action: Essays in Hermeneutics, II* (Evanston, IL: Northwestern University Press, 1991). Françoise Bastide and Jacques Bertin specifically have taken a semiotic perspective on graphing. See Françoise Bastide, 'The iconography of scientific texts:

principles of analysis', in Michael Lynch and Steve Woolgar (eds), *Representation in Scientific Practice* (Cambridge, MA: MIT Press, 1990), 187–229; Jacques Bertin, *Semiology of Graphics*, trans. W. J. Berg (Madison: University of Wisconsin Press, 1967/1983).

27. Triadic models of signs and sign use were established in different ways by Gottlieb Frege (Sinn, Zeichen, Bedeutung), I. A. Richards and C. K. Ogden (reference, symbol, referent), and Charles Sanders Peirce (interpretant, representamen, object). I follow Eco, who, with others, adopted the terms interpretant, sign, and referent. Umberto Eco, *A Theory of Semiotics* (Bloomington: Indiana University Press, 1976).

28. Whenever it is important, I use slashes (/atom/) and guillemets («atom») to distinguish signs and referents, respectively.

29. For example, John Law and Michel Callon made this point. In social studies of science, this move is used to establish a continuum between nature and word in order to get rid of the unfortunate binary approach that opposes nature and language. Any representation of nature can then be shown as being the result of a series of translation that turns nature into graphs and language. John Law and Michel Callon, 'Agency and the hybrid collectif', in Barbara Herrnstein Smith and Arkady Plotnitsky (eds), *Mathematics, Science, and Postclassical Theory* (Durham, NC: Duke University Press), 95–117.

30. Readers are advised that we are dealing here not with a mathematical function but a mathematical relation, that is, there are no unique mappings from one domain onto another, such as this would be the case for functions. Jon Barwise, 'On the circumstantial relation between meaning and content', in Umberto Eco, Marco Santambrogio and Patrizia Violi (eds), *Meaning and Mental Representations* (Bloomington: Indiana University Press, 1988), 23–39; Jon Barwise, *The Situation in Logic* (Stanford, CA: CSLI, 1989).

31. Such tasks have been used, for example, by Berg and Smith (op. cit. note 11) and Mokros and Tinker. Janice R. Mokros and Robert F. Tinker, 'The impact of microcomputer-based labs on children's ability to interpret graphs', *Journal of Research in Science Teaching*, Vol. 24, Vol. 4 (1987), 369–383.

32. Eco, op. cit. note 27.

33. Wolff-Michael Roth and Kenneth Tobin, 'Aristotle and natural observation versus Galileo and scientific experiment: An analysis of lectures in physics for elementary teachers in terms of discourse and inscriptions', *Journal of Research in Science Teaching*, Vol. 33, No. 1 (1996), 135–157.

34. Wolff-Michael Roth, Kenneth Tobin and Ken Shaw, 'Cascades of inscriptions and the representation of nature: How numbers, tables, graphs, and money come to re-present a rolling ball', *International Journal of Science Education*, Vol. 19, No. 10 (1997), 1075–1091.

35. Harold Garfinkel, Michael Lynch and Eric Livingston, 'The work of a discovering science construed with materials from the optically discovered pulsar', *Philosophy of the Social Sciences*, Vol. 11 (1981), 131–158.

36. Daniel Kennefick described the debate surrounding the detection of gravitational waves, which is based on analyses of data where the noise is of the same order of magnitude—or even larger—then the expected signal. The data analysis, then, involves a range of 'tricks' to get rid of much of the noise so that the expected signal, a sign that a gravitational wave has passed by the earth and has been detected. See Kennefick, op. cit. note 24.

37. On this point see, for example, the study by Françoise Bastide, op. cit. note 26.

38. Klaus Amann and Karin Knorr-Cetina, 'The fixation of (visual) evidence', *Human Studies*, Vol. 11 (1988), 133–169.

39. Howard Wainer, 'Understanding graphs and tables', *Educational Researcher*, Vol. 21, No. 1 (February 1992), 14–23.

40. Steve Woolgar, 'Time and documents in researcher interaction: Some ways of making out what is happening in experimental science', in Michael Lynch and Steve Woolgar (eds), *Representation in Scientific Practice* (Cambridge, MA: MIT Press, 1990), 123–152.

41. Bruno Latour, 'Drawing things together', in Michael Lynch and Steve Woolgar (eds), *Representation in Scientific Practice* (Cambridge, MA: MIT Press, 1990), 19–68.

42. See Amann and Knorr-Cetina op. cit. note 38; Wolff-Michael Roth and G. Michael Bowen, 'Digitizing lizards or the topology of vision in ecological fieldwork', *Social Studies of Science*, Vol. 29, No. 5 (October 1999), 719–764.

43. Michael Lynch and Samuel Y. Edgerton, 'Aesthetics and digital image processing: Representational craft in contemporary astronomy', in Gordon Fyfe and John Law (eds), *Picturing Power: Visual Depiction and Social Relations* (London: Routledge, 1988), 184–220.

44. See Lynch op. cit. note 37, at 163, emphases in the original.

45. On the differences between competencies that can be formalized and those that resist formalization see Hubert L. Dreyfus and Stuart E. Dreyfus, *Mind over Machine: The Power of Human Intuition and Expertise in the Era of the Computer* (New York: The Free Press, 1986).

46. Elinor Ochs, Patrick Gonzales and Sally Jacoby, '"When I come down I'm in the domain state": Grammar and graphic representation in the interpretive activity of physicists', in Elinor Ochs, Emanuel A. Schegloff and Sandra A. Thompson (eds), *Interaction and Grammar* (Cambridge: Cambridge University Press, 1996), 328–369, at 359.

47. See Alexandre Mallard, 'Compare, standardize and settle agreement: On some usual metrological problems', *Social Studies of Science*, Vol. 28, No. 4 (August 1998), 571–601, at 593.

48. See Charles Goodwin, 'Seeing in depth', *Social Studies of Science*, Vol. 25, No. 2 (May 1995), 237–274.

49. See for example Wolff-Michael Roth, 'Where is the context in contextual word problems?: Mathematical practices and products in Grade 8 students' answers to story problems', *Cognition and Instruction*, Vol. 14, No. 4 (1996), 487–527; Wolff-Michael Roth and G. Michael Bowen, 'Mathematization of experience in a grade 8 open-inquiry environment: An introduction to the representational practices of science', *Journal of Research in Science Teaching*, Vol. 31 (1994), 293–318.

50. The notion of reading graphs 'in captivity' is quite appropriate given that the participating scientists did not read my graphs for the sake of getting their own research done. This change in intentionality ('for the sake of which), as Martin Heidegger has shown, drastically changes actions and activity that can be observed. This change is also articulated in the activity theoretical analysis of scientists' reading graphs out of context, which I provide at the end of Chapter 5. See Heidegger, op. cit. note 17.

CHAPTER TWO

1. Marlene Scardamalia and Carl Bereiter studied expert writers and suggested that their performance was far from fluid and rapid. Sam Wineburg found that it was historians rather than students who were doubtful of their interpretations, second-guessed themselves, and extensively qualified their conclusions. See Marlene Scardamalia and Carl Bereiter, 'Literate expertise', in K. Anders Ericsson and Jacqui Smith (eds), *Toward a General Theory of Expertise: Prospects and Limits* (New York: Cambridge University Press, 1991), 172–194; Sam S. Wineburg, 'Historical

problem solving: A study of the cognitive processes used in the evaluation of documentary and pictorial evidence', *Journal of Educational Psychology*, Vol. 83, No. 1 (1991), 73–87.

2. See Sam S. Wineburg, 'Reading Abraham Lincoln: An expert/expert study in the interpretation of historical texts', *Cognitive Science*, Vol. 22, No. 3 (July–September 1998), 319–346.

3. The studies in this book were conducted in the context of other studies. Thus, I also interviewed 14 (current and future) science teachers and, more recently, 21 physicists. I do not draw on these data other than for comparisons made in this chapter. Furthermore, I recorded undergraduate ecology students in their second semester as they solved problems in their seminar courses related to the graphs that were used in my studies among scientists. The issues that arose during the analysis of the student and teacher data were structurally similar, but there were larger numbers of individuals who were unsuccessful. There were two reasons: science students and teachers (a) lacked relevant domain related knowledge that made for suitable referents and (b) articulated signs in the perceptual process that could not be related to relevant phenomena. Analyses of the student and teacher data can be found elsewhere. See G. Michael Bowen, Wolff-Michael Roth and Michelle K. McGinn, 'Interpretations of graphs by university biology students and practicing scientists: towards a social practice view of scientific representation practices', *Journal of Research in Science Teaching*, Vol. 36, No. 9 (1999), 1020–1043; Wolff-Michael Roth and G. Michael Bowen, 'Learning difficulties related to graphing: a hermeneutic phenomenological perspective', *Research in Science Education*, Vol. 30, No. 1 (2000), 123–139.

4. The non-scientists generally did not even realize that they might have problems constructing external referents to the graphs. Among the teachers, only two consistently constructed external referents (though not always appropriate ones), and one constructed external referents in the case of the isographs and population graph. All others were primarily engaged in structuring the graph qua sign, and establishing relations between graphical and verbal (caption) sign elements.

5. See, for example, Hermina J. M. Tabachneck-Schijf, Anthony M. Leonardo and Herbert A. Simon, 'CaMeRa: A computational model for multiple representations', *Cognitive Science*, Vol. 21, No. 3 (1997), 305–350.

6. See Robert Glaser, 'Education and thinking: The role of knowledge', *American Psychologist*, Vol. 39, No. 2 (February 1984), 93–104. A shortcoming of this approach was that it pertained to 'well-structured' domains rather than to the contexts that makes real-world problem solving so interesting.

7. When the participant reads aloud the text in the caption of the graph, I mark this reading by enclosing the text in quotation marks and preface it by some text in italics. '*Points to left intersection*', '*draws line*', '*writes "extinction"*', etc. are additional activities that are indicated in the transcript. For brevity sake, I use a notation such as *N[4]* to identify the value of the population size at the point indicated by gesture [4] in the associated figure.

8. See, for example, John R. Anderson, *Cognitive Psychology and its Implications* (San Francisco: Freeman, 1985).

9. This is a point about semiotic networks that was also made by Michel Callon and John Law. These authors note that 'Cats, catflaps, computers, and fax machines—not to mention X-ray sources, safety interlocks, and keys—*all* these order and organize, create paths and links. All of these signify, but they do so in their own ways'. See John Law and Michel Callon, 'Agency and the hybrid collectif', in Barbara Herrnstein Smith and Arkady Plotnitsky (eds), *Mathematics, Science, and Postclassical Theory* (Durham, NC: Duke University Press), 95–117, at 112.

10. See James D. Murray, *Mathematical Biology* (Berlin: Springer-Verlag, 1989), at 69 and 697–701.

11. See Warder Clyde Allee, *Animal Aggregations: A Study in General Sociology* (Chicago: University of Chicago Press, 1931); Warder Clyde Allee, A. E. Emerson, O. Park, T. Park and K. P. Schmidt, *Principles of Animal Ecology* (Chicago: University of Chicago Press, 1949).

12. P. W. Frank, C. D. Boll and R. W. Kelly, 'Vital statistics of laboratory cultures of *Daphnia pulex* De Geer as related to density', *Physiological Zoology* Vol. 30 (1957), 287–305.

13. See Robert E. Ricklefs, *Ecology* 3rd ed (New York: Freeman, 1990).

14. To the dismay of Canadian scientists and fishermen, other countries continued to fish the Atlantic cod on the tip of the Grand Banks outside of the 200-mile territorial waters of Canada. See, for example, Alan C. Finlayson, *Fishing for Truth: A Sociological Analysis of Northern Cod Stock Assessments for 1977 to 1990* (St. John's: Memorial University of Newfoundland, 1994).

15. Murray uses the term 'null cline'. The mathematics underlying the curves is quite complex and goes beyond what I can present here. Full discussions can be found in textbooks on mathematical biology. See Murray, op. cit. note 10, for example at 150.

16. On the mathematical formulation of this logistic curve see Appendix, particularly equation A.6.

17. See, for example, National Research Council Committee on Management of Wolf and Bear Populations in Alaska, *Wolves, Bears, and Their Prey in Alaska: Biological and Social Challenges in Wildlife Management* (Washington, DC: National Academy Press, 1997).

18. Palmer Morrel-Samuels and Robert M. Krauss, 'Word familiarity predicts temporal asynchrony of hand gestures and speech', *Journal of Experimental Psychology: Learning, Memory, and Cognition*, Vol. 18, No. 3 (May 1992), 615–622.

19. For the original article, see Stan Boutin, 'Predation and moose population dynamics: a critique', *Journal of Wildlife Management*, Vol. 56, No. 1 (1992), 116–127.

20. Greeno likened knowing mathematics to knowing once way around familiar environments. See James G. Greeno, 'Number sense as situated knowing in a conceptual domain', *Journal for Research in Mathematics Teaching*, Vol. 22, No. 3 (May 1991), 170–218.

21. On this point of rule following see, for example, Lucy A. Suchman, *Plans and Situated Actions: The Problem of Human-Machine Communication* (Cambridge: Cambridge University Press, 1987).

22. See Jill H. Larkin and Herbert A. Simon, 'Why a diagram is (sometimes) worth ten thousand words', *Cognitive Science*, Vol. 11, No. 1 (1987), 65–99.

CHAPTER THREE

1. As already indicated, predator-prey systems are modeled using second order linear differential equations that have become known as the Lotka-Volterra model that link predator (P) and prey (N) population dynamics in the following equations:

$$\frac{dN}{dt} = N(a - bP) \text{ and } \frac{dP}{dt} = P(cN - d)$$

where a, b, c, and d are constants. See, for example, James D. Murray, *Mathematical Biology* (Berlin: Springer Verlag, 1989), at 63–94.

2. In this, Nelson proceeded not unlike Brandon in the previous chapter. However, whereas Brandon abandoned his attempt to make the situation description (wolf-moose interaction) and the graph coherent, Nelson persisted.

3. See Ludwig Wittgenstein, *Philosophical Investigations* trans. G.E.M. Anscombe (New York: Macmillan, 1953/1968), at 193e.

4. See Martin Heidegger, *Sein und Zeit* [Being and Time] (Tübingen, Germany: Max Niemeyer, 1977); English translation by Joan Stambaugh (Albany: State University of New York Press, 1996.); Paul Ricœur, *From Text to Action: Essays in Hermeneutics, II* (Evanston, IL: Northwestern University Press, 1991).

CHAPTER FOUR

1. Craig A. Berg and Darrell G. Phillips, 'An investigation of the relationship between logical thinking structures and the ability to construct and interpret line graphs', *Journal of Research in Science Teaching*, Vol. 31, No. 4 (April 1994), 323–344, at 340.

2. Exceptions to this are the studies done by Moschkovich on graphing and similar studies by Izsák and Meira on other mathematical representations of scientific phenomena. See Andrew Izsák, 'Inscribing the winch: Mechanisms by which students defvelop knowledge structures for representing the physical world with algebra', *The Journal of the Learning Sciences*, Vol. 9, No. 1 (2000), 31–74; Judith N. Moschkovich, 'Resources for refining mathematical conceptions: Case studies in learning about linear functions', *The Journal of the Learning Sciences*, Vol. 7, No. 2 (1998), 209–237; Luciano Meira, 'The microevolution of mathematical representations in children's activity', *Cognition and Instruction*, Vol. 13, No. 2 (1995), 269–313.

3. See Lloyd C. Ireland, 'Crafting better charts', *Northern Journal of Applied Forestry*, Vol. 12, No. 4 (December 1995), 149–155.

4. See R. Breckenridge Church and Susan Goldin-Meadow, 'Using the relationship between gesture and speech to capture transitions in learning', *Cognition*, Vol. 23, No. 1 (1986), 43–71.

5. See Gaea Leinhardt, O. Zaslavsky and Mary K. Stein, 'Functions, graphs, and graphing: Tasks, learning, and teaching', *Review of Educational Research*, Vol. 60, No. 1 (1990), 1–64.

6. See G. Michael Bowen, Wolff-Michael Roth and Michelle K. McGinn, 'Interpretations of graphs by university biology students and practicing scientists: towards a social practice view of scientific re-presentation practices', *Journal of Research in Science Teaching*, Vol. 36, No. 9 (1999), 1020–1043.

7. See Berg and Phillips, op. cit. note 1.

8. Glenn D. Stone, 'Settlement ethnoarchaeology: Changing patterns among the Kofyar of Nigeria', *Expedition*, Vol. 33, No. 1 (1991), 16–23.

9. See, for example, Jenny Preece and Claude Janvier, 'A study of the interpretation of trends in multiple curve graphs of ecological situations', *School Science and Mathematics*, Vol. 92, No. 6 (October 1992), 299–306.

10. Berg and Phillips, op. cit. note 1, at 340.

11. Paul Cobb and Carrie Tzou, 'Supporting students' learning about data creation', *Cognition and Instruction* (in press).

CHAPTER FIVE

1. Jill Reimche, 'Group is a bridge over troubled waters', *Peninsula News Review*, (1998, December 16) at 9.

2. See, for example, Edward R. Tufte, *The Visual Display of Quantitative Information* (Cheshire, CT: Graphics Press, 1983).

3. Contrary to Glenn's statement, solubility is generally classified as a physical property, for it does not involve a chemical change in the substances involved. Chemical properties are related to the ways materials react with other materials to form new substances.

4. See, for example, Yrjö Engeström, 'Activity theory and individual and social transformation', in Yrjö Engeström, Reijo Miettinen and Raija-Leena Punamäki (eds), *Perspectives on Activity Theory* (Cambridge: Cambridge University Press, 1999), 19–38.

5. See Jean Lave and Etienne Wenger, *Situated Learning: Legitimate Peripheral Participation* (Cambridge: Cambridge University Press, 1991).

6. Yrjö Engeström and Virginia Escalante, 'Mundane tool or object of affection? The rise and fall of the postal buddy', in Bonnie Nardi (ed), *Context and Consciousness: Activity Theory and Human-Computer Interaction* (Cambridge, MA: MIT Press, 1996), 325–374.

7. William H. Sewell, 'A theory of structure: Duality, agency, and transformation', *American Journal of Sociology*, Vol. 98, No. 1 (1992), 1–29.

CHAPTER SIX

1. General, sociocultural analyses of the environmental group's activities can be found elsewhere. See Stuart Lee and Wolff-Michael Roth, 'How ditch and drain become a healthy creek: Representations, translations and agency during the re/design of a watershed', *Social Studies of Science*, Vol. 31, No. 3 (2001), 315–356; Stuart Lee and Wolff-Michael Roth, 'Of traversals and hybrid spaces: Science in the community', *Mind, Culture, & Activity*, (in press).

2. Valerie Walkerdine, *The Mastery of Reason* (London: Routledge, 1988), at 12.

CHAPTER SEVEN

1. Her ecological fieldwork was extensive and Sam had to collect data over three field seasons. Sam had to capture the animals by hand, augmented by some capturing with traps, keep female lizards in cages until they had given birth and, after having conducted all her measurements release the lizards in the same place that they had been captured. The offspring born in captivity had to be released in the same places where the mother was released.

2. Lizards are capable of tail autotomy, that is, they can shed their tail in a dangerous situation such as when a predator or scientist gets a hold of them at the tail but not the entire body. I know from personal experience in the field that lizards and skinks shed their tails, and considerable skill and speed is involved to catch the animals without damaging them, which, despite the lizard's ability to re-grow the tail, is the desired mode of capture.

3. Robert C. Stebbins, *A Field Guide to Western Reptiles and Amphibians* (Boston: Houghton Mifflin, 1966), quote at 17.

CHAPTER EIGHT

1. David Gooding presents a graphical way of representing actions, observations, and concepts in ongoing experiments such as these are available from historical records such as Faraday's laboratory notebooks. I retain here only his lexicographical scheme, A for apparatus and O for observation. However, rather than separating agency from the instrument or observation, following the convention in quantum mechanics, I denote each transformation or measurement as an operation. Because

these operators are in the form of matrixes, successive operations (operator multiplication) are denoted in a progression from right to left. As in quantum mechanics, the order of operations matters and the results are quite different or impossible to achieve in any other way. See David Gooding, *Experiment and the Making of Meaning: Human Agency in Scientific Observation and Experiment* (Dordrecht: Kluwer Academic Publishers, 1990).

2. Harold Garfinkel, Michael Lynch and Eric Livingston, 'The work of the discovery sciences construed with materials from the optically discovered pulsar', *Philosophy of the Social Sciences,* Vol. 11 (1981), 131–158.

APPENDIX

1. See Wolff-Michael Roth, G. Michael Bowen and Michelle K. McGinn, 'Differences in graph-related practices between high school biology textbooks and scientific ecology journals', *Journal of Research in Science Teaching*, Vol. 36, No. 9 (1999), 977–1019; G. Michael Bowen and Wolff-Michael Roth, 'Why do students find it difficult to interpret inscriptions?', *Research in Science Education*, Vol. 32 (2002), 303–327.

2. Robert E. Ricklefs, *Ecology* 3rd ed (New York: Freeman, 1990).

3. See Ricklefs, op. cit. note 2.

4. Sharon Kingsland provided an interesting history of the emergence of graphical models in ecology. Individuals who had come to ecology from the physical sciences and mathematics often conducted the modeling efforts; this has led to controversy in the field. See Sharon E. Kingsland, *Modeling Nature: Episodes in the History of Population Ecology* 2nd ed (Chicago: University of Chicago, 1995). Among the ecologists in this study, there were similar differences associated with quite varied responses regarding population graphs. Field ecologists often questioned the usefulness of such models for understanding 'real' systems. Theoretical ecologists (and physicists) found it quite interesting to explore the implications of the model itself regardless of their immediate usefulness for describing actual systems.

5. Hermina J. M. Tabachneck-Schijf, Anthony M. Leonardo and Herbert A. Simon, 'CaMeRa: A computational model for multiple representations', *Cognitive Science*, Vol. 21, No. 3 (1997), 305–350.

6. See, for example, *The American Heritage® Dictionary of the English Language* 3rd ed (Boston: Houghton Mifflin Company, 1992). Electronic version from INSO Corporation.

7. See David W. Stephens and John R. Krebs, *Foraging Theory* (Princeton, NJ: Princeton University Press, 1986).

References

Agre, Phil E., *Computation and Human Experience* (Cambridge: Cambridge University Press, 1997).

Allee, Warder Clyde, *Animal Aggregations: A Study in General Sociology* (Chicago: University of Chicago Press, 1931).

Allee, Warder Clyde, A. E. Emerson, O. Park, T. Park and K. P. Schmidt, *Principles of Animal Ecology* (Chicago: University of Chicago Press, 1949).

Amann, Klaus and Karin Knorr-Cetina, 'The fixation of (visual) evidence', *Human Studies*, Vol. 11 (1988), 133–169.

Anderson, John R., *Cognitive Psychology and its Implications* (San Francisco: Freeman, 1985).

Ballard, Dana H., Mary M. Hayhoe, Polly K. Pook and Rajesh P.N. Rao, 'Deictic codes for the embodiment of cognition', *Behavioral and Brain Sciences*, Vol. 20 (1997), 723–767.

Barwise, Jon, 'On the circumstantial relation between meaning and content', in Umberto Eco, Marco Santambrogio and Patrizia Violi (eds), *Meaning and Mental Representations* (Bloomington: Indiana University Press, 1988), 23–39.

_____, *The Situation in Logic* (Stanford, CA: CSLI, 1989).

Bastide, Françoise, 'The iconography of scientific texts: principles of analysis', in Michael Lynch and Steve Woolgar (eds), *Representation in Scientific Practice* (Cambridge, MA: MIT Press, 1990), 187–229.

Berg, Craig A. and Darrell G. Phillips, 'An investigation of the relationship between logical thinking structures and the ability to construct and interpret line graphs', *Journal of Research in Science Teaching*, Vol. 31, No. 4 (April 1994), 323–344, at 340.

Berg, Craig A. and Phillip Smith, 'Assessing students' abilities to construct and interpret line graphs: Disparities between multiple-choice and free-response instruments', *Science Education*, Vol. 78, No. 5 (1994), 527–554.

Bertin, Jacques, *Semiology of Graphics*, trans. W. J. Berg (Madison: University of Wisconsin Press, 1967/1983).

Bourdieu, Pierre, *Le sens pratique* (Paris: Les Éditions de Minuit, 1980).

_____, *Méditations pascaliennes* (Paris: Seuil, 1997).

Boutin, Stan, 'Predation and moose population dynamics: a critique', *Journal of Wildlife Management*, Vol. 56, No. 1 (1992), 116–127.

Bowen, G. Michael and Wolff-Michael Roth, 'Why do students find it difficult to interpret inscriptions?', *Research in Science Education*, Vol. 32 (2002), 303–327.

Bowen, G. Michael, Wolff-Michael Roth and Michelle K. McGinn, 'Interpretations of graphs by university biology students and practicing scientists: towards a social practice view of scientific re-presentation practices', *Journal of Research in Science Teaching*, Vol. 36, No. 9 (1999), 1020–1043.

Brown, John Seely, Alan Collins and Paul Duguid, 'Situated cognition and the culture of learning', *Educational Researcher*, Vol. 18, No. 1 (1989), 32–42.

Church, R. Breckenridge and Susan Goldin-Meadow, 'Using the relationship between gesture and speech to capture transitions in learning', *Cognition*, Vol. 23, No. 1 (1986), 43–71.

Cobb, Paul and Carrie Tzou, 'Supporting students' learning about data creation', *Cognition and Instruction* (in press).

Davidson, Donald, 'A nice derangement of epitaphs', in Ernest Lepore (ed), *Truth and Interpretation: Perspectives on the Philosophy of Donald Davidson* (Oxford: Blackwell, 1986), 433–446.

Dreyfus, Hubert L. and Stuart E. Dreyfus, *Mind over Machine: The Power of Human Intuition and Expertise in the Era of the Computer* (New York: The Free Press, 1986).

Eco, Umberto, *A Theory of Semiotics* (Bloomington: Indiana University Press, 1976).

_____, *Semiotics and the Philosophy of Language* (Bloomington: Indiana University Press, 1984).

_____, *The Name of the Rose* (Boston, MA: G. K. Hall, 1984).

Engeström, Yrjö, 'Activity theory and individual and social transformation', in Yrjö Engeström, Reijo Miettinen and Raija-Leena Punamäki (eds), *Perspectives on Activity Theory* (Cambridge: Cambridge University Press, 1999), 19–38.

Engeström, Yrjö and Virginia Escalante, 'Mundane tool or object of affection? The rise and fall of the postal buddy', in Bonnie Nardi (ed), *Context and Consciousness: Activity Theory and Human-Computer Interaction* (Cambridge, MA: MIT Press, 1996), 325–374.

Finlayson, Alan C., *Fishing for Truth: A Sociological Analysis of Northern Cod Stock Assessments for 1977 to 1990* (St. John's: Memorial University of Newfoundland, 1994).

Frank, P. W., C. D. Boll and R. W. Kelly, 'Vital statistics of laboratory cultures of *Daphnia pulex* De Geer as related to density', *Physiological Zoology* Vol. 30 (1957), 287–305.

Garfinkel, Harold, Michael Lynch and Eric Livingston, 'The work of a discovering science construed with materials from the optically discovered pulsar', *Philosophy of the Social Sciences*, Vol. 11 (1981), 131–158.

Glaser, Robert, 'Education and thinking: The role of knowledge', *American Psychologist*, Vol. 39, No. 2 (February 1984), 93–104.

Gooding, David, *Experiment and the Making of Meaning: Human Agency in Scientific Observation and Experiment* (Dordrecht: Kluwer Academic Publishers, 1990).

Goodwin, Charles, 'Seeing in depth', *Social Studies of Science*, Vol. 25, No. 2 (May 1995), 237–274.

Greeno, James G., 'Situated activities of learning and knowing in mathematics', in M. Behr, C. Lacampagne, and M. M. Wheeler (Eds.), *Proceedings of the 10th Annual Meeting of the PME-NA* (DeKalb, IL: IGPME, 1989), 481–521.

_____, 'Number sense as situated knowing in a conceptual domain', *Journal for Research in Mathematics Teaching*, Vol. 22, No. 3 (May 1991), 170–218.

Heidegger, Martin, *Sein und Zeit* (Tübingen, Germany: Max Niemeyer, 1977).

Henderson, Kathryn, 'Flexible sketches and inflexible databases: Visual communication, conscription devices, and boundary objects in design engineering', *Science, Technology, & Human Values*, Vol. 16, No. 4 (1991), 448–473.

Hutchins, Ed, *Cognition in the Wild* (Cambridge, MA: MIT Press, 1995).

Ireland, Lloyd C., 'Crafting better charts', *Northern Journal of Applied Forestry*, Vol. 12, No. 4 (December 1995), 149–155.

Izsák, Andrew, 'Inscribing the winch: Mechanisms by which students defvelop knowledge structures for representing the physical world with algebra', *The Journal of the Learning Sciences*, Vol. 9, No. 1 (2000), 31–74.

Kennefick, Daniel, 'Star crushing: Theoretical practice and the theoreticians' regress', *Social Studies of Science*, Vol. 30, No. 1 (February 2000), 5–40.

Kingsland, Sharon E., *Modeling Nature: Episodes in the History of Population Ecology* 2nd ed (Chicago: University of Chicago, 1995).

Knorr-Cetina, Karin D. and Klaus Amann, 'Image dissection in natural scientific inquiry', *Science, Technology, & Human Values*, Vol. 15, No. 3 (1990), 259–283.

Larkin, Jill H. and Herbert A. Simon, 'Why a diagram is (sometimes) worth ten thousand words', *Cognitive Science*, Vol. 11 (1987), 65–99.

Latour, Bruno, *Science in Action: How to Follow Scientists and Engineers through Society* (Milton Keynes: Open University Press, 1987).

_____, 'Drawing things together', in Michael Lynch and Steve Woolgar (eds), *Representation in Scientific Practice* (Cambridge, MA: MIT Press, 1990), 19–68.

_____, *Aramis ou l'amour des techniques* (Paris: Éditions de la Découverte, 1992).

_____, *La clef de Berlin et autres leçons d'un amateur de sciences* (Paris: Éditions la Découverte, 1993).

Lave, Jean, *Cognition in Practice: Mind, Mathematics and Culture in Everyday Life* (Cambridge: Cambridge University Press, 1988).

_____, 'The practice of learning', in Seth Chaiklin and Jean Lave (eds), *Understanding Practice: Perspectives on Activity and Context* (Cambridge: Cambridge University Press), 3–32.

Lave, Jean and Etienne Wenger, *Situated Learning: Legitimate Peripheral Participation* (Cambridge: Cambridge University Press, 1991).

Law, John and Michel Callon, 'Agency and the hybrid collectif', in Barbara Herrnstein Smith and Arkady Plotnitsky (eds), *Mathematics, Science, and Postclassical Theory* (Durham, NC: Duke University Press), 95–117.

Lee, Stuart and Wolff-Michael Roth, 'How ditch and drain become a healthy creek: Representations, translations and agency during the re/design of a watershed', *Social Studies of Science*, Vol. 31, No. 3 (2001), 315–356.

_____, 'Of traversals and hybrid spaces: Science in the community', *Mind, Culture, & Activity,* (in press).

Leinhardt, Gaea, O. Zaslavsky and Mary K. Stein, 'Functions, graphs, and graphing: Tasks, learning, and teaching', *Review of Educational Research*, Vol. 60, No. 1 (1990), 1–64.

Lemke, Jay L., 'Multiplying meaning: Visual and verbal semiotics in scientific text', in J. R. Martin & R. Veel (eds), *Reading science* (London: Routledge, 1998), 87–113.

Lynch, Michael, *Art and Artifact in Laboratory Science: A Study of Shop Work and Shop Talk in a Laboratory* (London: Routledge and Kegan Paul, 1985).

_____, 'Method: measurement—ordinary and scientific measurement as ethnomethodological phenomena', in Graham Button (ed), *Ethnomethodology and the Human Sciences* (Cambridge: Cambridge University Press, 1991), 77–108.

334   REFERENCES

_____, 'Extending Wittgenstein: The Pivotal Move from Epistemology to the Sociology of Science', in Andrew Pickering (ed.), *Science as Practice and Culture* (Chicago: University of Chicago Press, 1992), 215–300.

Lynch, Michael and Samuel Y. Edgerton, 'Aesthetics and digital image processing: Representational craft in contemporary astronomy', in Gordon Fyfe and John Law (eds), *Picturing Power: Visual Depiction and Social Relations* (London: Routledge, 1988), 184–220.

Mackenzie, Donald, 'Slaying the kraken: The sociohistory of a mathematical proof', *Social Studies of Science*, Vol. 29, No. 1 (February 1999), 7–60.

Mallard, Alexandre, 'Compare, standardize and settle agreement: On some usual metrological problems', *Social Studies of Science*, Vol. 28, No. 4 (August 1998), 571–601.

McGinn, Michelle K. and Wolff-Michael Roth, 'Assessing students' understandings about levers: better test instruments are not enough', *International Journal of Science Education*, Vol. 20 (1998), 813–832.

Meira, Luciano, 'The microevolution of mathematical representations in children's activity', *Cognition and Instruction*, Vol. 13, No. 2 (1995), 269–313.

Merz, Martina and Karin Knorr-Cetina, 'Deconstruction in a 'thinking' science: Theoretical physicists at work', *Social Studies of Science*, Vol. 27, No. 1 (February 1997), 73–111.

Mokros, Janice R. and Robert F. Tinker, 'The impact of microcomputer-based labs on children's ability to interpret graphs', *Journal of Research in Science Teaching*, Vol. 24, Vol. 4 (1987), 369–383.

Morrel-Samuels, Palmer and Robert M. Krauss, 'Word familiarity predicts temporal asynchrony of hand gestures and speech', *Journal of Experimental Psychology: Learning, Memory, and Cognition*, Vol. 18, No. 3 (May 1992), 615–622.

Moschkovich, Judith N., 'Resources for refining mathematical conceptions: Case studies in learning about linear functions', *The Journal of the Learning Sciences*, Vol. 7, No. 2 (1998), 209–237.

Murray, James D., *Mathematical Biology* (Berlin: Springer-Verlag, 1989).

National Research Council Committee on Management of Wolf and Bear Populations in Alaska, *Wolves, Bears, and Their Prey in Alaska: Biological and Social Challenges in Wildlife Management* (Washington, DC: National Academy Press, 1997).

Ochs, Elinor, Patrick Gonzales and Sally Jacoby, '"When I come down I'm in the domain state": Grammar and graphic representation in the interpretive activity of physicists', in Elinor Ochs, Emanuel A. Schegloff and Sandra A. Thompson (eds), *Interaction and Grammar* (Cambridge: Cambridge University Press, 1996), 328–369.

Preece, Jenny and Claude Janvier, 'A study of the interpretation of trends in multiple curve graphs of ecological situations', *School Science and Mathematics*, Vol. 92, No. 6 (October 1992), 299–306.

Reimche, Jill, 'Group is a bridge over troubled waters', *Peninsula News Review*, (1998, December 16).

Ricklefs, Robert E., *Ecology* 3rd ed (New York: Freeman, 1990).

Ricœur, Paul, *From Text to Action: Essays in Hermeneutics, II* (Evanston, IL: Northwestern University Press, 1991).

Ricœur, Paul, *From Text to Action: Essays in Hermeneutics, II* (Evanston, IL: Northwestern University Press, 1991).

Ricœur, Paul, *Oneself as Another* (Chicago: University of Chicago Press, 1992).

Roth, Wolff-Michael, 'Where is the context in contextual word problems?: Mathematical practices and products in Grade 8 students' answers to story problems', *Cognition and Instruction*, Vol. 14, No. 4 (1996), 487–527.

_____, 'Situated cognition and assessment of competence in science', *Evaluation and Program Planning*, Vol. 21 (1998), 155–169.

Roth, Wolff-Michael and G. Michael Bowen, 'Mathematization of experience in a grade 8 open-inquiry environment: An introduction to the representational practices of science', *Journal of Research in Science Teaching*, Vol. 31 (1994), 293–318.

_____, 'Digitizing lizards or the topology of vision in ecological fieldwork', *Social Studies of Science*, Vol. 29, No. 5 (October 1999), 719–764.

_____, 'Learning difficulties related to graphing: a hermeneutic phenomenological perspective', *Research in Science Education*, Vol. 30, No. 1 (2000), 123–139.

Roth, Wolff-Michael, G. Michael Bowen and Michelle K. McGinn, 'Differences in graph-related practices between high school biology textbooks and scientific ecology journals', *Journal of Research in Science Teaching*, Vol. 36, No. 9 (1999), 977–1019.

Roth, Wolff-Michael, G. Michael Bowen and Domenico Masciotra, 'From thing to sign and "natural object": Toward a genetic phenomenology of graph interpretation', *Science, Technology, & Human Values*, Vol. 27, No. 4 (2002), 327–356.

Roth, Wolff-Michael, Michelle K. McGinn and G. Michael Bowen, 'How prepared are preservice teachers to teach scientific inquiry? Levels of performance in scientific representation practices', *Journal of Science Teacher Education*, Vol. 9, No. 1 (1998), 25–48.

Roth, Wolff-Michael and Kenneth Tobin, 'Aristotle and natural observation versus Galileo and scientific experiment: An analysis of lectures in physics for elementary teachers in terms of discourse and inscriptions', *Journal of Research in Science Teaching*, Vol. 33, No. 1 (1996), 135–157.

Roth, Wolff-Michael, Kenneth Tobin and Ken Shaw, 'Cascades of inscriptions and the re-presentation of nature: How numbers, tables, graphs, and money come to re-present a rolling ball', *International Journal of Science Education*, Vol. 19, No. 10 (1997), 1075–1091.

Scardamalia, Marlene and Carl Bereiter, 'Literate expertise', in K. Anders Ericsson and Jacqui Smith (eds), *Toward a General Theory of Expertise: Prospects and Limits* (New York: Cambridge University Press, 1991), 172–194.

Sewell, William H., 'A theory of structure: Duality, agency, and transformation', *American Journal of Sociology*, Vol. 98, No. 1 (1992), 1–29.

Stebbins, Robert C., *A Field Guide to Western Reptiles and Amphibians* (Boston: Houghton Mifflin, 1966).

Stephens, David W. and John R. Krebs, *Foraging Theory* (Princeton, NJ: Princeton University Press, 1986).

Stone, Glenn D., 'Settlement ethnoarchaeology: Changing patterns among the Kofyar of Nigeria', *Expedition*, Vol. 33, No. 1 (1991), 16–23.

Suchman, Lucy A., *Plans and Situated Actions: The Problem of Human-Machine Communication* (Cambridge: Cambridge University Press, 1987).

Tabachneck-Schijf, Hermina J. M., Anthony M. Leonardo and Herbert A. Simon, 'CaMeRa: A computational model for multiple representations', *Cognitive Science*, Vol. 21, No. 3 (1997), 305–350.

The American Heritage® Dictionary of the English Language 3rd ed (Boston: Houghton Mifflin Company, 1992). Electronic version from INSO Corporation.

Traweek, Sharon, *Beamtimes and Lifetimes: The World of High Energy Physicists* (Cambridge, MA: MIT Press, 1988).

Tufte, Edward R., *The Visual Display of Quantitative Information* (Cheshire, CT: Graphics Press, 1983).

336   REFERENCES

Wainer, Howard, 'Understanding graphs and tables', *Educational Researcher*, Vol. 21, No. 1 (February 1992), 14–23.

Walkerdine, Valerie, *The Mastery of Reason* (London: Routledge, 1988).

Wineburg, Sam S., 'Historical problem solving: A study of the cognitive processes used in the evaluation of documentary and pictorial evidence', *Journal of Educational Psychology*, Vol. 83, No. 1 (1991), 73–87.

_____, 'Reading Abraham Lincoln: An expert/expert study in the interpretation of historical texts', *Cognitive Science*, Vol. 22, No. 3 (July-September 1998), 319–346.

Wittgenstein, Ludwig, *Philosophical Investigations* trans. G.E.M. Anscombe (New York: Macmillan, 1953/1968), at 193e.

Woolgar, Steve, 'Time and documents in researcher interaction: Some ways of making out what is happening in experimental science', in Michael Lynch and Steve Woolgar (eds), *Representation in Scientific Practice* (Cambridge, MA: MIT Press, 1990), 123–152.

Index